Impacts of the Media on African Socio-Economic Development

Okorie Nelson
Covenant University, Nigeria

Babatunde Raphael Ojebuyi
University of Ibadan, Nigeria

Abiodun Salawu
North-West University, South Africa

A volume in the Advances in Electronic Government, Digital Divide, and Regional Development (AEGDDRD) Book Series

IGI GLOBAL
DISSEMINATOR OF KNOWLEDGE

www.igi-global.com

Published in the United States of America by
IGI Global
Information Science Reference (an imprint of IGI Global)
701 E. Chocolate Avenue
Hershey PA, USA 17033
Tel: 717-533-8845
Fax: 717-533-8661
E-mail: cust@igi-global.com
Web site: http://www.igi-global.com

Library of Congress Cataloging-in-Publication Data

Names: Nelson, Okorie, 1981- editor, author. | Ojebuyi, Babatunde Raphael, 1974- editor, author. | Salawu, 'Biodun, editor.
Title: Impacts of the media on African socio-economic development / Okorie Nelson, Babatunde Raphael Ojebuyi, and, Abiodun Salawu editors.
Description: Hershey : Information Science Reference, 2017.
Identifiers: LCCN 2016046930| ISBN 9781522518594 (hardcover) | ISBN 9781522518600 (ebook)
Subjects: LCSH: Economic development--Africa, Sub-Saharan. | Information technology--Social aspects--Africa, Sub-Saharan. | Telecommunication--Social aspects--Africa, Sub-Saharan. | Social media--Africa, Sub-Saharan. | Mass media--Africa, Sub-Saharan. | Nigeria--Social conditions--21st century.
Classification: LCC HC79.I55 .I525 2017 | DDC 303.48330967--dc23 LC record available at https://lccn.loc.gov/2016046930

This book is published in the IGI Global book series Advances in Electronic Government, Digital Divide, and Regional Development (AEGDDRD) (ISSN: 2326-9103; eISSN: 2326-9111)

British Cataloguing in Publication Data
A Cataloguing in Publication record for this book is available from the British Library.

For electronic access to this publication, please contact: eresources@igi-global.com.

Advances in Electronic Government, Digital Divide, and Regional Development (AEGDDRD) Book Series

Zaigham Mahmood
University of Derby, UK & North West University, South Africa

ISSN:2326-9103
EISSN:2326-9111

Mission

The successful use of digital technologies (including social media and mobile technologies) to provide public services and foster economic development has become an objective for governments around the world. The development towards electronic government (or e-government) not only affects the efficiency and effectiveness of public services, but also has the potential to transform the nature of government interactions with its citizens. Current research and practice on the adoption of electronic/digital government and the implementation in organizations around the world aims to emphasize the extensiveness of this growing field.

The Advances in Electronic Government, Digital Divide & Regional Development (AEGDDRD) book series aims to publish authored, edited and case books encompassing the current and innovative research and practice discussing all aspects of electronic government development, implementation and adoption as well the effective use of the emerging technologies (including social media and mobile technologies) for a more effective electronic governance (or e-governance).

Coverage

- Emerging Technologies within the Public Sector
- Adoption of Innovation with Respect to E-Government
- E-Government in Developing Countries and Technology Adoption
- Current Research and Emerging Trends in E-Government Development
- Frameworks and Methodologies for E`-Government Development
- Knowledge Divide, Digital Divide
- Online Government, E-Government, M-Government
- Citizens Participation and Adoption of E-Government Provision
- E-Citizenship, Inclusive Government, Connected Government
- Issues and Challenges in E-Government Adoption

IGI Global is currently accepting manuscripts for publication within this series. To submit a proposal for a volume in this series, please contact our Acquisition Editors at Acquisitions@igi-global.com or visit: http://www.igi-global.com/publish/.

The Advances in Electronic Government, Digital Divide, and Regional Development (AEGDDRD) Book Series (ISSN 2326-9103) is published by IGI Global, 701 E. Chocolate Avenue, Hershey, PA 17033-1240, USA, www.igi-global.com. This series is composed of titles available for purchase individually; each title is edited to be contextually exclusive from any other title within the series. For pricing and ordering information please visit http://www.igi-global.com/book-series/advances-electronic-government-digital-divide/37153. Postmaster: Send all address changes to above address.

Titles in this Series

For a list of additional titles in this series, please visit: www.igi-global.com

Population Growth and Rapid Urbanization in the Developing World
Umar G. Benna (Ahmadu Bello University, Nigeria) and Shaibu Bala Garba (Qatar University, Qatar)
Information Science Reference • copyright 2016 • 358pp • H/C (ISBN: 9781522501879) • US $205.00 (our price)

Political Discourse in Emergent, Fragile, and Failed Democracies
Daniel Ochieng Orwenjo (Technical University of Kenya, Kenya) Omondi Oketch (Technical University of Kenya, Kenya) and Asiru Hameed Tunde (Umaru Musa Yar'adua University, Nigeria)
Information Science Reference • copyright 2016 • 412pp • H/C (ISBN: 9781522500810) • US $185.00 (our price)

International Perspectives on Socio-Economic Development in the Era of Globalization
Saurabh Sen (Sunbeam College for Women, India) Anshuman Bhattacharya (Sunbeam College for Women, India) and Ruchi Sen (Sunbeam College for Women, India)
Business Science Reference • copyright 2016 • 288pp • H/C (ISBN: 9781466699083) • US $175.00 (our price)

Comparative Political and Economic Perspectives on the MENA Region
M. Mustafa Erdoğdu (Marmara University, Turkey) and Bryan Christiansen (PryMarke, LLC, USA)
Information Science Reference • copyright 2016 • 386pp • H/C (ISBN: 9781466696013) • US $205.00 (our price)

Handbook of Research on Entrepreneurial Success and its Impact on Regional Development
Luísa Carvalho (Universidade Aberta, Portugal & CEFAGE, Universidade de Évora, Portugal)
Information Science Reference • copyright 2016 • 878pp • H/C (ISBN: 9781466695672) • US $430.00 (our price)

Trends, Prospects, and Challenges in Asian E-Governance
Inderjeet Singh Sodhi (St. Wilfred's P.G. College, India)
Information Science Reference • copyright 2016 • 410pp • H/C (ISBN: 9781466695368) • US $190.00 (our price)

Handbook of Research on Comparative Approaches to the Digital Age Revolution in Europe and the Americas
Brasilina Passarelli (School of Communications and Arts, University of São Paulo, Brazil) Joseph Straubhaar (The University of Texas at Austin, USA) and Aurora Cuevas-Cerveró (Complutense University of Madrid, Spain)
Information Science Reference • copyright 2016 • 530pp • H/C (ISBN: 9781466687400) • US $325.00 (our price)

Cloud Computing Technologies for Connected Government
Zaigham Mahmood (University of Derby, UK & North West University, South Africa)
Information Science Reference • copyright 2016 • 416pp • H/C (ISBN: 9781466686298) • US $195.00 (our price)

Editorial Advisory Board

Table of Contents

Detailed Table of Contents

Consequently, this paper considers the nature of risks and vulnerabilities affecting e-connectivity from a modernist theoretical perspective and contextualized this in the double edged implication affecting the use of the Internet. The first section of this paper is devoted to review on the nature of affinity between the Internet and crime while the final section engages the empirical analysis of secondary data on the consequences of cyber-crime to the global economy.

This chapter aims at analysing the effects of social media on African society by focusing on WhatsApp usage. The chapter will; one, show how usage of WhatsApp is affecting interpersonal relationships in Africa. Two, evaluate the common uses of WhatsApp in Africa. Three, scrutinise the reasons people in Africa prefer using WhatsApp. Four, add theoretical perspectives on how social media is impacting on the communication landscape of the region. The chapter will also introduce two concepts, one; medium veracity wherein different media have varying levels of trust or credibility and two; the social media domino effect, wherein social media platforms are mutually dependent and have a symbiotic relationship. By examining why people are using WhatsApp, this chapter seeks to add to the empirical conversation on futility and the transformative potential of social media.

Entrepreneurs all over seek ways of introducing their products to international markets, but international business environments pose a lot of opportunities and threats to foreign entrants. The cultural, political and technological environment has a lot to do in entrepreneurial success in global markets. The study examined the effect of social media on international business venturing. The general objective of the study is to critically determine the extent to which social media have contributed to the international venturing of entrepreneurs in Nigeria. The study concludes that since technological products provide ample benefits for international entrepreneurs, organizations should endeavour to provide indispensable technologies that support international transactions.

In an attempt to build a theoretical framework for new media within the African context, this chapter considers terminology that routinely surface around ICTs. Terms such as knowledge economy, information poverty, the digital divide, information, knowledge and data though used interchangeably or roughly to mean similar things, they are laden with value, meaning and reflect power relations within a global

24% higher than the females. These challenges should be put into consideration when formulating policies for education in developing countries. There is therefore the need to prioritize education, especially female education, as well as child rights in general in the local government through adequate funding, investment in teachers and creation of awareness about the rights of the child.

For ages, the African culture has limited the activities of women and conditioned, to a large extent, the mentality of most African countries about the place and positioning of women in the society. The mass media have been used as a tool in this. Promoting the abilities and achievements of women in the society is one of the important roles of the media in reducing the rate of gender discrimination. These achievements are becoming noticeable in politics and entrepreneurship; thus, setting a standard for other women in the society to build on. This chapter will explore the role of the media in promoting the woman entrepreneur in Nigeria. The study employed the survey research method for data gathering. Findings showed that women have equal opportunity to grow their businesses. Data shows that this growth is slow as there are no significant differences between respondents who agree or disagree with the availability of enabling environment for women to grow their business as compared to their male counterpart. The study recommends that the mass media need to give more voice to their businesses and activities to pave way and encourage the younger women in the society.

Discourse on the prevalent issues affecting the African mass media.

The Federal Government of Nigeria made several unsuccessful attempts at stopping petrol subsidy by arguing that the policy is unsustainable. Seven unsuccessful attempts were made between 1999, when the country returned to democratic rule, and 2012. Though the government might have good reasons for removing petrol subsidy, it was clear that there was an obvious disconnection between the government and the people as the people were not in support of the policy. This chapter examined coverage of petrol subsidy removal by three national dailies. Content analysis was adopted as the research method. It was discovered that 49.4% of the stories published by the selected newspapers in the study period were not in favour of petrol subsidy removal. It was therefore recommended that government should endeavour to carry Nigerian citizens along ever before introducing such kind of policy and that media practitioners should set an agenda for fighting corruption in the oil industry.

Section 4
Media, Health Behavior, and Development

The role of the media in promoting health issues in Africa.

The study investigated gender differences in attitudes towards sexuality in the context of HIV/AIDS among undergraduate students. The cross-sectional study was conducted among 1,498 randomly selected students. The majority of the students interviewed were females at 53.9% (M= 20.24, SD=2.98) compared to 46.1% males (M=21.67, SD=3.35). More Males (16%) responded in the affirmative with regard to having more than one sexual partner than female students (10%). Also, more male students (32.7%) than female students (18.6%) indicated that using condom reduces sexual pleasure and over 40% of both sets of respondents indicated that sexual intercourse should only be done between married couples. This study underscores the need to strengthen HIV/AIDS programmes to include issues on sexuality and encourage undergraduate students to attend these programmes.

Researches in clinical and medical science have shown that Human Immunodeficiency Virus (HIV), Hepatitis B, and to some extent Hepatitis C viruses, constitute a major public health challenge in the Sub-Saharan Africa. This is without prejudice to the Ebola virus disease (EVD) that is more deadly than the other three put together, and had affected some countries in West Africa- Liberia, Sierra Leone and Guinea. There is little awareness and education via the mass media on some common viral infections in Nigeria like Hepatitis compared to HIV. Therefore, there is dire need for information, sensitisation and education on the viruses, means of transmission, preventive measures and also therapy. Edutainment is the fusion of education into entertainment programming which can come in form of drama, music, poetry and lots more, and it had been used as a platform to create awareness for positive reproductive health and HIV/AIDS by leading health organisations like WHO, USAID, DFID, SFH, and UNFPA. This paper therefore, examines why edutainment should be used and how it can be used to educate media audience in Nigeria on some viral infections that pose serious health risks and how they can live healthy lives.

Preface

The idea of using communication for the purpose of advancing the goals of development received a boost when it was realised that development has to do with the generation of psychic mobility and changing of attitude. It was also realised that the failures of certain development projects had been due to the absence of communication component in the development calculus (Salawu, 2008).

Since this realisation had dawned on development agencies, communication has remained an integral but indispensable part of the development process. And, this has been made manifest in the various Information, Education and Communication (IEC) documents that have been produced for the various development projects.

Sustainable Human Development is development that is participatory and pro-people. The question is how do we get the people informed and, therefore, mobilize their informed capacities? The answer lies in communication. Development communication is the application of the process of communication to the development process (Moemeka, 1991). In other words, development communication is the use of the principles and practice of exchange of ideas to fulfill development objectives. It is further said that development requires that a mass of people with a low rate of literacy and income and the socio-economic attributes that go with it, first of all be informed about and motivated to accept and use a sizeable body of hitherto unfamiliar ideas and skills in very much less time than that process would normally take. Thus, the job of development communication is to inform and motivate at the national, sectoral and project levels. The job of development communication is the process of development itself. The ultimate aim of communication for development is the facilitation of the exchange of information for increased productivity and development (Goldhaber, 1983).

In concise terms, the purpose of development communication is to understand the needs and social realities of the people and mobilise them towards the development goals. In other words, development communication is one with a "social conscience", having the people as its focus. Development communication involves the generation of psychic mobility or empathy, raising of aspirations, teaching of new skills and encouragement of local participation in development activities. The point to note here is that no amount of investment or provision of technology and inputs will bring any lasting improvement in the living standards of people unless they, themselves, change their attitude and behavioural patterns. In essence, it is people who will bring about development, and there can be no change for the better without their informed participation, without mobilising their capacities and energies, and without increasing their knowledge and skills. This is where communication comes in because it caters to these human dimensions.

Precisely, this is what this book, *Impacts of Media on Socio-Economic Developments*, has done in its various chapters. Contributions in this book have variously considered the use of different kinds of media – print, broadcast, digital and edutainment – for development purposes.

The book is divided into four sections, namely, Section 1: ICT and Development in Africa – The role of ICTs in promoting socio-economic development in Africa; Section 2: Gender, Children, and the Media – The role of the media in promoting gender and children issues in Africa; Section 3: Issues in African Mass Media – Discourse on the prevalent issues affecting the African mass media; and Section 4: Media, Health Behaviour and Development – The role of the media in promoting health issues in Africa.

Section 1 contains seven chapters. In Chapter 1 (Role of ICT in Socio-Economic Development in Kenya: Demystifying Gender Empowerment Through Mobile Telephony), Juliet Macharia discusses the magic revolution of mobile telephony in Africa and specifically in Kenya. She argues that innovations such as MPESA have transformed people's lives in Kenya and in other emerging economies. She also discusses the role of gender socializationin relation to how the ICTs have eliminated the long held belief that females are not participants in societal development.

In Chapter 2 (Modern Information Technology, Global Risk and the Challenges of Crime in the Era of Late Modernity), Ajibade EbenezerJegede looks at the other side of the coin of ICTs. The chapter considers the nature of risks and vulnerabilities affecting e-connectivity from a modernist theoretical perspective. It specifically dwells on cybercrime. Brian Pindayi Rusangu's Chapter 3 (Social Media Uses and Effects in Africa: The Case of WhatsApp in Africa) analysed effects of social media on African society by focusing onWhatsApp. In the chapter, he also introduced two concepts: *medium veracity* and social *media domino effect*. Ogbari, Ejimofor, Akanbi and Suleiman were also concerned with social media effect, but on international business venturing in Nigeria. In Chapter 4 (Social Media Effect on International Business Venturing in Nigeria: A Conceptual Review), they concluded that since technological products provide ample benefits for international entrepreneurs, organizations should endeavour to provide indispensable technologies that support international transactions.

Anusharani Sewchurran said his chapter (5) was an attempt to build a theoretical framework for new media within the African context. The author asserted that the digital divide offered a key entry point into ICT discourses and the opportunities and challenges presented by new media technology within the context of Africa. In the chapter titled "Digital Divide, Data Trash, and the Commodification of Information: Discourses around the Digital Divide," Anusharani attempted a review of the theoretical concepts of the digital divide within the larger context of global constructions of information wealth and information poverty. Democratisation of communication and media space engendered by Web 2.0 was the focus of Chapter 6 (Discourse and Multimodal Analysis of Netizen Reactions to the Nigerian 2015 Presidential Elections: Case Study of Osun Political Parrot Facebook Site) authored by Aretha Oluwakemi Asakitikpi and Samuel Oluwafemi Adeyeye. They specifically looked at readers' comments on the Facebook page of the *Osun Political Parrot*. Focusing on the #NoToSocialMediaBill Twitter protest against the introduction of The Frivolous Petitions (Prohibition, etc.) Bill 2015, in Nigeria. Chapter 7 (Nature and Outcome of Nigeria's Twitter Protest against the Frivolous Petitions Bill 2015), authored by Oyewole Adekunle Oladapo and Babatunde Raphael Ojebuyi, examined the composition of the loudest voices that framed the protest, explored the level of engagement of those voices with the content of the bill, and analysed the nature and outcome of the protesters' reactions to the bill. The authors came up with the finding that a combination of hashtag and retweet features of Twitter functioned effectively as collective framing tools in the #NoToSocialMediaBill Protest. Though the frames drew global media and human rights activists' attention in support of the protesters' goal of resisting an attempt to restrict freedom of speech, they did not achieve the same with the recall of the initiator of the bill.

Section 2 of the book begins with Chapter 8 (Portrayal of Women in Nollywood Films and the Role of Women in National Development). In this chapter, Suleimanu Usaini, Ngozi Chilaka and Nelson Okorieaimed show how the images of women were reflected in Nollywood films, with a focus on investigating the influence of such portrayals on their role in national development. They discovered that over two-third of major female characters in the films analysed were portrayed as dependent, 80% were depicted in situations of physical, sexual, and emotional abuse, while only 30% of the major female characters were portrayed as career professionals and intellectuals. They concluded that portrayals and representations of women have negative influences on their contributions towards national development. Chapter 9 (Education, Gender, and Child-Rights: Salient Issues in SDGs Years in Ado-Odo/Ota Local Government Area of Ogun State, Nigeria) examined the place of education, gender and child rights within the United Nations' Sustainable Development Goals (SDGs) in Ado Odo/Ota local government of Ogun State, Nigeria. Taiwo Abioye, Kehinde Oyesomi, Esther Ajiboye, Segun Omidiora and Olusola Oyero found out that there was a limited awareness about child rights among children in primary schools and secondary schools; teaching materials and instructors were grossly inadequate in many of the schools sampled and basic needs such as water and electricity were unavailable. They also observed that enrolment of male children in the schools was 24% higher than the females. They recommended that these challenges should be put into consideration when formulating policies for education in developing countries.

Chapter 10 concludes Section 2 of the book and it is titled "Women Entrepreneurs in Nigeria: Where Is the Mass Media?" In the chapter, Tolulope Kayode-Adedeji, Oyinkansola Ige and Thelma Ekanem attempted to identify instances of gender inequality against women, and explore the role of the media in promoting woman entrepreneur and achievements in Nigeria. They recommended that the mass media should promote women entrepreneurs in Nigeria; give more voice to their businesses and activities to pave way and encourage the younger women in the society.

Section 3 takes off with Chapter 11 (Nigerian Government's Attempts at Removing Fuel Subsidy and the Attendant Mass Protests: A Content Analysis of National Newspapers' Coverage). Coverage of petroleum subsidy removal in three national dailies of Nigeria was the concern of Adeyemi Aderogba and Olanipekun Gbenga in this chapter. They observed that over years attempts by Nigerian governments to remove petroleum subsidy had been resisted by Nigerians. In their study, they observed that a considerable number of the stories sampled were not in support of the subsidy removal. They therefore recommended that government should endeavour to carry Nigerian citizens along ever before introducing such kind of policies.

The chapter by Nelson Okorie focused on Press Freedom and Socio-economic issues in Nigeria and Uganda Democratic Landscape. Using Nigeria and Uganda as examples, the chapter explored press freedom and socio-economic issues in a democratic media landscape. Nelson concluded that press freedom promotes transparency and flourishes more in countries with democratic principles and ideals. Nneoma Anozie's Chapter 13 is "A Discourse on Mass Media and Society." The chapter examines the relationship between mass media and society and suggests ways in which mass media can facilitate a peaceful society. In Chapter 14, Seamogano Mosanako examined Broadcasting Policy in Botswana: The Case of Botswana Television. She observed that there is a broadcasting policy vacuum in Botswana in that there is no formal media policy to guide the operations of the media. She argued that this lack of media policy in Botswana and possibly other developing countries is generally responsible for poor performance of the broadcast media regarding its role in national development. According to her, such poor performance of the media is reflected by domination of government originated content, domination of foreign content on local television, state control of the media as well as increasing pressure on the media to be more accessible to opposition political parties.

Section 4 contains the rest of the chapters (15, 16, and 17). Paul Bigala, in Chapter 15 (Examining Sexuality in the Context of HIV/AIDS Knowledge among Students of the North-West University Mafikeng Campus) examined gender differences in attitudes towards sexuality in the context of HIV/AIDS among undergraduate students. The study underscored the need to strengthen HIV/AIDS programmes to include issues on sexuality and encourage undergraduate students to attend these programmes. In Chapter 16 (Awareness and Education on Viral Infections in Nigeria Using Edutainment), Suleimanu Usaini, Tolulope Kayode-Adedeji, Olufunke Omole and Tunji Oyedepo examined why edutainment should be used and how it can be used to educate media audience in Nigeria on some viral infections that pose serious health risks and how they can live healthy lives. The last chapter (An Assessment of Media Contribution to Behaviour Change and HIV Prevention in Nigeria) written by Olugbenga Elegbe explored media influence on behaviour change and its implication for HIV prevention programmes in Nigeria. The chapter concluded on the positive influence of media intervention in sexual behaviour change programmes but notes that behaviour change still is a personal decision. Therefore it recommended other forms of communication like interpersonal communication, community mobilization and advocacy to support media strategies for effective HIV prevention initiatives in Nigeria.

Chapters in this book have looked at development in its socio-economic dimensions and have outlined how media of all sorts have been applied to advance the purposes of development.

Abiodun Salawu
North-West University, South Africa

REFERENCES

Goldhaber, G. M. (1983). *Organisational communication*. Dubuque, IA: C.B. Company Publishers.

Moemeka, A. (1991). Perspectives on development communication. In K. Boafo (Ed.), *Module on development communication* (pp. 15–30). Nairobi: ACCE.

Salawu, A. S. (2008). Development communication: The preliminaries. In Globalisation and development communication in Africa. Ibadan: Ibadan University Press.

Acknowledgment

We thank God Almighty for the speedy and successful completion of this book. He has been the source of our inspiration. We acknowledge, generally, that this book project is, without doubt, a product of collective and assiduous efforts of various individuals and organisations. They, at various phases, and in different capacities, ensured that our dream to collaborate and produce this book did not fizzle.

Specifically, we express our gratitude to the Managements of IGI-Global, North-West University, South Africa and Covenant University, Nigeria for creating the enabling platform for us to successfully execute this book project. We cannot quantify the contributions of our various reviewers: Prof. Juliet W. Macharia; Dr. Ebenezer Oluseun Ogungbe; Dr. Olugbenga Elegbe; Dr. SulemanuUsaini; Dr. Oyesomi Kehinde; and Dr. David Igbokwe. Your critical comments, cogent corrections and recommendations assisted us to select the best from numerous papers submitted to the editorial board. Also, we thank Mr Oyewole Oladapo and Mr Obasanjo Oyedele, both of the Department of Communication and Language Arts, University of Ibadan, Nigeria. Your rigorous language editing added to the quality of the chapters. We are forever indebted to executive members of the African Council for Communication Educators (ACCE) and South African Communication (SACOMM) for their significant contributions towards the success of this book. Finally, our sincere appreciation goes to all scholars in the field of Communication, who made significant contributions to this book. We acknowledge the time you spared, and the valued intellectual resources that you invested to produce your respective chapters.

Okorie Nelson
Covenant University, Nigeria

Babatunde Raphael Ojebuyi
University of Ibadan, Nigeria

Abiodun Salawu
North-West University, South Africa

Section 1
ICT and Development in Africa

The role of ICTs in promoting socio-economic development in Africa.

Chapter 1

Role of ICT in Socio-Economic Development in Kenya:

Demystifying Gender Empowerment through Mobile Telephony

Juliet W. Macharia
Karatina University, Kenya

ABSTRACT

The growth of Information Communication Technology (ICT) has empowered communities in diverse ways. Both males and females use ICTs to transact businesses and carry on relationships. Access and use of mobile phones have changed people's lives. This chapter discusses the magical revolution of mobile telephony in Africa and specifically in Kenya. Innovations such as MPESA have transformed people's lives in Kenya and in other emerging economies. Both males and females have used the innovations to bring about development that has enriched their lives. Secondly, the role of gender socialization is discussed in relation to how the ICTs have eliminated the long-held belief that females are not participants in societal development. Thirdly, the role of the media in development is examined because they socialize people into the gender roles. Through mobile telephones, females perform diverse roles. Positive portrayal of men and women in development acknowledges that men and women must be active members of the process.

INTRODUCTION

The development of ICTS especially the mobile telephony in Africa- has been faster than any other innovation that came before. The access and use of mobile phones in Africa and specifically in Kenya have shown that gender empowerment is possible. A study conducted in Machakos, Kenya, by Wandibba, Nagendo and Mulemi (2014) showed that both men and women have equal access and use of mobile phones and that more women than men used MPESA. Also, it has been observed that the mobile phones have enabled participants with low level of education to transact businesses that they would not have done before. They have opened and ran mobile telephony related businesses; at the same time, some of them have used the mobile telephony innovations to empower others through small and medium enterprises.

DOI: 10.4018/978-1-5225-1859-4.ch001

As business owners and employers, they have proved that given the right environment, males and females equally contribute towards socio economic development of their countries.

The media in Kenya have consistently been voted as the most trusted institutions therefore the massages that are broadcast or written in print media carry a lot of weight in matters of credibility. When the media positively set the agenda for the nation and positively communicate development messages, the listeners or views take the messages seriously. The media portray men and women differently and this has lead many gender scholars to argue that women's contribution is towards social economic development in African countries is ignored (Macharia, 2016a).

In this chapter, the writer discusses the access and use of ICTs-the mobile phones- and the role that mobile money services such as MPESA have played in gender empowerment in Africa and specifically in Kenya. Studies, such as the one conducted by Wandibba, et al (2014) on Machakos in the region, demonstrate that when women access and use mobile money, their communities have been empowered to a higher level. This is evident in the field of business, agriculture, education, health and others. Projects that rely on mobile telephony are identified and examined in details to see how they have contributed to gender empowerment in Africa. There are more female-headed households who are economically empowered in East Africa than before. Technology as a double-edged sword has brought about advantages that have positioned ICT access and use in some African countries with those in the developed world. However, the challenges that ICTs and specifically the mobile telephony has introduced in the social and economic contexts are discussed. Recommendations are made on how to mitigate the challenges brought by access and use of ICT in Kenya.

Access and Use of ICTS in Africa

The African ICT landscape shows that Information and Communication Technologies have enhanced communication. United Nations (2014) and World Bank Group Strategy (2012-14) demonstrate that they are available in most African countries and the only difference is the level at which rural and urban populations access them. The way communicators used to disseminate information has changed in recent times, though the traditional channels of mass communication, the radio, film, television, newspapers and magazines are still in use. However, new communication technologies such as: computers, mobile phones, internet, radio, television, websites, messaging services, satellite systems have changed the communication field.

According to African Partnership Forum (2008), ICTs contribute to economic development of African countries in the following ways:

- Increasing productivity across all sectors;
- facilitating market expansion beyond borders to harvest economies of scale;
- lowering costs of and facilitating access to services, notably in administration, education, health and banking;
- providing access to research products and services;
- contributing to better governance,

Information Communication Technologies can enhance networking, participation and advocacy within a society and can also improve interaction between governments and the citizens (UNESCO, 2003). That is, men and women need to have access to ICTs but we find that in Africa and in Kenya, women's access to ICTs is constrained by factors such as level of education, access to funding and cultural beliefs that go beyond the technological and socio -economic environment (UNESCO, 2003). According to World Bank (2003c), ICTs can be used as tools of development to positively change people's attitudes (including women), and to enable them to participate in development.

When countries such as Ghana, Kenya, and Nigeria make the ICT policies, gender should be a concern because ICTs facilitate commerce by increasing the flow of information from the sender to the receiver, thus reducing the cost of interaction and transactions (World Bank, 2003c). For example, the radio is a medium particularly suited to promoting economic development and empowerment, as it is affordable and accessible to listeners and demands few specific skills. A recent study established that radio plays a prominent role in tackling gender issues and promoting women's visibility in African countries such South Africa, where the average person listens to radio for more than four hours every day; and Mali is an example of a country with a vibrant radio landscape, including but not limited to community radio (African Development Forum, 2008:5).

The African landscape shows an increasing convergence across different technologies. For example, ICT such as the Internet is not only a mechanism for the dissemination of in-formation, but are also used as broadcasting system, a platform for individuals to interact and a marketplace. In addition to this, ICTs such as the mobile phones are not only used for voice services, but they also facilitate internet-access/data collection and even financial services. Before the advent of mobile phones, people used fixed telephone lines for communication (voice) and the computers to access the internet. In practical terms, the smart phones have easily replaced the need for computers because one can perform multiple functions using the mobile phones (African Development Forum, 2008:7; United Nations, 2014).

The ICT of interest in this paper is the mobile phone that is used for varied purposes. Aker and Mbiti (2010) posit that mobile phones bring economic development to Africa by:

- Using mobile phones to improve access to and use of information by improving coordination among agents and increasing market efficiency.
- Communicating to improve firms' productive efficiency by allowing them to better their supply chain
- Using mobile phones to create new jobs to address demand for mobile-related services, thereby providing income generating opportunities to rural and urban areas.
- Using mobile phones to facilitate communication among social networks in response to shocks, reducing households exposure to risk.
- Using mobile phone applications and development projects that have potential to facilitate the delivery of financial, agricultural, health and educational services.

Gender and Information Communication Technology in Africa

In terms of development, Information Communication Technologies (ICTs) are helping emerging economies to sort out various problems, so it is important for the majority of members in the population to have access and be able to use them. Ling and Horst (2011:370) argue that ICTs do not empower, but it is access and use that empower. They note that "seen through the lenses of power, the mobile phone changes the rules regarding who can interact with whom". Research conducted in ten countries (Kenya, Uganda, Mozambique, Nigeria, Cameroon, Egypt, Colombia, India, the Philippines and Indonesia) showed that web-based ICTs contribute to the re-adjustment of social rules for interaction, providing new channels for accessing and avoiding relationships as well as forming new ones (World Wide Web Foundation, 2014). Castells, Fernandez-Ardevol, Linchuan Qiu and Sey (2007) observe that the way people use ICTs is closely related to the construction of their culture. Seen this way gender is also a construction of the society so in essence, the society influences how men and women access and use ICTs.

Studies conducted in African countries (Kenya, Uganda, Zimbabwe, Nigeria, Cameroon, Egypt, Senegal, Zambia, Mozambique and South Africa) showed that men and women access and use ICTs differently and that their ability to use ICTs is dependent on many variables among them are their level of education, and the location where they are situated-rural or urban (United Nations, 2014; World Bank Group Strategy, 2012-14). According to UNESCO (2003), women and girls may lack basic skills and are likely to be left out of local initiatives. What the authors are saying is that ICTs are necessary for development but women are likely to miss participation because they lack literacy and language skills that are required of them. Those dealing with ICTs and especially the mobile phone have come up with applications and locally relevant content that play a crucial role in linking people to ICT -- they make ICT infrastructure come alive and accessible to people. Applications include e-mail, SMS, electronic transactions, Internet-based telephony and multimedia (African Development Forum, 2008:12). However, for those in Kenya, the problem is minimized by the fact that the mobile phone and the internet have information that is presented in Kiswahili, the East African Lingua franca which three quarters of the population use.

Gender Roles and Socialization

Every society is mandated to socialize its young ones on what is expected of them as girls and boys, men and women. Most African societies are patriarchal and the way children are socialized reflects what the societies value and frown at. According to Gamble and Gamble (2013), gender is a society construct of the roles of men and women. It is dynamic though culturally determined. Wood (2011) argues that the way societies socialize their children demonstrate how men and women are viewed and whether their gender roles are respected or not. These differences are seen in both developed and the emerging economies.

Socialization shapes our aspirations and experiences. More often than not what is communicated borders on gender stereotypes about men and women. According to Tannen (2013) and Wood (2011), the following are agents of socialization:

- The media-television, newspapers, internet, magazines, radio, new media,
- Family-parents, siblings and significant others,
- Church,
- School,
- Peers,
- Society.

The family is the primary agent of socialization. The church and the school build on the foundation laid by the family. The media are the most influential agents of socialization and they impact the concepts 'gender' and 'gender roles' (Wood, 2011). This is because almost all the regions have a gendered media, which are male-owned and perpetuate male ideals. Gamble and Gamble (2003:351) argue that "the media both reflect, refract and reshape our perceptions, cultural values and ideals about gender… and what it means to be a man or a woman."

The above is achieved through advertising in the media, both print and electronic. The media send messages that use images and representations of men and women which capture our attention and persuade us about their intentions.

The portrayal of men and women in America, Europe, Nigeria and in Kenya depict men and women differently. Women are portrayed as passive, dependent, powerless, as decorative or as sexual objects in Africa. That is why in advertisements, they are seen in roles such as mothers, nurses, caretakers, cooks, and wives in kitchens preparing meals or washing clothes for the family while men are presented in offices as managers, pilots, lawyers or as powerful professionals carrying out various activities (Macharia, 2016 b; Nwalikpe, 2014; Tannen, 2013; Wood, 2011). The three authors are in agreement on how women are portrayed in the media. What happens in Kenya and Nigeria is a reflection of what is happening in Europe, America and Australia. One would expect the women in developed world to be in advantaged positions but media images flashed on the screens across the continents show negative portrayals of most of them, geographical region notwithstanding. What this shows is that the media messages are a powerful force in the socialization process in that they are repeated many times. For example, televisions power and the social media out-do what the parents have done in years.

In Kenyan media, one sees advertisements whereby men sit in big offices with laptops, smart phones, tablets and other digital gadgets while in another office, the secretary types letters on a big old desktop computer. Images such as these create the impression that technology and especially the trendy and expensive mobile phone- belong to the men and that these symbols of power are not shared with others. However, there are some adverts of young beautiful young female CEOs who own businesses that make a lot money and create jobs for fellow Kenyans. In my opinion, we need to see a mix of these portrayals to give the correct and the real picture in the country (Macharia, 2016b).

The socialization process should be dynamic to show the changes that have taken place in Africa and specifically in Kenya where access to ICTs and use have developed to such a level that one can claim gender representation although not on 50/50 percent basis. Therefore, the young ones should be socialized in a way that prepares them for roles that are not stereotypical and to be exposed to role models and media images that convince them that there are no roles that are meant for men and not for women. Today, men and women are empowered, have access and use mobile phone for personal, regional and national development.

Mobile Money and Development in Kenya

The mobile phone penetration in Africa is high. Out of the 735 million mobile subscribers in Africa, 37 million are Kenyans. (Communication Commission of Kenya, 2015; GSMA, 2014). Most of the subscribers are also accessing the internet through their mobile phones. For ICT such as the mobile phone which is available in Kenya (more than 80 percent of the population has mobile phones), women can participate in many activities using mobile phones (Oteri, 2015).

The mobile phone has revolutionized the Kenyan society. The immediate feedback provided by mobile phone operations empower many users. It was in 2007 that the first mobile money transfer service was unveiled in Kenya. The main purpose then was to enable the customers (the un-banked population) to send and receive money. The innovation known as M PESA –M stands for mobile and P for Pesa, the Kiswahili name for money (Aker & Mbiti, 2010). Safaricom service provider, the owners of the innovation, did not immediately realize that the mobile application would facilitate a variety of transactions for the users.

In the early stages of MPESA, it was assumed that the application was going to be useful to the unbanked population. However, it was later realized that it was useful to all, the unbanked and the banked population in Kenya. Most Kenyan banks put stringent measures as requirements for opening bank accounts and running them. Some insisted on minimum balances that the citizens could not afford and MPESA changed all this because all what one needs to register is a phone, sim card and identification card.

The mobile phone is one of the developments in technology. Recent studies in Kenya show that most of the people who access the internet do it using the modern sophisticated mobile phone. The availability of ordinary phones that are internet-enabled and the affordable smart phones have made sending and receiving emails easier than before. People are able to check for information in the privacy of their homes, offices and even when travelling. Mobile internet subscriptions has continued to dominate the internet market with 98.8 percent of the total internet subscriptions being from the mobile internet sub-market.

Most of the smart phones owners have no need of a computer because the mobile phones have features that enable them to perform many functions such as watching TV, listening to the radio, listening to music, playing games, photography, commerce, products branding and marketing. In some African countries especially Kenya, the development in ICT and especially the mobile phone is alleviating poverty (Ndunge & Mutinda, 2011).

The mobile money field has grown globally to such a level that the service providers such as Safaricom's MPESA deal with bigger volumes of money than banks. According to GSMA (2015), mobile money accounts for 60 percent in world's developing market. They also noted that in 2014, there were 255 mobile money services in 89 markets and that mobile money agents outnumbered bank branches in developing economies. In Kenya, the Nation of 19th March, 2016 reported that currently, mobile money service providers transactions have grown in volume from 911 million to1.114 billion and that most banks are relying on mobile phone services such as MPESA to sustain their growth. What this means is that the bank customers, male and female can access their bank accounts through MPESA and transact business without going to the bank.

According to GSMA (2015:2), mobile money is the cornerstone of the economy because it empowers people and society. It contributes through:

- Direct contribution to mobile operators,
- Contribution to mobile ecosystem,
- Indirect impact on the broader economy, and
- Increase in productivity brought about by use of mobile technologies.

Gender, Mobile Money, and Development in Kenya

The mobile phone penetration in Kenya is more than 80 percent and can be equated with that of the developed world. Both men and women have access to mobile phones and those who do not have one can share with those who have. A study conducted in Machakos, Kenya, found that equal numbers of men and women owned mobile phones. (Wandibba et al, 2014). The argument in this chapter is not whether they have smart phones or ordinary phones, the issue is about access and use for personal, social or economic development. That is, they can call, send text messages, receive calls, take photos, access the internet and carry out mobile money transactions. To demonstrate this, see the following examples:

Rahab Kamau in Muranga County is 90 years old and owns a Nokia phone that has a camera and can use internet if need be. She communicates with her many grand-children who are scattered in many towns and cities in Kenya. She has basic education and is able to send and receive money through MPESA. She can discuss the Bible reading tasks with members of her church group and she is comfortable listening to Christian music one of the grandchildren uploaded for her.

Florence Anyango in Kisumu County is 65 years old and owns a Techno mobile phone. She is a retired primary school teacher and a part time dress maker. She can access the internet, instagram and WhatsApp through her mobile phone. Her clients can send photos of the dresses they want her to make for them and they pay her through MPESA. Just like Rahab, she uses her phone for other basic things such calling, sending and receiving messages.

Jane Njeri in Nairobi County is 30 years' old and a graduate of one of the public universities in Kenya. She owns two phones (private and business) and an iPad. She uses technology to carry out her online design business from the comfort of her apartment in Nairobi. The three women named above are of different ages, education background and capabilities are all empowered in different ways.What they have in common is that all own mobile phones.

James Mkulima in Mombasa County is a 28 –year old entrepreneur from Mombasa county who has two smart phones and a tablet that he uses to carry out his business transactions. He is available every hour because some of his clients are international. Because of access to ICT (mobile phone), he does not need to travel from Kenya to meet his clients because he does all that through his internet-enabled mobile phones.

The Africa landscape shows that women are left behind as consumers of mobile services but also as employers and leaders in the mobile industry. According to research by GSMA (2014), women are 14 percent less likely to own a mobile phone than a man because of finances, security and social norms that bar women from owning and using mobile phones. The situations in Kenya, Nigeria and South Africa are different in that men and women access and use mobile phones for personal and business purposes without a big gender gap as in other African countries.

While this may be correct in many African countries, the situation in South Africa is different because equal numbers of men and women have mobile phones (GSMA, 2015). In Kenya, the mobile phone penetration is more than 80 percent and what this means is that both men and women, young and old have access and use mobile phones; and out of the 37 million mobile phone subscribers, 30 million have mobile money accounts (Oteri,2015; Wandibba, et al, 2014).

What the mobile phone and the mobile money platform have allowed men and women to do on matters of empowerment is immense. There are different ways that the Kenyan societies especially the women have benefitted from mobile telephony. The women are able to freely communicate or transact business with others without supervision by the men. This is important in that in patriarchal societies, men may still want to limit access and usability of the mobile phone. Available research findings show that "although women are the primary bread winners, males continue to have an upper hand in technology access and use" (Bukesha et al, 2009' Ekine,2011; Wamala, 2010).

The other type of empowerment is that women have set up their own businesses that they run themselves. The mobile money has enabled them to join other citizens in getting involved in mobile phone related businesses. A look at Kenyan media reveal that the country has a robust landscape populated with young, well-educated entrepreneurs who own ICT-related businesses, while there are others who are in small and medium enterprises. Then there are those who own or run multi- million businesses. Whatever women are involved in results in personal, social or economic development (Lusike and Macharia, 2016).

The Kenyan landscape is different from what obtains in other African countries in that men and women have access and can use the mobile phones for whatever business they choose. This confirms research reported in Okello (2012) "that male-domination does not exist in some farming households, and access and use of mobile phones reflect the gender relationships existing within each household." The findings are important in that it is assumed that because the males have more access to mobile phones, then they have to dominate in everything ICT. According to GSMA (2010), gender gap 'suggests that the mobile phones in particular are helping women to save time, money, improve productivity, increase returns on investments and maximize household resources.' What this demonstrates is that some of the women have overcome some cultural norms that hinder them from participating in issues related to the ICT field. It shows that whatever women have they are making use of it.

Mobile Money Projects in Kenya

The mobile phone industry in Kenya has grown big because of its applicability to other industries thereby diversifying its functions. The largest of this is MPESA-platform that is owned by Safaricom service provider. The platform has grown from one function of offering money transfer services to diverse uses that are carried out through the mobile phone. According to the Daily Nation newspaper of 19th April, 2013, internationally, the mobile money transfer service has grown to 226 billion from 205 billion deposits from the previous period. Total African mobile money transfers are expected to exceed $200 billion in 2015, accounting for approximately 18% of the continent's GDP (GSMA, 2015).

For one to use MPESA service, one has to register with MPESA as a customer. One can send money to those who did not register with MPESA but have mobile phones. Because of ICT-Mobile telephone, Safaricom is able to conduct business with business organizations, schools, hospitals, airlines, restaurants, and even farmers. The following services are transacted through MPESA:

- Money transfer Service-This is transferring to and receiving money from other people.
- M-Shwari-Saving and borrowing money.
- M karo (mobile fees, karo is Kiswahili word for fees) service which is used by schools and colleges for parents to pay school fees. Parents do not have to go to school since they can use their mobile phones at home or work place.
- M –Health-this helps in facilitating health services. One can call the doctor, get a diagnosis, a prescription and pay the bills through MPESA.
- M-Commerce-people use MPESA to pay for goods and services. According to Aker and Mbiti (2010), MPESA has transitioned from pure money transfer service into platforms that allow hospitals, NGOS and other firms to send and receive money. For example, in various restaurants, shops, supermarkets and gas stations in Kenya, people can pay their bills using MPESA. The innovation has enabled Kenyans to pay utility bills, such as water and electricity bills through MPESA.
- E-Banking- Most customers use MPESA to deposit and withdraw money from their accounts. The banks are encouraging all mobile phone holders to have the mobile accounts that they can run from their phones.
- M-Agriculture- Kenyan men and women use the opportunities provided by MPESA to buy and send money to suppliers and markets in Nairobi and other parts of the country.

The mobile phones are used to render other services to people in Kenya and in Africa. Some of the unique services are:

Table 1. Other mobile phone services in Africa

Service	Country	Uses
M-agriculture	Kenya	Empowering small scale farmers
Internet	Most of African countries	Access internet
M-Kilimo	Kenya, Tanzania	Empowering small scale farmers
Equitel	Kenya	Money transfer and loan services
Stereo me	Nigeria	For education content
M Health	Kenya, Malawi, Tanzania, Rwanda, Malawi	Health delivery service
Mom Connect	South Africa	Pregnant mothers engagement with healthcare providers
M-Pedigree	Ghana	Determine authenticity of medication
M-Pewa	Tanzania	Save and borrow money
Mobile midwife	Nigeria	Sends messages to women and family
Hello doctor	South Africa	Discussion, diagnosis and treatment
Duma	Kenya	Connects job seekers with employers
I-Hub	Kenya	Assists businesses and offers financial support
CKW	Uganda	Connects farmers with experts
M-Education	African countries	To deliver education content
E-Learning	African countries	To deliver education content
E-Ticketing	Kenya	Provide E tickets to airline customers.

Source: Constructed by Author.

The services listed in the table above benefit both men and women. However, there are services that are meant only for women such as the issues of babies and pregnant women (m-Health). The fact that women have access to mobile phones is important enough because this gives them the opportunity to get information and to have their voices heard. (GSMA, 2014). A study reported by GSMA (2013) indicated that when women have mobile phones, it makes them feel" safer, connected and independent". The study also posited that their owning mobile phones unlocked economic opportunities for them.

To the women, mobile phones are more accessible that other channels in terms of cost, geographical region, coverage and usage (Aker & Mbiti, 2010). While research shows that both men and women have benefitted from mobile phone access and use, Oteri et al (2015) and Wakunuma (2012) argue that mobile phones have enabled women to establish and run small and medium enterprises. This is because the phones provide financial inclusion via mobile services. The women can transact their businesses through mobile phones without the hassle previously experienced by most of them who could not access banking services. Public opinion may be concerned about availability, but what is most important in Africa is what the women do with the ICTs (mobile phones) available to them. Therefore, the big picture is as important as the details of usability.

A study conducted in Machakos, Eastern Kenya by Wandibba et al (2014) found that financial services through mobile money have had a positive impact on women empowerment in rural areas. The study also established that when women were financially empowered, they were able to take up the roles which were culturally meant for men, which seemed to threaten men's position in the community. The most encouraging finding from this study was that more women than men sent and received more money in this county. However, the researchers also established that more men used more M-Banking than women. This can be explained by the fact that the study was carried out in rural Kenya.

The access and use of mobile phones and mobile money have contributed more to gender empowerment than other ICTs. For example, in Kenya, mobile phones have helped the other mass communication channels -television and radio- to communicate their messages better by having interactive radio and television programs where listeners and viewers can participate in a program in real time. Some radio and television stations allow their audiences and viewers to either call or send short messages during the programs.

This type of participation has encouraged such radio and television station owners to have live programs where experts in the studio provide knowledge and skills to the listeners and the viewers. The interaction through the platform of mobile phones is an inclusive method of sharing information with all interested parties. The most favoured method of communication is short text messages commonly known as SMS.

The mobile money MPESA has empowered communities by allowing men and women carry out projects that have had big impact on their communities. For example, in most Africa countries, food security is an issue that is always at the centre of gender discourse. Women in these countries are expected to feed the nations and their families yet the patriarchal systems of the communities have not empowered them to do so. They have less education than men and have little access to finances. According to Maree et al (2013), the use of mobile money has benefitted communities in Kenya and Uganda because of its products that have empowered women.

Wandibba et al (2014) and Iteri et al (2015) have demonstrated that the agricultural sectors in these two countries have benefited from mobile phone communication because information, knowledge and skills can be communicated through the mobile phone. Products such as KickStart 2013) and (Musoni 2012) have lifted communities out of poverty and more so women. Musoni is a cashless microfinance institution based in Nairobi, Kenya whereby members borrow money and act as guarantors for each other. Disbursements and repayments are made through mobile phones through the M-Pesa platform (Maree, 2013:11). On the other hand, Kickstart project helps members of the community to buy pumps and start businesses thus empowering the community to be in charge of their development.

The mobile money services have placed women in positions that have changed the lives of their communities by enabling them to be active participants in development in their regions. For example, Women of Uganda Network (WOUGNET) use mobile phones to create opportunities for themselves. The rural women-farmers in Uganda have received knowledge and skills through mobile phones in their own local languages. They have also used mobile money to transact agricultural businesses in the community (Maree, 2013). The WOUGNET members have made their communities food secure thanks to mobile money services.

With Kenya's mobile phone penetration of more than 80%, the Mininistry of Agriculture has decided to package information for use in the industry. The communication channel chosen is the mobile phone and language experts have been busy translating the content into Kishwahili, Kenya's national language. The mobile phone industry has empowered communication in ways that were unimaginable in the 20th century. According to GSMA (2015), mobile phone is addressing social, economic and environmental issues. This is done by delivering digital inclusion to the unconnected populations. However, African countries need to empower women more by providing more mobile phones to accelerate social and economic growth (GSMA, 2014:12).

Media, Agenda Setting, and Development

The media in many African countries, Kenya included, are male-owned and more often than not set male oriented agenda (Macharia, 2016a). Studies such as the one by the Media Council of Kenya (2012, 2011 & 2005) depict them as the most trusted channels by the receivers of the messages. What this means is that the media have performed their function as media researchers; Maxwell MCCombs and Donald Shaw envisaged that "the mass media are… stunningly successful in telling us what to think about" (Gamble & Gamble, 2003:351).

What the quotation above demonstrates is that even when we are dealing with matters of development in relation to gender empowerment, the media will have played an important role in communicating messages that give the perception that men are active participants while women are passive receivers of development. This is achieved in the way the media report news, events and other activities. Studies conducted in Kenya by Media Council of Kenya (2012), AAJA (2008) and AWC (2014) reveal that most of the stories covered by broadcast media, newspapers, magazines, and the internet were about men, what they did, said or achieved in the public domain. Women appeared on the periphery and were not the central figures in the story, event or the lead character in the movie. Majority of men were depicted as sources of information while women were seen as mere receivers of the media messages.

With the expansion of the ICTs in Africa, one would have expected this to provide more room for both men and women to equally participate in communicating development messages, but we are reminded that even with availability of the ICTs, women have less access than men and that is why most of the messages communicated through television or the internet will not reach them.

The media have a tendency to make important issues that have to do with the males more important than others. For example, there are many programs aired on TV in Kenya that educate the public on the diagnosis and treatment of prostate cancer than there are about cervical and ovarian cancers that kill more women than men (INOORO TV, 2016). The media in their educative function have the power to provide information that can lead to gender empowerment in Kenya and this can only happen when they set the agenda to pursue the development agenda that concerns itself with both males and females in the countries (Dominick, 2005; Macharia, 2016).

The media in their diverse roles have to consider the diversity of the countries' needs by taking into consideration that men and women have different needs, viewpoints, knowledge and concerns (UNESCO, 2003). More often than not the media messages portray half of the population that feeds the nations negatively. The images of women and children in hunger-stricken, war prone countries boldly grace our newspapers, magazines, television screens, the internet, social media and radio stations. When countries are involved in conflict or have been destroyed by natural disasters such earth quakes, both men and women are affected but the media hardly depict men as victims.

The media's educative role are expected to without discrimination engage in communication for development whereby both men and women participate in development. However, most of the times, the media hardly consult women as experts or stakeholders in their respective countries especially in Africa. One would be tempted to think that Africa has no female professionals because they are not visible(AMWIK,2015 & Macharia,2016a).Kenya despite her vibrant media still perpetuate the view that women should concern themselves with social issues while men remain in the public domain, engaged in development activities.

UNESCO (2003c:29) acknowledges that "women are excluded from opportunities to acquire knowledge and skills, access to resources, enhance social status and obtain status and power" The media should be the ones to make men's and women's contribution to social economic development visible to all.

As the media involve themselves in communicating positive messages, the focus should be on the role women play when they have access and use ICTs-the mobile phone and mobile money services. The development of mobile money and its empowering role are important factors to enable us to change the social relations that generate inequality without marginalizing the men (UNESCO, 2003).

Challenges of Access and Use of the Mobile Phones in Kenya

- Despite the gender empowerment demonstrated by access and use of mobile phones, the African media continue to portray women as powerless and voiceless victims. They fail to communicate the positive aspects of development brought about by ICTs and mobile telephone.
- Every technology is a double-edged sword. The ease with which mobile phone communication takes place has made crooks and other criminals to take advantage of it. Some people have lost money to criminals who called or sent a text message to their victim informing them of winning some amount of money. The receivers are directed on how to claim the money they have won. In the process, they reveal their personal identification details which the crooks use to channel money from their phones.
- Both men and women are using the mobile phone not for development but for harassing or threatening others. In one private university in Kenya, staff precipitated the resignation of a Vice Chancellor by spreading malicious information about his job performance through social media.
- Employees in companies and other institution are spending a lot of their employers' time searching the internet or chatting on social media with their friends from their smart phones.
- Frauds in the banking Industry-E banking –The mobile phone technology has empowered criminals to use the same technology to steal money directly from customers or the banks. Between 2011 and 2015, many banks lost millions of Kenya shillings to these frauds. As a mitigating factor, most of them changed the Automated Teller Machine cards to make it difficult for the crooks to steal money from ATMs. However, the super hackers evolve everyday and the problem is still disturbing the banks and other financial institutions.
- The other industries have also been affected by these undesirable behavior. Many MPESA operators have also lost money to criminals who give fake money to unsuspecting business men and women, leading to closures of some businesses. Despite enhanced security in the country, MPESA and other mobile money agents are targeted by armed robbers.
- Kenya has had her share of accidents caused by drivers talking on phones or reading or writing text messages while driving. The penalties are very high for this kind of crime but by the time the results of this behavior are seen; people will have died or been maimed for life.
- Wandibba et al's (2014) study in Machakos Kenya found that the mobile phones have impacted the family both positively and negatively. There have been cases of family relationships broken or weakened because women have been empowered by access and use of mobile money, thus threatening the gender roles in the community. The situation in Machakos may be an isolated case where the men feel threatened when both groups participate in development.
- The mobile phone's high penetration is having negative effects on low income groups who choose to buy air time for their mobile phones instead of spending the money on basic needs.

SUGGESTIONS FOR RESEARCH

With the growth of the telecommunication field, the service providers and the academic institutions have yet to partner to conduct research in this field to enable the consumers access more benefits than is the case currently.

Findings of research conducted in universities remain in the schools and are not shared with the consumers or with those innovators who can come up with more innovative ways of improving mobile phone use. The service providers in my opinion should partner with scientists, the governments and the universities to fund research on finding new ways of enhancing mobile phone use and methods of curbing crime committed using mobile phones. This way, Africa will achieve more economic, political and social development in the 21st century.

CONCLUSION

The access and use of ICTs, specifically the mobile phones, have brought about both personal and national development. The mobile phones have also played a positive role by empowering communities, making it possible for men and women to start and operate businesses through the MPESA platform and other mobile money services, and to transact businesses with individuals and organizations.

The mobile telephone is helping to fuel growth on the continent by equipping citizens with the tools they need to improve their lives in ways never possible before. For example, parents with children in schools pay fees using M-KARO;other organizations such as Kenya Power (supplying electricity) and Nairobi Water have benefitted from the mobile phone technology because their customers pay all their bills through MPESA, making it easy for them to collect revenue. Communication between people without computers has improved because with their mobile phones, they can access the internet and other services such as E-banking and E commerce, and save and borrow money through M-Shwari.

This chapter has demonstrated the centrality of the media to gender socialization and development. The way the media portray men and women is important in that the media set the agenda for the society. When media messages encourage participation of both men and women in social economic development, it will be because in most countries, people trust the media more that they trust politicians and other leaders.

The mobile money services offered by Safaricom in Kenya and others in East Africa have empowered women to participate in development. In other words, mobile telephony has demystified the role of both women and men in development. When people use mobile money services, it does not matter whether they are men or women. This is demonstrated in many projects in Kenya that are mobile money related.

RECOMMENDATIONS

- The media should present a balanced view of men and women as they participate in social and development activities in their countries.
- The media also have a role to play in positively socializing the young people because negative strategies only damage the future of nations.

- Countries should improve on their gender policies to streamline access and use especially in middle income economies.
- To safe guard the benefits of mobile telephony specifically mobile money, Kenya should enact laws that protect people from criminals and other negative elements.
- African countries should support the small and medium enterprises started and run by women because these are the bedrock of development.
- The Kenya government has put in place rules and regulations to minimize fraud, mobile phone threats and what is termed as offensive messages or hate speech.

Mobile phone users should cultivate appropriate phone use etiquette. A study conducted by Pew et al, quoted in Tubbs and Moss (2008), found that people are likely to lie about their whereabouts and that "cell phone use is changing expectations about when and how others are available to us." It becomes a challenge when users demand immediate feedback even when time is not appropriate.

Finally, African countries should encourage more innovations to spearhead development in the middle income countries such as Kenya and others.

REFERENCES

Accenture, V. (2011). *Connected agriculture: The role of mobile in driving efficiency and sustainability in the food and agriculture value chain.* Berkshire: Vodafone.

Africa Partnership Forum. (2007). *Gender and economic empowerment of women, APF Support Unit and NEPAD, Briefing Paper 3.* Retrieved from http://www.oecd.org/dataoecd/16/42/39921766.pdf

Aker, J. C., & Mbiti, I. M. (2010). Mobile phones and economic development in Africa. *The Journal of Economic Perspectives, 24*(3), 207–232. doi:10.1257/jep.24.3.207

Amobi, I. (2013). *Portrayal and participation of women in Nigerian media.* Retrieved from https://profreri.wordpress.com/2013/07/27

Association of Media Women in Kenya. (2015). Media monitoring analysis of the coverage of the 2/3rds gender principle. Nairobi, Kenya: Author.

Bakesha, S., Nakafeero, A., & Okello, D. (2009). ICTs as agents of change: A case of grassroots women entrepreneurs in Uganda. In I. Buskens & A. Webb (Eds.), *African Women & ICTs: Investigating Technology, Gender and Empowerment.* Pretoria: Unisa Press.

Barrientos, S., Gereffi, G., & Rossi, A. (2010). *Economic and social upgrading in global production networks: Developing a framework for analysis.* Capturing the Gains Working Paper 2010/03.

Bowen, H. (2010). *Information at the grassroots: Analyzing the media use and communication habits of Kenyans.* Washington, DC: InterMedia Survey Institute. Retrieved from http://audiencescapes.org/sites/default/files/AudienceScapes%20/Kenya%20Survey%20Research%20Report%202010.pdf

Bratton, M. (2013). Briefing: Citizens and ccll phones in Africa. *African Affairs, 112*(447), 304–319. doi:10.1093/afraf/adt004

Buskens, I., & Webb, A. (2009). *African women and ICTs: Investigating technology, gender and empowerment*. Zed Books.

Castells, M., Fernandez-Ardevol, M., Linchuan Qiu, J., & Sey, A. (2007). *Mobile communication and society: A global perspective*. Cambridge, MA: MIT Press.

CCK (Communications Commission Kenya). (2012). *Quarterly sector statistics report: Third quarter of the financial year 2011/12 (January-March 2012)*. Available at http://www.cck.go.ke/resc/downloads/SECTOR_STATISTICS_REPORT_Q3_11-12.pdf

Chen, D. H. C. (2004). *Gender Equity and Economic Development. The Role for ICTS*. Washington, DC: World Bank Working Paper 3285.

Daly, J. A. (2003). *ICT, gender equality, and empowering women*. Retrieved from www.development-gateway.org/node/133831/sdm/blob?pid=5233

Dominick, J. A. (2005). *Dynamics of Mass Communication. Media in the Digital Age*. Boston: Mc Graw Hill.

Galvin, K. M., & Terrell, J. (2009). *Communication works. Communication applications in the workplace*. London: National Text Book Company.

Gamble, T. K., & Gamble, M. W. (2003). *The gender communication connection*. Boston: Houghton Mifflin Company.

Githui, D. M. (2011). Mobile money transfer in Kenya: An ethical perspective in research. *Journal of Finance and Accounting*, *2*(2).

Global Information Society Watch. (2013). *Report on Women's Rights, Gender and ICTs*. Association for Progressive Communication (APC) and Humanist Institute for Cooperation with Developing Countries (Hivos). Retrieved from http://www.giswatch.org/2013-womens-rights-gender-and-icts

Global Media Monitoring Project. (2010). *Who makes the news, 2010 report highlights*. Toronto: The World Association for Christian Communication. Retrieved from http://www.whomakesthenews.org/images/reports_2010/highlights/highlights_en.pdf

Griffin, E. M. (2012). *A first Look at Communication Theory*. McGraw Hill Higher Education.

GSMA. (2014). *Women and Mobile: A Global Opportunity: A Study on the Mobile Phone Gender Gap in Low and Middle Income Countries*. Retrieved from http://www.gsma.com/mobilefordevelopment/wpcontent/ uploads/2014/01/GSMA_Women_and_Mobile-A_Global_Opportunity.pdf

GSMA. (2014). *The Role of Mobile Economy in Sub Saharan Africa*. GSMA.

GSMA. (2015). *The Mobile Economy. Empowering People and Society*. GSMA Intelligence.

Horst, M., & Miller, D. (2006). *The Cell Phone: An Anthropology of Communication*. Oxford, UK: Oxford International Publishers.

Hughes, N., & Lonie, S. (2007). M-Pesa: Mobile money for the "unbanked" turning cell phones into 24-hour tellers in Kenya. *Innovations*, 66–81.

Inooro TV. (2016). *Dagitari* [Doctor]. Author.

Jahannessan, J. (2006). *The Role of the Media in the Cultural Struggle of Gender Transformation in Tanzania* (PhD Thesis). Norwegian University of Science and Technology.

KickStart. (2013). Retrieved from http://www.kickstart.org/

Kiiti, N., & Mutinda, J. (2011). *Mobile Money Services and Poverty Reduction: A study of Women's Groups in Rural Eastern Kenya*. Academic Press.

Lee, J., & Gereffi, G. (2013). *The co-evolution of concentration in mobile phone global value chains and its impact on social upgrading in developing countries*. Capturing the Gains Working Paper 2013/25.

Ling, R., & Horst, H. (2011). Mobile Communication in the Global South. *New Media & Society*, *13*(3), 363–374. doi:10.1177/1461444810393899

Lusike, L., & Macharia, J. (Eds.). (2016). *Political Influence of the Media in Developing Countries*. IGI Global Publishers.

Macharia, J. W. (2016a). *Gendered African Media: The Kenyan Experience. In Political Influence of the Media in Developing Countries*. IGI Global Publishers.

Macharia, J. W. (2016b). The Portrayal of Men and Women in the Media: The African Landscape. *Journal of Communication and Media Research, 8*(1).

Maree, J., Piontak, R., Omwansa, T., Shinyekwa, I., & Njenga, K. (2013). *Developmental uses of mobile phones in Kenya and Uganda.* Manchester, UK: Capturing the Gains.

Mathenge, D. G. (2011). Mobile Money Transfer in Kenya: An Ethical Perspective. *Research Journal of Finance and Accounting, 2*(1). Retrieved from www.iiste.org

Mbogo, M. (2010). The impact of mobile payments on the success and growth of micro-business: The case of M-Pesa in Kenya. *Journal of Language Technology and Entrepreneurship in Africa*, *2*(1), 182–203.

Media Council of Kenya. (2005). *Baseline Survey of Media in Kenya*. Author.

Media Council of Kenya. (2011). *The Performance of Vernacular Radio Stations in Kenya Monitoring Report*. Nairobi: Media Council of Kenya.

Media Council of Kenya. (2012). *The Pre-election Coverage by Kenyan Newspapers Monitoring Report*. Nairobi: Media Council of Kenya.

Musoni. (2012). *Musoni next generation microFinance*. Available at http://www.musoni.eu/

Muthiora, B. (2015). Enabling mobile money policies in Kenya: Fostering a digital financial revolution. In GSMA, 2015.

Okello, D. (2012). *Mobile phones breakout group discussion*. Capturing the gains East African Workshop, Nairobi, Kenya.

Oteri, O. M., Kibeti, L. P., & Ndungu, E. N. (2015). Mobile Subscription, penetration and coverage trends in Kenya telecommunication sector. *International Journal of Artificial Intelligence*, *4*(1).

Research I. C. T. Africa. (2013). *Lifting the Gender Veil on ICT Indicators in Africa.* Accessed on 25 May 2014. http://www.cprsouth.org/wpcontent/uploads/2013/07/Presentation_CPRsouth-8%EF%80%A2CPRafrica-2013_23_08_2013.pdf

Safaricom. (2012). *Celebrating five years of changing lives.* Available at http://www.safaricom.co.ke/mpesa_timeline/timeline.html

Safaricom. (2013). *M-Pesa agents.* Available at http://www.safaricom.co.ke/personal/m-pesa/m-pesa-agents

Tubbs, S., & Moss, S. (2008). *Human communication: Principles and contexts.* Boston: Mc Graw Hill.

Ugangu, W. (2012). *Normative media theory and the rethinking of the role of the Kenyan media in a changing social economic context* (Doctoral dissertation). University of South Africa. Retrieved from http://uir. unisa.ac.za/bitstream/handle/10500/8606/thesis_ugangu_w.pdf?sequence=1

UNESCO. (2003C). *Gender Issues in information society.* Paris: UNESCO.

United Nations. (2014). Measuring ICT and gender: An assessment. New York: UN.

Wakunuma, K. (2012). Mobile phones for development: Are we in danger of losing sight of the bigger Picture. *Standpoint*, 8.

Wamala, C. (2010). *Does IT count? Complexities between access to and use of information technologies among Uganda's farmers* (PhD thesis). Luleå University of Technology, Sweden.

Wandibba, S., Nangendo, S. M., & Mulemi, B. A. (2014). *Gender empowerment and access to financial services in Machakos County.* Academic Press.

Wood, J. (2011). *Communication, gender and culture.* Belmont: Thomson Wadsworth.

World Bank. (2003a). *Engendering ICT: Ensuring gender equality in ICT for development.* Washington, DC: Author.

World Bank. (2003b). *ICT and MDGs: A World Bank Group Perspective.* Author.

World Bank Group. (2006). *2006 Information and communications for development. Global trends and policies.* Washington, DC: World Bank.

World Bank Group Strategy. (2012-15). *ICT for greater development impact.* Washington, DC: Author.

KEY TERMS AND DEFINITIONS

Development: Growth, changing from one state to a higher state.

Empowerment: The situation where one can participate in, economic activities, decision making and have power over economic resources.

Gender: Societal construct of the roles of men and women in the society.

Media: Communication channels such as radio, newspapers, Tv, internet.

Mobile Telephony: Provision of telephone services from phones that are not fixed.

Socialization: The process through which the society passes cultural values to the young ones.

Chapter 2
Modern Information Technology, Global Risk, and the Challenges of Crime in the Era of Late Modernity

Ajibade Ebenezer Jegede
Covenant University, Nigeria

ABSTRACT

The rapidity and efficient use of most communication technologies remain the driver of accelerated developments across the major societies of the world today. This is quite exciting when compared to what existed in the past. However, current happenings indicate that the contribution of these technologies to sporadic development of nations is fraught with recordable socio-economic risks whose effect is unprecedented and affecting the nature of trust required for social continuity in human environment. Consequently, this paper considers the nature of risks and vulnerabilities affecting e-connectivity from a modernist theoretical perspective and contextualized this in the double edged implication affecting the use of the Internet. The first section of this paper is devoted to review on the nature of affinity between the Internet and crime while the final section engages the empirical analysis of secondary data on the consequences of cyber-crime to the global economy.

INTRODUCTION

Crime in the modern society is identified as a social problem of a major concern. There has always been a number of humans that are bent in their ethics, morals, socio-psychological makeup, or sense of justice, and are ready, willing, and able to commit crimes of all types (Singleton & Singleton, 2010). Apparently, Graham (1993) posits that the world is thickly populated with minds and taking cognizance of the thickness of both "good" and "bad" minds in human environment, significant efforts and resources are generally committed by government and other stakeholders toward crime control or eradication and equally devoted to ordering human relationship (NCPI, 2001). Therefore, since the beginning of recorded

DOI: 10.4018/978-1-5225-1859-4.ch002

history, violations of the social and legal code by individuals or groups have been formally proscribed (Hughes, 2004; Garland, 2001; Muncie & McLaughlin, 2001; Sarri, 1995). Despite the formal proscription of criminal acts across national boundaries, the magnitude and the character of old and newly evolving crime remain alarming (Stark, 2004; Packer, 1968). Many factors can conveniently be attributed to this development but the most central which is the focus of this paper is the role of communication technology in the promotion of criminality. In the global arena, rapid advances in technology and consumer demands have created greater access to cyber communications (McShane & Williams III, 2007:147). Modern breakthroughs in the technological world is not completely insulated from the shortcomings that hitherto attended traditional technologies. Discoveries from time immemorial have been fraught with one anomaly or the other both helping to solve human challenges and substituting same for higher if not tougher challenges. The advent of diverse technologies into societies worldwide is laden with both positive and negative impacts entrenched in socio-economic related challenges existing in all known society. Consequently, current reflection is closely linked to the duality of developments attendant to information and business technologies that are jointly enhancing global economy and breeding new crimes. Global experiences have basically show that the ICTs allow legitimate global economies and knowledge networks to exist, they also allow for a darker side that sees the illegitimate use of information networks (Castells, 1998; Williams, 2006:17). Put succinctly, just as the ICT revolution facilitates international and local businesses, it also creates a threat to social/corporate trust and transactional relationships. In line with these known attributes therefore, the focus on ICTs in this analytic discourse is restricted to the examination of the Internet, a technology with a capacity viewed as both enabling, disenabling, fostering human progress and generating vices. Despite the Internet's instrumentality to the spread of globalization, its existence has simultaneously nurtured existing crimes and as well intensify the evolution of yet to be known crimes. Cyber-crimes are growing and by 2017, the global cyber security market is expected to skyrocket to $120.1billion from $63.7billion in 2011(cigionline, 2015). Consequently, this paper reflects on the role of the Internet in rising profile of crime and evaluates the resultant risks affecting interpersonal relationship in the global community with the view of arresting the situation.

Brief History and Locus of Use of the Internet

It is important to know that technology in distant past has historically been shaped by governmental, military and commercial interests or requirements, but the development of the internet is increasingly driven by its users (Green, 2010). The Internet is conceived as a rapidly expanding system of networks linking millions of people in new spaces and changing the way we think, the nature of our sexuality, the form of our communities, our very identities'(Turkle, 1995: 9). Five billion devices currently access the Internet and by the end of 2020 it is projected that there will be fifty billion users that will hook to the Internet for the information requirements (New Zealand Management, 2013). In the context of the massive web underpinning Internet use, Turkle posits that the Age of the Internet 'hasushered in the opportunity to build virtual communities in which all will be able to participate with people from all over the world, people with whom we exchange information daily, people with potentials of business partnership with the appearance of whom are capable of generating interactive session devoid of physical contact (Turkle, 1995:10). As at the end of 2001 the number of Internet users in the world has crossed the threshold of 500 million (up from 16 million in 1996) and in North America and Scandinavia over 60 percent of the population has access to the Internet (Nua, 2002). The data on the use of the Internet followed the same pattern for most advance countries such as the USA, Britain and Canada with the utilization of e-mail topping the list of use. Tracing the history of the current development that necessitated the adoption and simplicity of the use of e-mail as one of the derivatives of the Internet, Winston (1998:330) argues that the creation of designated file where messages between users could be left had been developing since the early 1960's.

Reception of messages on a relatively large scale took off around 1972. Ray Tomlinson became the first programmer that added @ to e-mail address thereby making today web addresses to be so patterned (Winston, 1998:330). Commercial Internet exchange was established in the year 1991 and was later followed by Tim Banners Lee protocol that created World Wide Web which entered into full business in 1992 giving the Internet wider coverage and multitude of users (Dufour, 1995:39). The advent of browsers, beginning with Mosaic in 1993 (then Netscape), transformed the Internet from a platform for geeks to a new tool for the world. In 1995 there were approximately 5 million users. In the autumn of 1995 Microsoft Explorer came out. By 1996 there were 50 million users. By 2000, there were 211 million of beneficiaries. The so-called dot.com bust did not change this trend. By December 2004, there were 812 million users. By the end of December 2005, there were 1,018,057,389 users, according to World Internet Usage Statistics. Jordan (1999: 38) rightly pointed to the fact that the e-mail was the first sub media to spring up with the APRANET thus instantiating the domestic usage of the Internet. Similarly, Winston asserts that from the very beginning it has been clear that the most unambiguously valuable facility provided by the net is e-mail (Winston, 1996:335). Connectivity was registered to involve person to person interaction at the commencement of Internet use by the general public. People connect to people using computers, a development which gave rise to the over-arching term computer mediated communication. It is within the context of the general concept of mediation that the illicit manipulations of the Internet for crime pursued in the current paper rest solidly. By July 1995 there were supposedly anything from 6.5 million machines worldwide to 10.3 million in the US alone connected to the net (Winston, 1998:334). In 2011, more than one third of the world's total population had access to the Internet while over 60% of all Internet users are in developed countries, with 45% of all Internet users below the age of 25 years (UNODC, 2013).

Wellman and Haythornwaite (2002) report that the e-mail is rapidly more used in the global environment than the telephone. This is obvious since the monetary cost of sending an e-mail is near zero. People now spend more time online and use Internet more often (Macaluso, 2001). Research indicates that e-mail and searching for information take high priority in Internet time (UCLA CCP, 2001). In its estimate, over 80% of Internet traffic is implicated in sending e-mail messages and users rank e-mail as number one reason for being online (Katz and Aspden, 1997). Two third of users also consider the information function of the Internet as extremely important. Statistics shows that well over 80% of users engage web surfing and browsing on daily basis (UCLA CCP, 2000). Information search linked with youth's activities online consists of catching the fun and accessing game. Large proportion of Internet users also hook on the net for e-commerce purposes. Statistics report that e-commerce users range between 36-51% of users in the United States alone, 33% in Britain and 19% in Canada between 1999-2000 (Nie & Erbring, 2000; UCLA CCP, 2000; National Statistics Omnibus, 2000; Statistic Canada, 2000). Just as the users of in the arena of e-commerce is increasing so also the threat posed by cyber-crimes and the nature of risks. Rotich et al, (2014:13) assert that the threats posed to organizations by cyber-crimes have increased faster than the rate which potential victims or cyber security professionals can deal with and thus placing targeted organizations at considerable risk. With the high intensity of the application of the Internet to the business arena, more and more violation of business ethics became a common place and in the same way, risks of doing business was heightened and the level of victimization rose astronomically. Correspondingly, there is a disconnect between the hitherto known institutionalized trust in the public area and the rising artificial trust in the Internet driven world. The nature of risk inherent in this age of late capitalism calls to question the basis of our coexistence and the continuity of our collective agreement, and thus, generating an overtly non predictable modernity. The modernity far beyond our comprehension and both challenging us and threatening the reason for our existence.

Theoretical Position: Giddens and Beck Reflexive Modernity

The advent of the Internet technology ushered in both globalization and dis-embeddedness of social relation. Giddens analyzed disembedding in the context of time-space distanciation and that which involves the “lifting out” of social relations from local context of interaction. This allows people to relate to and interact with others who do not live in the same area (Haralambos et al, 2004:980). Due to the contraction of time and space which necessitated unimpeded socio-economic interactions globally, life in the stage of modernity then rest solely on absolute trust. This environment of trust operates under a capitalist milieu that is restless and considerable unstable in terms of human capacities to predict with some level of exactitude their stake in the scheme of things. Modernity then became a lived experience for individuals and which open up new opportunities for the pursuit of any avowed area of economic interests. The flexibility promoted by modernism also entrenches high consequence risks for socio-economic interaction and which threatens the basis of human corporate existence. Risk in the period of modernity has been co-opted as a term reserved for a negative or undesirable outcome and hence synonymous with the term danger or hazard (Fox, 1999:12). In analyzing the synergy between modernity and enormity of risks in today world therefore, one would require to focus on a specific area of research interest. The necessity of doing this had earlier been reported by Dean (1999) when he posited that the significance of risk does not lie with either observable or calculable risk itself but what risk gets attached to. Most central is the risks resulting from the increasing interdependence between local and global processes and institutions which defines globalization (Dodd & Hutter, 2000). In the current discourse, risk is closely enmeshed with the discovery and the mass usage of the e-mailing facility linked with the Internet technology. And toeing the line of thought of Beck (1992) in this context, there are proliferation of risks as a consequence of technological innovation and such scientific breakthroughs most often makes the risks encounterable by man get out of control. Although Beck asserts that most risks only exist in terms of the knowledge about them and their anticipated effects are open to social definition and construction, his summation nevertheless failed to capture the calculable monetary and material costs of risks linkable to displacement of trust and other cyber related crimes. Risks in the context of crimes are commoditized, made tangible and with overt consequences. Risk in today world generates its material existence in fear, socio-psychological emotions and in disincentiveness to morality and engagement in the global business. With the lifting out of social relationship from the scene of event of economic activities, vulnerabilities become high and the outcome has the potential of proving debilitative to both personal and group expectations. This corresponds to what Beck aptly observed that the society becomes a problem for itself; a problem to the extent that it attempted to make life easier (through scientific breakthroughs fashioned out in technologies and unrestricted socio-economic interaction) but in the process putting a lot of people into jeopardy of either losing their moral acumen, lives or dispossessing them of their scarce resources as in the case of fraud. The age of the Internet ushered in conveniences in time and space distanciation and went ahead to implant monetary implicated insecurities. It questions the basis of our continued interaction in the face of myriad of uncertainty. Enclosed in the world of risk are the challenges of crime defying legal comprehension that was ever engaged in human history. The Internet technology has single handedly promoted and expanded the web of crime and its attendant risks thus as it were the dislocation of all known social agreement. In part, its existence portends progress and in the other, quantifiable threat to socio-economic interaction and hence the requirement for an analytic attempt of its dual contribution to human civilization.

The Dual Character of the Internet in the Global Interactive Environment

The contribution of the Internet technology to the robustness of human interactive environment have been examined by diverse scholars in the social sciences. The medium functions fo social, political and economic purposes. While some researchers eulogized its existence as advantageous to social and modern trade cum global interactional flows, few others have chronologically exposed its negative effects on human relationship. But in summary, what appears outstanding in this era is the uses into which the Internet technology is adapted which determines its effect on human interaction. On a positive note, the Internet is vast becoming part of the fabric of our lives, not only in advanced societies but in the core activities and the dominant social groups in most of the world (Castells, 2002:xxix). Life without the internet seems almost unimaginable for most people globally. The Internet stands out as a comprehensive techno-science world which exemplifies cyberspace'(Homes, 2005: 50). Castel (2001) also describes the Internet as galaxy of regulated sub media and a model for cyber space relations. Friedman also posited that the modern world consists of flattering forces empowering more and more individuals today to reach farther, faster, deeper and cheaper than ever before thereby generating equalization of power and opportunities and also giving so many more people the tools and ability to connect, compete and collaborate. Clough (2010:184) equally gave the market implications of the Internet and he describes it as a large marketplace. He proceeded to give an account of the volumes of trade that took place through the Internet. In his account, in 2003, 17.6% of adults in the United States conducted banking online and 32.3% purchased a product or service via the Internet. In Canada, 44% of adults had purchased goods or services over the Internet in 2003, while in the UK the figure in 2009 was 80%, with 55% banking online. In excess of 3.8 billion transactions were made on Australian issued cards in the twelve months to 30 June 2008. This activity translates to a lot of virtual 'money being transferred online. In 2003, Canadians spent an estimated CDN$3,034,000,000 in electronic commerce, while in the UK, the estimated value of Internet sales by business in 2005 was £103.3 billion. According to the US Census Bureau, the seasonally adjusted estimated value of US retail e-commerce sales for the fourth quarter of 2007 was US$36.2 billion. Finally, Winston and Walton (1996:82) attributed the increase in the use of the Internet to the instantaneous effects deducible in messaging. They posited that data transmission times have dropped so that it takes only 2/3 of a second, for example for e-mail message from the US to Antarctica. Such connections, competition, collaborations and time economy are not without cost. Regarding the negative effects of the Internet, Clough (2010:3) juxtaposed between the positive and negative dimension of the global connectivity when he asserts that "we live in a digital age, the proliferation of digital technology and the convergence of computing and communication devices, has transformed the way in which we socialize and do business.

The pervasive existence of the digital technology in its global context is overwhelmingly positive but it however possesses qualitative dark side to these developments. This further reinforces the maxim that crime follows opportunity and that every advancement will sundry times be accompanied by a corresponding niche to be exploited for criminal purposes. "Many of these cybercriminals see the Internet as a job opportunity," says Thomas. "With low employment, they can use their technical skills to feed their family" (Thomas, 2006). Technological innovations across ages have been fraught with illicit applications and the recent development represented in Internet crime conforms with what is known in research. Today the internet is intrinsic to media and communication, entertainment, politics, defense, business, banking, education and administrative systems as well as to social interaction.

Internet and Fraud in the Late Modernity

Varied reasons have been adduced for the advent of the Internet. The successes recorded in the field of Internet fraud are made possible by the opportunity the medium has for users especially that of operating under concealed identity. It is important to note that the Internet offers its users tremendous opportunities to experiment with their identities and that several scholars have pointed out this fact (Katz & Rice, 2002; Rheingold, 1993; Smith & Kollock, 1999; Turkle, 1995; Wallace, 1999). The executive Vice Chairman, Nigerian Communications Commission Ndukwe (2009) asserts that the structural embodiment of the Internet possesses the defective attributes of openness, inventiveness and over reliance on assumption of the goodwill it offers the world. Popular discussions of the Internet, for example, veer between celebration and paranoia; on the one hand, the technology is seen to create new forms of community and civic life and to offer immense resources for personal liberation and empowerment, while on the other, it is believed to pose dangers to privacy, to create new forms of inequality and commercial exploitation, as well as leaving the individual prey to addiction and pornography. cyberspace in general, and the Internet in particular, are notoriously vulnerable to a frightening and expanding range of accidents and attacks by a spectrum of hackers, criminals, terrorists, and state actors who have been empowered by unprecedented access to more people and organizations than has ever been the case with any infrastructure in history (Goodman & Lin, 2007:xi). Ass et al (2009:1) also argue that inherent duality and dialectic exist between our search for security latently (arising from technological uses) and the manifest tendency to generate simultaneously insecurity (crime and anti-human relationship) in the process. There are two major areas with security implications in Internet use. The first involves the scares over the abuse of credit card details, identity theft and other forms of scamming which are all facilitated by a lack of physiological grounding of on-line interactions and exchanges. Meridien research (2000) has earlier reported that fraud loss accruing to credit card industry amount to $1.5 billion annually, of which $230 million is often traceable to online transactions. With the growth in online banking also, there is a correspondent increase in economic related crimes. Electronic mediated embezzlement became unprecedented and account take over by fraudulent users also increased. And the second being the concerns about contracting computer viruses, Trojans and worms, which lurk beneath deceptive programmes, which lure unsuspected Internet users into rendering their computers completely vulnerable to destruction, spyware or other forms of infiltration (Van Loon, 2008:122). Perhaps the most insidious type of attack prone crime that exists today is the *botnet*. In short, a botnet involves remote control of a collection of compromised end-user machines, usually broadband connected PCs. The controlled end-user machines, which are referred to as *bots,* are programmed to attack some target that is designated by the botnet controller. The attack is tough to stop because end-user machines are typically administered in an ineffective manner. Furthermore, once the attack begins, it occurs from sources potentially scattered across geographic, political, and service provider boundaries. Perhaps worse, bots are programmed to take commands from multiple controller systems, so any attempts to destroy a given controller result in the bots simply homing to another one (Amoroso, 2011:7). It is vividly true that the need for social comfort and the drive towards ease in meeting life challenges occasioned the emergence of the Internet; the technology is fashioned to foster socio-economic interaction on a global scale. The technology is seen as having the potential of liberating individuals from constraint and from narrowly hierarchical ways of working. Contrary to this, the Internet is regarded as a false substitute for the supposedly authentic values it is seen to be replacing (Buckingham, 2008:11). Yar (2009:190) rightly puts it that ——the growing societal dependence upon and use of computerized systems, and not least the centrality of electronic communication networks such as the internet, has brought in its wake a host of (real or perceived) insecurities. The Internet has brought diverse crimes such as pornography, cyber stalking, intellectual property theft and privacy, hate speech, identity theft and fraud. Internet remains vulnerable in recent times because it has become so central to the development and delivery of goods and services. The nature of modern economy promotes acute exclusion of many people and in the process producing the class of economic destitute.

Metaphorically, modern destitute are left bowling alone under the clouds of economic uncertainty; a climate consistently nurturing individualistically created meanings centered on the quest for survival at the expense of the collective. In realism, the Internet remained the saving grace of the oppressed group for expression and the mediator of the chaotic economic field for the downtrodden population who opted for crime and fraud as a means of making ends meet. When rephrased and drawing a departure from what Castells initially conceived therefore, the Internet represents a new form of connection outside the institutional realm of modern failing states. A succor for modern cyber criminals, terrorists, disillusioned people and other activists. It supplies the missing link between what constitute the needs of the economically battered or yearning populace on one hand and the limitation of failing state to deliver on the other hand.

Reviewing the major indicators explaining the prevalence of cyber-crime or fraud for instance in the global context, it is clearly evident that significant response points to a single direction, joblessness, destitution, falling standards of living and disillusionment with the modern capitalist arrangement. Significant states in the Third World is completely implicated and in a way they are to be prosecuted internationally peradventure if they are not willing or ready to address the plights of their citizens. Noticeably, this stance is not only applicable to a singled out country but to several other nations whose citizens are barely existing. In fact, if most developing nations would continue to run their citizens as machines and not mind the implication of this for their existence, the world is yet to harvest greater consequences than what is being experienced now. Nations and governments of the world should be made to be responsive to the material needs of their citizens or else we will have more problems to face.

Locus of Attack in the Internet Configuration

Internet has several characteristics that often makes it amenable to deceptive usage. These characteristics include reduced auditory and visual cues, something that affords cyber criminals and fraudsters the opportunity to change or conceal personal features. For instance, internet between the fraud expert and the victim is characterized by anonymity throughout the duration of the transaction.

The atmosphere of anonymity often predisposes major victims to disclosing vital private or secret information since the adverse repercussions for such disclosure in real life are reduced (McKenna & Bargh, 2000; Spears et al. (2000).

Furthermore, it may be observed that Internet communication often takes place in social communities that are isolated from those in real life. Such remote communities always involve limited commitments and tend to encourage identity faking or experiment (Turkle, 1995). In the analysis of human potential assuming different identity, research has shown that while individuals have only one self, they possess the capacity to assume different identities when they have the intentions of victimizing or defrauding others and in the most recent time, through the manipulation of the Internet (Brinthaupt & Lipka, 2002; Harter, 1999; Hogg et al. 1995). The successful orchestration of cyber-crime/fraud is guided by the ability of criminal assuming different identities within a specific time frame.

It involves the ingenuity of the criminal taking on the victim with a convincing identity akin to identity faking. Research has shown that adolescents and modern day youths mostly use Internet to experiment identity faking (Calvert, 2002; Lenhart et al, 2001; Maczewski, 2001; Turkle, 1995; Valentine & Holloway, 2002). Considering Internet crimes directed at national infrastructures, vulnerabilities entail malicious physical attacks, such as equipment tampering, cable cuts, facility bombing and asset theft. For the fraudsters, cyber space creates or provides an enterprising environment and secures participation in fraud business via Internet. However, this type of business requires the ability to predict the environment with a sufficient level of accuracy in order to forestall detection and arrest. The benefits accruable through cyber fraud, crime and risk concerned in the arrest of fraudsters are often disseminated to potential trainees by already established net fraudsters or other cyber criminals. The ingenuity of a significant number of people operating undetected and amassing wealth becomes a reference point upon which potential fraudsters or would be cyber criminals base their decision or justify their participation in cyber related deviance.

Consequence of Cyber Related Crimes on Modern Global Environment

Research have shown that the Internet provides the opportunity for relational expansion in all field of human endeavor and it simultaneously induces a form of apprehensive approach when put to use due to the likelihood of risks and vulnerabilities attached to the Internet environment. Cybercrime and cyber espionage has been discovered as a fast-growing area of crime thus carrying monetary loss implications for individuals, firms, economic sectors, national economies, and global markets. Cyber enabling environment afford more and more criminals the luxury of exploiting the speed, convenience and anonymity of the Internet to commit a diverse range of criminal activities that know no borders, either physical or virtual (Interpol, 2015). As a result, enforcing sanity in the Internet environment becomes problematic to both state actors and enforcement agencies. The problem is made worse by the borderless nature of the crimes involved and the difficulties in tracking perpetrators. Ponemon report (2014) indicates that cyber-crime is growing more common, more costly and taking longer to resolve due to the elusiveness of the actors in crime. Consequently, the reach of cyber-crime is on the leap on a daily basis and its effect appearing more dastard than any hitherto existed crimes. The web of victimization is equally wide and having the potential of affecting at least six of every ten users of the Internet. The Norton Cybercrime Report: The Human Impact," showed that two-thirds (65 percent) of Internet users globally, and almost three-quarters (73 percent) of U.S. users have been a victim of cybercrime (Lennon, 2010). Similarly, the Center for Strategic and International studies (CSIS) describes cyber-crime as a growth industry with the capabilities of damaging trade, innovation and competitiveness. The effect of the damages is quite substantial in the light of the volumes of trade in the global community promoted by the flatness of the world postulated by Caircross. Statistically, IC3 (2014) reported that the damage accruable from cyber-crime is on the increase and cutting across financial, computer infrastructural and informational concern of significant Internet users in its recent impact. Analyzing the data on the amount of monetary damage linkable to reported cases of cyber-crime from 2001 to 2014 that was released by IC3 therefore, the cost generatable from cyber-crime is incontestably rising. In 2001, the monetary cost was puts at $17.8 million and in 2002, it stood at $54 million. By 2003, it rose to $125. 6 million and in 2004, it was estimated to be within the border of $68.1. In the year 2005, the amount loss to cyber-crime totaled $183.1 million and the following year 2006, $198.4 million was documented as accruals from cyber-crime. From 2007 to 2014, an upward steady increase was documented. In 2007 alone, $293.1 million was found to be the cost of damage and by 2008, $264.6 million indicating a surge was reputed to be cost of cybercrime. In subsequent years, the cost of damage stood at $559.7 million, in 2009, $485.25 million, in 2011, $581.44 million, in 2012, $781.84 million, in 2013 $800.49 million in 2014 respectively (IC3, 2015). The net returns from e-crime perpetration is not only great but on the increase while the cost and risk of apprehension calculable by cyber criminals remain drastically low (CSIS, 2014). The opposite is the case for businesses suffering skyrocketing cost to ward of the predatory effect of cyber-crime. Chunk of organizational or corporate finance is often lost to cyber-crime thus resulting to reduced profitability. The center also puts the annual cost of cyber-crime and fraud to the global economy at more than $400 billion. Cyber-crimes have the potential of extracting between 15% to 20% of the total amount realizable from the Internet economy globally which was puts at between $2 trillion and $3 trillion. In the US. for instance, 85% of the direct financial cost to cyber-crime include, 42% to fraud, 26% to repairs to network damages,17% theft or loss and 15% to others. The cumulative effect on the victims can also be substantial. Yearly victim count estimate is puts at 556 million and on daily basis over 1.5 million while victimization rate per second is puts at 18 persons. In relation to the percentage of cyber-crime victims by gender, 71% were found to be male and 63% female. The possible explanation to non-parity in this regard is connected to findings on gendernet habit earlier recorded in research. Male gender are more glued to the Internet technology than their counterpart, the female gender. This is also a function of socio-economic variables determining the rate of the Internet technology use. In relation to age category of victims, Statistita (2013) reported that 50 percent of online adults have been victim of cyber-crime or negative online experiences such as unwanted nude images or cyber stalking. The rate of victimization for other categories should be relatively higher since significant number of young people are more often found in the Internet environment than the adults due the prevalence, usability and mobility of social media networks.

Another dimension of victimization affects information compromise. In 2013 alone, the victimization rate involving information compromise touched upon 40 million people in the US., 54 million in Turkey, 20 million in Korea, 16 million in Germany, 20 million in China and with a total of 800 million people on a global scale. In 2014, U.S. victimization figure was puts at 69% representing a rise from 60% in 2011, Ukraine 63% up from 39% in the 2011, Russia 60% from 37% previously, and Australia 57% a rise from 47% in 2011 respectively. Statistita (2014) equally reported on the trend in Germany for 2014, the breakdown of victimization shows that 40% reported malware infection of PC, 19% login data spied out, 16% involuntary sending of spam mail using own name, 14% concerning online shopping, online auctioning or private sale online, 4% fraud in connection with online banking and 2% financial loss as a result of malware or data theft. Many more people are likely to become victims on a global scale in several years to come. In addition to the data examined on the rate and types of victimization, the prevalence level of economic crime in part affecting cyber-crimes across regions of the world was given by GECS (2014). The rate as listed indicates 50% in Africa, 41% in North America, 39% in Eastern Europe, 35% each in Latin America and Western Europe, 32% in Asia pacific and 21% in the Middle-East. In the figures provided, 25% of survey respondents claimed to have been victims of cyber-crime at one point or the other (European Business News, 2014). The impact of cyber driven crimes is astronomical and with grave consequences for social and economic relationship. These impact ranges from loss of online businesses and consumer interest; compromise of critical infrastructure; loss of financial resources often leading to emotional conflict; loss of jobs and business asset; incremental cost required in arresting e-fraud and crime; growth of other subsidiary crimes having linkages with cyber-crimes; pressure on law enforcement agency relating to the cost of apprehending cyber-crimes. In terms of loss to the economy, Africa suffered the least in the reviewed year, while U.S.(0.64%GDP), China (0.63%GDP) Europe and Asia were hit the hardest, Germany (1.5%GDP) and Netherland (1.6%GDP) suffer average losses based on percentage of their GDPs while Japan and Australia had lower than average losses to cyber-crime in a survey conducted in 2013 (Brinded, 2014; Lawrence, 2014). In the analysis of cyber-crimes, it should be known from the onset that this form of crime pose business strategy problems. In a Global Economic Crime Survey conducted in (2013), it was disclosed that 30% of businesses in Australia reported a cyber-crime incident and 16% within this figure have lost more than 5 million dollars. Cyber related crimes cost New Zealand enterprises an estimated $625 million in 2011. Average loss or damage to business in the U.S. was $12.69 million, Germany $8.13 million, Japan $6.91 million, France $6.38 million, UK. $5.98 million, Australia $3.99 and Russia $3.33 million as at June 2014. All this were attributed to increasing reliance on technology and technology enabled processes in all aspect of business and the growing of economic energy towards emerging markets. Talking more about the cyber driven economic crimes, there is a gap in capturing the totality of its effects. It was aptly noticed that a precise single figure for the cost of cyber-crime and cyber espionage is unattainable but optimistically a more accurate estimate of the range of potential losses can be developed (CSIS and McAfee, 2013). This gap is created by reason of non-availability of data in most developing countries on the prevalence rate of cyber-crimes, the exact cost of these crimes to many nations' economies, losses in human capita utilization and absence of modality for capturing losses that cannot be easily quantified in a tangible and monetary terms. Apart from the cost emanating from loss to tangible product or material that were afore listed, there were several other cost affecting information theft mostly resulting from attacks from espionage activities against organizations and nations both of whom generate negative consequences to economic growth and safety. The combined cost of financial and espionage cost emanating from cyber-crime affecting U.S. alone stood at $140 billion (Amber Corin, 2013) and to the UK. $11.4 billion. This aspect in most cases encompasses intellectual property, loss of business information, employment disruption, reduced trust to online activity, increasing cost required to secure networks, insurance and recovery from cyber-attacks, and reputational damage that are pretty difficult to calculate. Intellectual property and business information are so vital and needed to strengthen the economic base of nations and the losses from them may be running to billions of dollars (CSIS and McAfee, 2013). Many nations and organizations are finding it difficult to capture, evaluate the cost implication and control the impact of cyber-crime and cyber espionage. The awareness of this threat is more real in advanced economies while in the developing economies, the awareness and concerted steps required to nip its effects on economic growth is quite unknown. This in part may be due to the level of industrialization attained so far or by reason of the component of their economies which in most cases is mainly agrarian. .Irrespective of the divide among the major economies, it is expedient to know that cyber-crime and cyber espionage pose debilitative effects to the sustainability and growth of most economies and in the wisdom of Ashford (2015) halting cyber-crime could have a positive impact on the global economy. Certainly this will be cost effective if concerted effort is made toward checkmating its rampaging effects on the wellbeing of the global community.

FUTURE RESEARCH DIRECTIONS

It is quite important for both technology systems developers, criminologists and other stakeholders in the ICT industry to act spontaneously on how best to arrest the state of emergency created bythe illicit uses of communication devices in the age of modernity. Research focus should be directed at how best to reconfigure ICT devices to enable automatic censorship of interactional flows in the cyber space. Users alongside information tracking system related research should be embarked upon to help determine the appropriate or relevant elements needed for the detection of both the location and identities of Internet users. ICT technologies should be embedded with "on the spot garget" that will make anonymity difficult to impostor. To achieve this feat, the requirement of interdisciplinary research is long overdue. The collaboration between social scientists and computer hardware/software professionals will help promote the production of systems that will be resistant to fraudulent and other unconventional uses. The expert environment are expected to synergize between human/machine interfaces and ultimate, the robust research results derivable from these efforts will help safeguard users from being victimized on one hand and invariably, a modest and secured society will eventually be built on the other hand.

CONCLUSION AND RECOMMENDATIONS

It is substantially established in this discourse that the Internet is relatively unsafe and threat ridden by reason of the activities of cyber criminals. The trend is getting worse by the day and the coverage of its negativity results in acute vulnerability, victimization and economic loss to individuals, organizations and national entities. Several scholars equally expresses the need to correct this unwholesome behaviours prevalent in the Internet environment. It then becomes imperative to galvanize collective effort towards its reduction on a short run and probably its complete eradication on the long run. The solutions to the challenges relating to crime-internet affinity will require an amalgam of strategies. In this wise, concentration of strategies to achieve this is will expected in the following areas: increment in enforcement funding, improvements in training of cyber experts on cyber technology know-how, intra and international data sharing on coverage of threat and possibly measures at curbing the probable and actual incidences of cyber-crimes, institutionalization of laws, regulations and policies regime against the surge of e-crimes. To check espionage, there is the increasing need to police the capabilities of individuals, corporations and nations from prying and spying on other users information or strategic data. This can only be done by instituting convention on a global scale to prevent its occurrences. Action needed will include data protection and computer misuse legislation that will be formulated and made effective across the major boundaries of the world. The growing of the cyber expert community is also essential to arresting the surge of cyber-crime and cyber espionage. Their nature of competence will involve timely spotting of cyber infringement at all level which may be through crowd sourcing via the cyber powered media, capturing the trends and patterns, clearly demarcated line of reporting infringements affecting individuals, organizations and national entities. Basically, this study recommend further studies into the affinity that exists between the Internet and organized crime an interesting area which is completely omitted in this discourse.

REFERENCES

Amoroso, E. G. (2011). *Cyber Attack: Protecting National Infrastructure*. Butterworth-Heinemann.

Ashford, W. (2015). *Cyber-crime is a threat to global economy*. Retrieved from: http://www.computerweekly.com/news/2240239300/Cyber-crime-is-a-threat-to-global-economy-says-researcher

Beck, U. (1992). *Risk Society: Towards A New Modernity*. London: Sage.

Brinded, L. (2014). Cybercrime Costs $445bn Worldwide and Hits 150,000 EU Jobs. *International Business Times*. Retrieved from: http://www.ibtimes.co.uk/cybercrime-costs-445bn-worldwide-hits-150000-eu-jobs-1451895

Brinthaupt, T. M., & Lipka, R. P. (2002). *Understanding Early Adolescent Self and Identity: Applications and Interventions*. Albany, NY: State University of New York Press.

Buckingham, D. (Ed.). (2008). *Youths, Identity, and Digital Media*. MIT Press.

Cairncross, I. (1997). *The death of distance: how the communications revolution will change our lives*. Cambridge, MA: Harvard Business School Press.

Calvert, S. L. (2002). Identity Construction on the Internet. In S. L. Calvert, A. B. Jordan, & R. R. Cocking (Eds.), *Children in the Digital Age: Influence of Electronic Media on Development* (pp. 57–70). Westport, CT: Praeger.

Castells, M. (2001). *The Internet Galaxy: Reflections on the Internet, Business, and Society*. Oxford, UK: Oxford University Press. doi:10.1007/978-3-322-89613-1

Castells, M. (2002). The Internet and the Network Society. In The Internet in Everyday Life. Blackwell Publishing Company.

CERT. (2013). *Australia Cyber-crime and Security Survey Report*. Retrieved from https://www.cert.gov.au/

Center for Strategic and International Studies (CSIS). (2014). *Net Loses: Estimating the Global Cost of Cyber-crime*. Retrieved from http:///csis.org/

cigionline. (2015). Retrieved from https://www.cigionline.org/publications/global-cyberspace-safer-you-think-real-trends-cybercrime?gclid=COrAyIbP7sYCFS7KtAodmTENRQ

Clough, J. (2010). *Principles of Cybercrime*. Cambridge, UK: Cambridge University Press. doi:10.1017/CBO9780511845123

Corin, A. (2013). *Tracking the Cost of Cyber-crime*. Retrieved from http://fcw.com/articles/2013/07/23/cyber-crime-cost.aspx

CSIS and McAfee. (2013). *Cyber-crime costs global economy $445 billion a year: report*. Retrieved from http://www.reuters.com/article/2014/06/09/us-cybersecurity-mcafee-csis-dUSKBN0EK0SV20140609

Dean, M. (1999). Risk, Calculable and Incalculable. In Risk and Socio- cultural Theories: New Directions and Perspectives. Cambridge, UK: Cambridge University Press.

Dodd, N., & Hutter, B. M. (2000). Geopolitics and the Regulations of Public Life. *Law & Policy*, *22*(2), 1–24. doi:10.1111/1467-9930.00083

Dufour, A. (1995). *Internet*. Paris: Presses Universitaires de France.

European Business News. (2014). *PwC survey finds economic crime rising globally, all business sectors, regions suffer from impact*. Retrieved from www.european-business.com/news/business_economy_news/economic_crime_rising_globally/

Fox, N. J. (1999). Postmodern reflections on 'risk' 'hazard' and life choices. In Risk and socio-cultural theories: New directions and perspectives. Cambridge, UK: Cambridge University Press.

Garland, D. (2001). *Culture of control*. Oxford, UK: OUP.

Global Economic Crime Survey (GECS). (2014). *The benefits—and –risks –of a networked world are here to Stay*. Retrieved from: http://www.pwc.com/gx/en/economic-crime-survey/cybercrime.jhtml

Graham, G. (1993). *Philosophy of mind: An introduction*. Oxford, UK: Blackwell Publishers Ltd.

Green, L. (2010). *The Interne: An introduction to new media*. New York: Berg.

Haralambos, M., Holborn, M., & Heald, R. (2004). *Sociology: themes and perspectives* (6th ed.). London: HarperCollins Publishers Limited.

Harter, S. (1999). *The construction of the self: A developmental perspective*. New York: Guilford Press.

Hogg, M. A., Terry, D. J., & White, K. M. (1995). A tale of two theories: A critical comparison of identity theory with social identity theory. *Social Psychology Quarterly*, *58*(4), 255–269. doi:10.2307/2787127

Homes, D. (2005). *Communication theory: Media, technology and society*. Sage Publications Inc.

Hughes, G. (2004). Crime prevention, community safety and crime and disorder reduction. In Student Handbook of Criminal Justice and Criminology. Cavendish Publishing Ltd.

Interpol. (2015). *Cybercrime*. Retrieved from: http://www.interpol.int/Crime-areas/Cybercrime/Cybercrime

Katz, J. E., & Aspden, P. (1997). Motivations for and barriers to Internet usage: results of a national public opinion survey. *Internet Research: Electronic Networking Applications and Policy*, *7*(3), 170–88.

Katz, J. E., & Rice, R. E. (2002). *Social consequences of Internet use: Access, involvement andinteraction*. Cambridge, MA: MIT Press.

Lenhart, A., Lewis, O., & Raine, L. (2001). *Teenager life online*. Retrieved from: http://www.pewinternet.org/2001/06/21/teenage-life-online/

Lawrence, D. (2014). *The global cost of cybercrime: More than $400 billion per year.* Retrieved from: http://www.bloomberg.com/bw/articles/2014-06-09/the-global-cost-of-cybercrime-more-than-400-billion-per-year

Lennon, M. (2010). *New Report Gauges Emotional Impact of Cybercrime*. Retrieved from http://www.securityweek.com/new-report-gauges-emotional-impact-cybercrime

Macaluso, N. (2001, August 13). Nearly 60% of homes are online. *E-Commerce Times.*

Maczewski, M. (2001). Exploring identity through the Internet: Youths experiences online. *Child and Youths Care Forum, 31*(2), 111-129.

Meridien Research. (2000, March 8). Shopnow.com unveils PCcharge transaction processing alliance, an innovation in credit card fraud protection. *PR Newswire.*

McKenna, K. Y. A., & Bargh, J. A. (2000). Plan 9 from cyberspace: The implications of Internet for personality and social psychology. *Personality and Social Psychology Review, 4*(1), 57–75. doi:10.1207/S15327957PSPR0401_6

McShane, M. D., & Williams, I. I. I. (Eds.). (2007). Youth violence and delinquency: Monster and myths (vols. 1-3). Westport, CT: Praeger Publishers.

Muncie, J., & McLaughlin, E. (Eds.). (2001). *The problem of crime.* London: Sage Publication Ltd.

National Crime Prevention Institute (NCPI). (2001). *Understanding crime prevention* (2nd ed.). Woburn, MA: Butterworth-Heinemann.

New Zealand Management. (2013, April). Cyber-crime: Cyber threats and data privacy. *Magazine of the Institute of Management.*

Ndukwe, E. C. A. (2009). *Combating cyber-crime in 21st century.* Paper presented at the 1st annual conference of National assembly anti-money laundering and cyber-crime coalition.

Nie, N. H., & Erbring, L. (2000). *Internet and society: A preliminary report.* Stanford Institute for the Quantitative Study of Society (SIQSS), Stanford University, and InterSurvey. Available online at http://www.stanford.edu/group/siqss/

Nua.com. (2002). *How many online*? Retrieved from: http://www.nua.ie/surveys/how_many_online/world.html

ONS Omnibus Survey. (2000). Retrieved from: http://discover.ukdataservice.ac.uk/Catalogue/?sn=4528&type=Data%20catalogue

Packer, L. H. (1968). *The limits of the criminal sanction.* Stanford, CA: Stanford University Press.

Ponemon. (2014). *Global report on the cost of cyber-crime.* Retrieved from: http://www.ponemon.org/library/2014-global-report-on-the-cost-of-cyber-crime

Rheingold, H. (1993). *The virtual community: Homesteading on the electronic frontier.* Addison-Wesley.

Rotich, E. K., Metto, S. K., Siele, L., & Muketha, G. M. (2014). A survey on cyber-crime perpetration and prevention: A review and model for cybercrime prevention. *European Journal of Science and Engineering, 2*(1), 13–28.

Sarri, C. R. (1995). Criminal behavior overview. Encyclopedia of Social Work, 19(1), 637-646.

Singleton, T. W., & Singleton, A. J. (2010). *Fraud auditing and forensic accounting* (4th ed.). John Wiley & Sons, Inc. doi:10.1002/9781118269183

Smith, M. A., & Kollock, P. (1999). *Communities in cyberspace.* London: Routledge. doi:10.5117/9789056290818

Spears, R., Lea, M., & Postmes, T. (2000). One side: purview, problems and prospects. In T. Postmes, M. Lea, R. Spears, & S. Reicher (Eds.), *SIDE, issues center stage: recent developments in studies of de-individuation in groups* (pp. 1–16). Amsterdam: KNAW.

Stack, C. (2004). Self esteem and psychological impact of employment. *Social Psychology Quarterly, 49*(6), 67–72.

Cobb, S. (2015). *Cyber-crime deterrence: 6 important steps.* Retrieved from: http://www.welivesecurity.com/2015/01/20/cybercrime-deterrence-6-important-steps/

Statistita. (2013). *Percentage of online adults who have fallen victim to cyber-crime as of August 2013.* Retrieved from: http://www.statista.com/statistics/294684/online-adult-cyber-crime-victimization/

Statistita. (2014). *What types of cyber-crime have you encountered personally during the last 12 months*? Retrieved from: http://www.statista.com/statistics/429689/internet-users-experiences-with-cyber-crime-germany/

Statistita. (2015). *Average costs of cyber-crime in selected countries as of June 2014 (in million U.S. dollars).* Retrieved from http://www.statista.com/statistics/293274/average-cyber-crime-costs-to-companies-in-selected-countries/

The, U. C. L. A. CCP Internet Report. (2001). *Surveying the digital future.* Retrieved from: http://www.worldinternetproject.net/_files/_Published/_oldis/ucla-internet-2001.pdf

Turkle, S. (1995). *Life on the screen: identity in the Age of the Internet.* New York: Simon and Schuster.

UNODC. (2013). *Comprehensive study on cyber-crime—draft.* Vienna, Austria: UNODC.

Valentine, G., & Holloway, S. L. (2002). *Cyberkids? Exploring children's identities and social networks in on-line and off-line worlds.* Retrieved from: http://knowledgepublic.pbworks.com/f/cyberkids.pdf

Van Loon, J. (2008). *Media technology: Critical perspectives.* New York: Open University Press.

Wallace, P. (1999). *The psychology of the Internet.* Cambridge, UK: Cambridge University Press. doi:10.1017/CBO9780511581670

Wellman, B., & Haythornwaite, C. (Eds.). (2002). *The Internet in everyday life.* Oxford, UK: Blackwell. doi:10.1002/9780470774298

Williams, M. (2006). *Virtual criminal: Crime, deviance and regulation online.* New York: Routledge.

Winston, B., & Walton, P. (1996). Netscape: Virtually free. *Index on Censorship, 25*(168).

Winston, B. (1998). *Media technology and society: A history from the telegraph to the Internet.* London: Routledge.

Yar, M. (2009). Computer crime control as industry: Virtual insecurity and the market for private policing. In Technologies of insecurity: The surveillance of everyday life. Routledge-Cavendish.

KEY TERMS AND DEFINITIONS

Cyber-Crime: This involves a range of illegal activity often perpetrated through the use of computers. It represents any form of crime directed at computer network or devices and also covering advances into other crime activities bearing cost in both psychological and financial aspect of human life.

Economy: This encompasses trade and industry, productive activity, wealth creation and gamut of efforts directed at making a living in a given society. It is inclusive of the medium for promoting the processes of production of goods and services, management of resources and the structure of economic life at either local, national or international scale.

E-mail: This involves messages transmitted by electronic means from one computer/internet user to one or more interactors within a network system. It involves messages transmitted through telecommunication.

Global: Global relates to issues encompassing the whole of humanity and that which are likely to be affecting the whole world. It connotes the use of ICT or Internet worldwide.

ICT: An umbrella term often used in the description of communication devices or applications. It represents the gamut of technology mostly used to facilitate interaction across spectrum of media gadgets encompassing telecommunication, broadcast media, network based control and communication monitoring devices.

Internet Technology: Communication infrastructure equipped with the ability to transmit information and data resulting from interaction between dual or multiple users through different servers and computer systems. It consist of network of autonomous computers linked together by wires, cable or fiber links and other devices enabling the exchange of information or data within a wide range of socio-scientific concerns.

Risk: It represents exposure to danger, harm or loss as a result of using the Internet technology. It is a state of being vulnerable to attacks, financial loss or data theft. It is a probability of something least expected happening through the use of computer infrastructure.

Space: This involves the non-physical terrain created by computer systems. It represents a notional environment in which communication over computer networks.

Chapter 3
Social Media Uses and Effects:
The Case of WhatsApp in Africa

Brian Pindayi
Rusangu University, Zambia

ABSTRACT

This chapter aims at analysing the effects of social media on African society by focusing on WhatsApp usage. The chapter will; one, show how usage of WhatsApp is affecting interpersonal relationships in Africa. Two, evaluate the common uses of WhatsApp in Africa. Three, scrutinise the reasons people in Africa prefer using WhatsApp. Four, add theoretical perspectives on how social media is impacting on the communication landscape of the region. The chapter will also introduce two concepts, one; medium veracity wherein different media have varying levels of trust or credibility and two; the social media domino effect, wherein social media platforms are mutually dependent and have a symbiotic relationship. By examining why people are using WhatsApp, this chapter seeks to add to the empirical conversation on futility and the transformative potential of social media.

INTRODUCTION

"We become what we behold … we shape our tools and afterwards our tools shape us" (McLuhan, 1964). New media technologies have transformed media ecosystems in the sense that conventional media professionals no longer have a monopoly on the generation of news content, as people now generate and share their own content. User-generated content creates numerous possibilities which affect how people relate towards one another and construct meaning or everyday reality (Gauntlett, 2011). As shown by the popularity of web 2.0, new communication technologies have affected the communication landscape in numerous ways ranging from political effects to economic and social effects. For example, the Arab Spring of 2011 showed how new communication technologies can provide a platform for political mobilisation (Acemoglu, 2015). In South Africa, the #Rhodes must fall and #Fees must fall movements of 2015 were successful in part because of the communicative platforms afforded by new communication technologies such as WhatsApp (Luescher, 2016). In Kenya, the #StopPoliceBrutality movement of 2016 saw many people participate in protests because of the information that was shared through social

DOI: 10.4018/978-1-5225-1859-4.ch003

media platforms (Buchanan, 2016).These effects of new communication technologies have led some researchers (Newman, 2012) to talk of the emergence of a fifth estate as the public appears more empowered through user-generated content to gather and share content without relying on the fourth estate or conventional media houses. Likewise, the impact of new communication technologies has led some researchers to suggest that an alternative public sphere and in some instances a counter public sphere is constituted by the internet and by inference social media, as users deliberate on a less restricted platform and raise divergent views or anti-establishment sentiments (Aslama, 2009).

Social media have been growing in popularity in Africa at an exponential rate (Balancing-Act, 2014). It is estimated that around 9% of the population in Africa is using social media (Park, 2016). Also, 9% is a substantial number when one considers the uneven technological distribution on the continent and the big divide between urban and rural Africa.

(Kalyango, 2013) notes that social media have mobilized people in both political and social ways in East and West Africa. As a way of exploring the question of the effects of social media in Africa, this chapter focuses on WhatsApp and examines how the messaging platform is affecting interpersonal relationships, political movements, civil society, citizenship journalism andcultural, educational and technological issues etc..

As a communicative space, WhatsApp is a social media platform that is shaping or affecting the lives of African citizens. The word development in the broadest possible sense means "a multi-dimensional phenomenon which encompasses economic, political, social and cultural aspects," this chapter seeks to investigate how WhatsApp is affecting development in Africa. Specifically, this chapter intends to:

- Highlight the major strengths and weaknesses of WhatsApp with regards to its societal, economic, cultural and political impact.
- Discuss the Uses and Effects of WhatsApp in Southern Africa.
- Analyse how communication is impacting development in Africa.
- Discuss theoretical developments and possibilities that are created by Social media through WhatsApp.

BACKGROUND

Social media have affected the way people interact with one another and deliberate over pertinent issues. The way people interact through social media generates meaning and sociality through user-generated content; users create and share their own text, photos, videos audio files and much more. User - generated content creates a sense of empowerment as people who are communicating with each other have a sense of ownership and consume the information without the trepidation that comes with consuming news or information that has been sifted by gatekeepers or professional organisations (Gauntlett, 2011).

As a typical example of a platform for user-generated content, WhatsApp creates endless possibilities as people are able to discuss issues that are topical to them and are able to generate a momentum of mutual influence. User- generated content eliminates certain barriers to entry as people are able to generate and share their own video files and movies, or their own news stories. A good example of this is the citizen journalism experience of South African citizens which helped them to exert pressure on President Jacob Zuma to address the issue of alleged funding irregularities in the building of his personal home in Nkandla, Kwazulu - Natal province (Shapshak, 2016).

For civil societies, WhatsApp provides an opportunity to disseminate messages to a very large audience at a very cheap rate and without the scrutiny or censorship of the state or government. Thus WhatsApp as part of social media has encouraged political mobilisation and several protests in Africa. Again in South Africa, students organised and executed protests in 2015 using *WhatsApp, Facebook, Twitter* and other forms of social media (Holmes, 2016).

There is extensive literature on the strengths of technology in spreading development. There are two main polarised positions regarding the potential of social media in transforming lives or instigating development. The first is the optimistic view which postulates that social media creates opportunities and networks that empower people and result in economic, social, political and cultural advancement. The optimistic view credits social media with the ability to instigate change in social, economic and political structures. For example, social media can be said to have played a pivotal role in mobilising students in South African for the protest in 2015(Jenkins H., 2015).

The second is the critical view which contends that social media is not really empowering because it lacks the material resources and ability to transform lives or the status quo in society since the means of production or companies behind social media are private owned and elitist. This school of thought thus holds that the ideas generated through social media can at best cause minor disturbances rather than a revolution or drastic economic, political, social or cultural change. The critical view asserts that there are several factors which inhibit the transformative power of social media. These range from limitations of technological adaptation as many people in Africa do not have the means purchase smartphones; to literacy barriers as many Africans will not be able to operate the smartphones, and ideological deficiencies as the smartphones are meant for communication that is incapable of posing critical questions. In the case of the South African students' protests, proponents of the critical view would submit that the protests where superficial and did not really succeed in attaining what the protesters were agitating for. Thus, rather than propagate change in South Africa social media, it simply caused a minor disturbance (Fuchs, 2013).

Despite the polarity on the transformative potential of social media between the optimistic view and the critical view, proponents of both views agree on the importance of the extensive utility of social media. In the case of WhatsApp, users are able to send and receive texts, share pictures, videos, make voice calls and in some cases video calls. The ability of users to send countless messages cheaply has made WhatsApp the most popular and widely adopted social media platform in South Africa (Shambare, 2014).

Social media can be defined as a communicative platform or space in which there is sociality as categorised by cognition, communication and co-operation (Trottier, 2015). Cognition refers to sending and receiving information that engages and affects the user's mind-set. Communication refers to the sending and receiving of messages that cement relations and build a sense of community, while co-operation refers to co-production as users construct texts, narratives, images and videos in a collaborative manner. This triple C model qualifies WhatsApp as part of social media in that there is cognition as users send and receive texts, communication as users send texts that maintain relationships, and cooperation as users of a group forum exchange ideas and views about various subjects. It can therefore be said that WhatsApp plays an integral role in the sociality of Africa as users send messages to maintain relationships, post videos as a way of sharing their views, asserting their creativity and engaging in pertinent issues.

WhatsApp Messengers is a cross messaging platform which allows users to send and receive messages cheaply. It can be used on most smartphones including IPhone, Blackberry, Nokia, Samsung, Sony and many more. Once the software is installed on one's phone, the application synchronises with the contacts, enabling one to send - receive texts, audio files, video files and to make either voice or video calls. When the receiver accesses a message, two ticks appear to alert the sender that the message has been delivered and the colour of the ticks change from black to blue. WhatsApp has an ideal age limit of 16 but anyone with a smartphone can create a WhatsApp account and begin communicating. Each user has a prolife picture and status information which can be updated frequently and fellow users are able to know when someone was last online by viewing one's profile. The application has a provision to block any contact one does not wish to communicate with. WhatsApp is an Over- The -Top (OTT) messaging service because it operates independently from both smartphone and service provider which makes it difficult to regulate its content and use. As of February 2016, the Facebook-owned WhatsApp Messenger boasted of having one billion users (Titcomb, 2016).

As has been the case worldwide, WhatsApp has been growing in popularity in Africa at an exponential rate. This growth in popularity has seen WhatsApp become a tool of communication with effects on various aspects of life.

In business, people have created group chats to enable them to advertise garage sales and companies have created WhatsApp forums as a way interacting with clients. For example, in Zambia, MTN launched a WhatsApp forum as a way of improving customer services (Lisulo, 2016). In interpersonal relationships, people have created new ways of communicating and maintaining relationships (Langat, 2015); and in politics, people have used WhatsApp as a source of information and organising protests. For instance, in Burundi, WhatsApp was among the social networks that were used to mobilise people for protests in 2015 (Caulderwood, 2015). In education, students and teachers have been engaging in academic discussions through WhatsApp, (Zanamwe). In the Information and Communication Technology (ICT) sector, WhatsApp has had a profound impact as a digital disruptor, reshaping the business of text messaging from conventional SMSes and forcing service providers to come up with bundles for smses, as a competitive measure against the WhatsApp potential to send countless SMSes cheaply (Bradely, 2016). As a digital disruptor WhatsApp has arguably increased the sales of smartphones as people have been buying smartphones as a way of getting on WhatsApp (Shambare, 2014). Even in law enforcement, WhatsApp has had an impact as nude pictures shared via WhatsApp resulted in a police officer being reassigned in Zimbabwe (Zaniest-Zimbabwean, 2016). The many uses of WhatsApp explain in part why it is such a popular phenomenon, an application with such diverse utility and relative affordability is bound to be popular. It has been suggested that apart from the various uses and benefits it offers, WhatsApp is popular because it fulfils the condition of enabling people to exercise faithfulness in maintaining and nurturing relationships. That WhatsApp enables people to connect across space and time to such an extent that they remain in touch or in sync despite being far away from each other. Thus, they are able to engage in sociability by interacting in a digital space. WhatsApp enables people to communicate in time and across different contexts or realities and situations to such an extent that the concept of dwelling materialises within WhatsApp as one communicates in a manner that fulfils the need to maintain relationships, be part of a society, be part of a community (O'Hara, 2014).

THE USES AND EFFECTS OF WHATSAPP

The effects of WhatsApp are diverse and can be categorises as follows:

Social and Interpersonal Effects

WhatsApp is a communicative forum where people largely maintain relationships rather than establish new ones. It therefore has phatic communication attributes as people maintain relationships across space and time (Shih-Hsiung Liu., 2013). At an interpersonal level, therefore, WhatsApp has been instrumental in enabling people to communicate more extensively by cheapening the cost of sending messages. Arguably, this implies that people find it easier to maintain relationships because of WhatsApp (O'Hara, 2014).

WhatsApp also improves one's ability to produce person-centred messages since it allows one to create and send countless messages. This in turn enables one to create more social capital. An illustration of the benefit of social capital is the case of a WhatsApp-based program called Brighter Futures in Kenya. Youths under this program have been using WhatsApp to discuss issues relating to sexual health, contraceptives and positive choices; the results suggest that the youths are making commendable choices because of the cheap communicative platform created by WhatsApp (Devezin, 2014).

Some scholars have cited an addictive feature of WhatsApp and other mobile-based applications as causing serious anxiety and social withdrawal symptoms when one loses a smartphone. Based on a quantitative study in South Africa (Dlodlo, 2014), a sizeable number of South African youths have become addicted to mobile phones and cannot function normally without them because of messaging platforms such as WhatsApp. This phenomenon has been described by Shamabare (2012) as twenty-first century non-drug addiction.

After t a comparative case study of tertiary institutions, Yeboah (2014) infers that WhatsApp keeps students fiddling on phones while texting personal messages to friends and family members, and this distract them from their studies to such an extent that their academic performance is negatively affected.

Since WhatsApp has been known to be very involving in as far as texting and receiving messages is concerned. This is because one may appear anti-social and may not reach out to make new friends when in a public environment. For example, when in a public place, one might prefer to chat through the phone rather than to the person seated next to them in a bus (Drago, 2015).

As a social media platform which enables people to share pictures and videos, WhatsApp has been used to send nude pictures and videos as users try to conform to sexualised stereotypes and to portray a likeable image of self to others (Boyd, 2014; Mascheroni, 2015). The prevalence of sending nude pictures and graphic videos through WhatsApp in African countries is exemplified by the predominance of revenge porn in Zimbabwe. Revenge porn occurs when someone disseminates a nude picture of another person without their consent and with the intention of humiliating or embarrassing them. Revenge porn normally occurs when a jilted lover posts nude pictures or a graphic video on social media (Arimoro, 2015). The prevalence of revenge porn in Zimbabwe has become rife that a civil society organisation called Katswe Sisterhood was compelled to petition the Zimbabwean Parliament in March 2016 to consider enacting a law to curb the practice of revenge porn (Kabweza, 2016). In Nigeria, revenge porn is prevalent but many cases go unreported because people fear social stigmatisation (Arimoro, 2015).

By virtue of enabling people to create group forums which members are determined by the group administrator, WhatsApp has allowed people to create their own pseudo-communities wherein people can share information and receive information in an exclusive manner. These groups have the effect of making people feel that they are part of a community which can discuss issues in a relatively secretive manner. Group chats are however not without problems, for example in Zimbabwe, some police officers were the subject of investigations in February 2016 when they sent messages on a group forum complaining about non-payment of their salaries (Harare-Bureau, 2016).

Technological Effects

A profound effect of WhatsApp has been its digital disruption of the mobile service producers' market. Digital disruption can be defined as "a phenomenon whereby a company or market is radically disrupted as a result of a fundamental change in the competitive landscape due to the innovative application of a technology-based solution" (Prinsloo, 2015).

As a digital disruptor, WhatsApp revolutionized the SMS services by forcing service providers to introduce message bundles as a way of competing with the cheap messaging capability of WhatsApp. The introduction of WhatsApp has boosted demand for smartphones which has resulted in a steep increase in sales of smartphone coupled with a reduction in smartphone prices (Goldstuck, 2014).

The increase in demand of smartphones has enabled service providers to be more innovative about the prices of data as they have had to accommodate users who may want WhatsApp but not the internet and vice versa. In South Africa for example, service providers have had to reduce the cost of sending SMS and voice calls in order to compete with the cheap rates offered by WhatsApp (Vermeulen, 2015). WhatsApp has also enabled business ventures like Ongair in Kenya to benefit from its popularity by offering customers a multiple user interface where employees can communicate instantaneously via the WhatsApp platform (Mulupi, 2015).

Stork (2016) conducted comparative case studies of some African countries and identified six interventions or reactions by mobile service providers to OTT's (such as WhatsApp) as companies battle to remain profitable. One, blocking OTT's wherein some mobile operators have resorted to blocking or throttling peer to peer traffic or VOIP in countries such as Morocco and Egypt. Two, regulating OTT's whereby regulation is aimed at either preventing customers from accessing OTT's or protecting the market interests of mobile operators. For example, in South Africa, some mobile phone operators like MTN have lobbied for regulation against messaging services such as WhatsApp in a bid to protect their revenue base (Grootes, 2016). Three, embracing OTT's whereby operators have incorporated OTT's in their pricing policies to offer either bundles for social media or free basic packages as is the case in Namibia, Kenya and South Africa. In fact, South Africa's Cell C has embraced WhatsApp by offering a zero-rating package which has enabled it to gain a bigger market share (Chen, 2016). Four, Simulate OTT's through bundling; mobile operators have resorted to bundling voice, text and data services in a manner that resembles OTT like services. The number of SMS is very high or unlimited in order to resemble OTT texting. In Zambia, every mobile operator has embraced this strategy of blended bundles which is popular in major countries. Five, offering Non Telco OTT's, mobile service operators have teamed up with OTT's to offer popular services such as video streaming and music file sharing in order to protect their market share. For example, Vodacom and MTN in South Africa have discounted video streaming services for specific partners. Six, offering own OTT's; some mobile operators have developed their own OTT services but this intervention has not been widely adopted in Africa. These six interventions or reactions show that WhatsApp has had a profound technological effect on the African mobile phone market.

In some instances, WhatsApp has been used to install malware on users' phones thus perpetuating cybercrime. WhatsApp users have been scammed into installing malware on their phones by responding to a request to install a WhatsApp premium or gold version, or being invited to click and listen to a WhatsApp voicemail message; or being invited to install a WhatsApp spy software which ends up loading malware rather than monitoring other users; or being invited to click a gift voucher which says the user has been randomly selected to win a specific prize, or being invited to invest or buy stock in a company whose share prices are sky rocketing (Fin24, 2016). In May 2016 in South Africa, some WhatsApp users received texts inviting them to install a so-called gold version through clicking on a link which resulted in malware being installed on phones which stole user information and invited other friends to install the malware (ICT-Security-Monitoring-Services-Report, 2016).

Economic Effects

The group chat function has enabled people to conduct garage sales by advertising their products through WhatsApp. Likewise, where people are buying material on behalf of others in distant places, photos of products such as cars have been sent through WhatsApp to enable the potential buyer to have an accurate image of what they are planning to buy. In Zimbabwe and Zambia, people have made extensive use of WhatsApp as they go to buy and transport cars from the ports of Tanzania and Namibia. People who travel across countries have been able to avoid costs of buying sim cards and airtime from local operators by using WhatsApp through Wi-Fi at hotels and other public places.

As a communication platform, WhatsApp can be used to promote business in several ways (Ogunlade, 2015). One, using WhatsApp for business support and engagement in which SMS and video chat are used to dialogue with customers and ensure customer satisfaction. For example, in South Africa, some small-scale businesses have been using WhatsApp to market their products and inform customers of new prices (Hyde-Clarke, 2013). In Kenya, the Ongair platform has enabled businesses to enhance customer satisfaction using WhatsApp (Mulupi, 2015). The second is promotions and advertising using WhatsApp. Some businesses have been using WhatsApp to advertise their products and services; for example, BBC Africa is using WhatsApp to reach out to people in Africa and to tell the story of Africans using WhatsApp (Morrison, 2016). Likewise, in South Africa, IOL (Independent Online) is using WhatsApp to interact with customers and to disseminate news items (IOL, 2016). The third is the fact that WhatsApp is a platform that can be used for market surveys. Through the success of its anti-Ebola campaign in West Africa, BBC Africa was able to ascertain the potential of WhatsApp as a communication tool, hence the official partnering of BBC with WhatsApp in Africa. Four, WhatsApp group chats have been used to coordinate workers and enhance teamwork. For example, health workers in Kenya used WhatsApp to coordinate and disseminate vital work-related information through a group forum (Henry et al, 2016). Five, WhatsApp can be used as a social sharing tool for publishers and bloggers. For example, in Nigeria, the Nairaland forum uses WhatsApp to reach out to people (Publishers-Posts, 2016). Six, WhatsApp is being used for commercial services and business transactions. For example, Jovago, a hotel booking service is making use of WhatsApp and other social media platforms to book rooms in Nigeria, Kenya and Tanzania (Jovago, 2014). Seven, WhatsApp can be used to provide a concierge service such as Russsh in India which uses WhatsApp as a booking platform, this service is relatively unexplored in Africa but has potential to be launched. Eight, WhatsApp can be used to order food deliveries as exemplified by restaurants like Spicy Snacks in Randburg, South Africa, which enables customers to order food and free delivery through voice calls on WhatsApp (Spicy-Snacks, 2016). Nine, WhatsApp can be used for medical consultation and health services, as shown by the BBC Africa which used WhatsApp to disseminate information to help fight against the Ebola outbreak in West Africa in 2015.Similarly an NGO called Life-Line Western Cape is enabling South Africans to access free counselling services through WhatsApp voice calls (Fin24, 2016).Ten, WhatsApp can be used by journalists to source and disseminate information or news items as is the case in Tanzania where some journalists have joined several group forums so as to source and distribute news (Mohammed, 2016).

Cultural Effects

The existence of group chats has arguably resulted in some form of collective values being maintained. Most group chats are constituted by members who are either already acquainted or share some common interests. By enabling people to share explicit content and discuss sensitive issues, WhatsApp has the effect of making culture more dialogic as people are able to discuss issues that they would not otherwise have been be able to discuss. For instance, issues of sharing nude photos, issues of sharing certain advice pertaining to sexual issues which would not otherwise be sharable in a real public gathering are important here. The provision of a group chat creates a sense of belonging which has the potential of unleashing effects of groupthink in a profound way as people share information and ideas in a relatively secret environment. The ability to send information to a group also enables some users to boost their self-esteem by occasionally sending positive messages to group members (Bouhnik, 2014).

By helping people to share information directly, WhatsApp distorts cultural barriers and issues of respect as people begin to communicate directly as peers. For example, a daughter who is communicating with her father via WhatsApp may not be as respectful when asking for pocket money as would have been the case if the father was there in person. Similarly, the sharing of jokes and customised videos and pictures has led to information being shared which would not otherwise be shared if the WhatsApp group did not exist.

WhatsApp has enabled users to text selfies and other explicit photos in an unprecedented way (Phippen, 2016). This has in some way fanned a culture of revenge porn as exemplified by the case of the Midlands State University's (MSU) student in Zimbabwe whose video was leaked and extensively circulated through WhatsApp (MusvoZimbawe, 2016). Likewise, several celebrities and common folks alike have been humiliated by having nude photos circulated through WhatsApp, making Zimbabwean musician Alick Macheso to vow that he will never use Facebook or WhatsApp as a way of protecting his image (Nehanda-Radio, 2013). Similarly, social media platforms such as WhatsApp are said to be exposing youths to sexually explicit content with detrimental effects in Nigeria (Amoo, 2013). This culture of revenge porn seems to have grown in prominence because of the ease with which one can upload and share sexually explicit content via WhatsApp.

The cheap cost of sending messages via WhatsApp has enabled people to send devotional materials and other inspirational materials to friends and family members. With this function in mind, several Christian groups have formed thriving communities through WhatsApp as group members interact with and support each other.

Political Effects

By allowing people to share information on a platform that is not easily monitored by governments or state agents, WhatsApp enables users to share information that results in political views being crystalized. For example, the Magamba Network distributes short anti-Mugabe satirical videos through social media channels like WhatsApp as a way of evading harsh censorship laws in Zimbabwe (Msonza, 2015). Likewise, in 2015, the opposition political party 'Transform Zimbabwe' used WhatsApp to distribute messages aimed at mobilising political support (Young, 2015).

During the 2015 xenophobic attacks in South Africa, citizens used WhatsApp to circulate graphic videos of the assaults on foreigners, while foreigners used WhatsApp to circulate warnings to each other in order to avoid violent hotspots (Naidu, 2015). In regional countries of Malawi, Zambia, Mozambique and Zimbabwe, protests against xenophobia in South Africa ensued partly because of the graphic videos that were circulating via WhatsApp showing South African citizens assaulting foreigners. This instance exemplifies how WhatsApp has changed the dynamics of citizen mobilization in Africa. Little wonder then that the Ugandan government blocked WhatsApp and other forms of social media during the disputed presidential elections of 2016. In Burundi, WhatsApp has been used by citizens to keep track of political activities and mobilise protests (Vircoulon, 2016). In Rwanda, Kizito Mihigo, a popular singer, was imprisoned for allegedly conspiring to overthrow the Rwandan government by sharing information through WhatsApp and Skype (Freedom-House, 2016).

It should be noted that the use of WhatsApp in politics is not always antagonistic. In Kenya, the government has been using various forms of social media, including WhatsApp, to engage citizens and to create a dialogue with common people with an aim of improving social amenities and services (Wangui, 2014).

Educational Effects

By enabling people to share information cheaply, WhatsApp has empowered teachers and students to share academic information. In some instances, university lecturers have been able to clarify academic issues and students have been able to have their own discussions via WhatsApp. Even at the high school level, some students have been able to use WhatsApp to communicate academic issues (Rambe, 2013). As far as health education is concerned, some NGOs or non - governmental organisations in Ghana and Zimbabwe have been using WhatsApp to disseminate messages about sexual health, sanitary practices in times of cholera outbreaks and warnings about certain products (Mavura, 2015). Likewise, in South Africa, the Right to Know campaigners have been using WhatsApp to send messages to group members in order to improve sexual health practices (Waldman and Stevens, 2015). Similarly, NGOs- USAID and UNICEF- have utilised WhatsApp to enhance health awareness among the youth in Liberia (Avis, 2015).

Furthermore, WhatsApp has been identified as being disruptive to students' study habits by making them attend to a barrage of messages at a time when they are supposed to be concentrating on either lectures or studies. A peculiar case of enhancement academic dishonesty was when examination papers where leaked to thousands of South African students through WhatsApp, prompting the South African Department of Higher Education to launch an investigation into what they termed group copying (The-Hispian, 2015). However, in other instances, WhatsApp has been hailed as playing a positive role in enhancing the quality of education by creating interactive platforms and study groups for students (Kahari, 2013).

Media Landscape Effects

WhatsApp has affected the media landscape by becoming an alternative source of news and boosting the reach of citizen journalists. Stories that are disseminated through WhatsApp are affecting the way news is consumed as people quickly share stories and photos before traditional news sources are able to get the story. For example, in Nigeria, it is estimated that at least one in every seven stories originates from social media such as WhatsApp (Africapractice, 2014).

By spurring on citizen journalism, WhatsApp has in somewhat contributed to a decline in ethics of journalism as some stories which are untrue are peddled and shared to a point where it becomes difficult to determine whether a story is true or false. For instance, in February 2016, it was rumoured through

WhatsApp that the Eritrean Government was encouraging men to marry two wives. Sharing of the story via WhatsApp led to jokes and pictures being modified to emphasise the story to such an extent that it became difficult to ascertain whether the story was true or false (Ndlovu, 2016).

Medium Veracity and the Domino Effect

The WhatsApp content is user-generated and this affects the way a receiver interprets or reacts to messages sent through WhatsApp. User-generated content has varying levels of veracity or reliability depending on the content that is being disseminated. Thus, a story that appears on WhatsApp needs to be corroborated by the user through cross-checking it either on other social networks or on television or in a newspaper. By necessitating cross-checking or verification of content, it can be said that WhatsApp generates a social media domino effect.

The foregoing suggests that a medium's content is not taken as wholly genuine by everyone. This is reminiscent of Stuart Hall's, (1973/1980) concepts of dominant reading, negotiated reading and oppositional reading. With regards to television viewing, a viewer can agree with ideological assumptions and mainstream suggestions of a story or content (dominant/hegemonic reading). Even, a viewer can add or subtract from the ideological assumptions and suggestions of a story or content (negotiated reading). Again, a viewer can disagree with the ideological assumptions and suggestions of a story or content (oppositional reading). In other words, when one interacts with a text, the interaction results in a reading or reaction which can either be in line with what the writer/producer of the text intended (dominant reading) or partly in line with what the producer intended (negotiated reading), or not in line with what the producer intended (oppositional reading).

By extension, we can analyse how someone reads or interacts with a text via WhatsApp by inferring that they know that the content has been generated by peers and not experts. This peer-generated content necessitates that the content be cross-checked or verified, especially if it relates to a news item or an innovation. This cross-checking or verification implies that WhatsApp texts predominantly elicit a negotiated reading because of the low medium veracity predicated upon user-generated content.

Medium veracity refers to the extent to which one holds information conveyed through a medium as either true or false or genuine versus fabricated. A headline news story in a newspaper is regarded as true because the reader believes the producer of the content is a professional. The same is true for television content which one believes is generated by professionals albeit the content may be subjected to the various kinds of readings as noted above. Thus, it can be said that newspaper and television contents have a high medium veracity rating as compared to user-generated content circulated through WhatsApp.

Medium veracity is contingent on two main variables, authorship or the producer of the content and the reader's mind-set. In the instance of authorship or the producer, if someone believes that God wrote the Bible, then they will read the entire Bible as flawless and inspired and will not questions its claims. In this case, the medium veracity of the Bible is high. If someone reads a newspaper title, they will believe the content partially because they believe the story was written by a professional and the same is true for a television news story. However, a story shared through WhatsApp is subjected to scrutiny and verification from other media sources because the author or producer is not necessarily a professional. An illustration we can take is the example of a tsunami warning. If either a newspaper or television station warns of a tsunami in Botswana, the reader or viewer will pause to digest the information; however, if the same story is disseminated on WhatsApp, it will be briskly dismissed as a hoax because of the issue of authorship or source.

Thus, medium veracity means that different media channels have different levels of reliability or attract different levels of trust. The kind of content that is shared through a medium attracts different levels of medium veracity. For example, a video shot by eye witnesses of a xenophobic attack is treated as genuine via WhatsApp because the person who has shared it is authenticating what they have seen. Using and generating content create a level of credibility and sense of ownership which makes WhatsApp more authentic for certain content like personal videos and selfies, but not for content concerning hard news commentary.

The issue of medium veracity results in people using various channels of social media to cross check information. In other words, one cannot take a news story shared via WhatsApp at face value, it can therefore be said that WhatsApp has an effect of encouraging readers or audiences to develop a habit of cross checking information with other news platforms and sources.

By necessitating cross checking or verification of stories, user-generated content encourages a social media domino effect. Information on traditional forms of media such as radio and television can be accepted by the audience without verification, for example a state of the nation address by South African president Jacob Zuma on national television is accepted as genuine. A video clip of the same address on WhatsApp is normally subjected to verification by going online or switching on the television to ensure that the story is real. This domino effect makes social media and WhatsApp for instance highly dependent on second opinions or verification. Thus, a story shared via WhatsApp is seen as credible only after the user has seen the same story on other media channels or social media platforms. So the more social media platforms a story occupies the more credible it appears, hence the term social media domino effect as social media platforms are in a symbiotic relationship.

Media Literacy Effect

Media literacy refers to "the ability to access, analyse, evaluate and create messages across a variety of contexts" (Livingstone, 2004). The importance of media literacy is underscored by how many governments have been creating legislations to enhance media literacy. As a relatively cheap resource, WhatsApp enables people to gain access to and improve media literacy. By virtue of allowing everyone who has the means to post content, WhatsApp presents users with a barrage of information and messages which enables one to become more analytical as they sift through large quantities of messages. Since it presents information from various sources, it can be argued that WhatsApp makes users more analytical as they have to sift through scores of information and opinions in a dialogic manner. As a user-generated content platform, WhatsApp enhances media literacy by involving users in the production of content. Some researchers have gone so far as to suggest that WhatsApp can help improve writing skills and language acquisition (Alsaleem, 2014). By typing texts, taking and sharing pictures and videos, WhatsApp enables users to become more media literate.

An increase in media literacy can arguably empower citizens and enable them to participate actively in media culture (Jenkins H., 2015). Little wonder then WhatsApp has been used to improve sexual health preferences among adolescents in Kenya, or to organise a boycott against steep DSTV prices hike in Zambia, or to organise a demonstration by women in Zimbabwe. Even in Uganda after the government blocked WhatsApp and Facebook during the 2016 presidential elections, people still found creative ways of getting on to social media platforms by using alternative service providers.

The fact that WhatsApp is used by people of all age groups shows how it has improved media literacy levels as even old people who were born before computers (BBC) have had to learn how to navigate smartphones in a bid to keep up with the young people or so-called digital natives when communicating.

SOLUTIONS AND RECOMMENDATIONS

The exponential growth in popularity of WhatsApp has made it a messaging platform which mobile operators, users and governments continue to harness and adopt. Given the boom in its use, five recommendations are made for its improvement.

WhatsApp has affected interpersonal relationships in ways that vary from an increase in revenge porn to providing a forum for free counselling services. As such, it is important for appropriate legislations to be put in place to ensure that Whatsapp messaging is not abused. African governments need to introduce legislations which deter people from seeking to inflict emotional and psychological harm on fellow citizens.

The popularity of WhatsApp has compelled mobile operators to introduce blended bundles in a bid to compete with the OTT. This era of cheaper prices and data should be maintained rather than interrupted by inhibitive legislation or regulations which are now looming in countries such as South Africa.

The common uses of WhatsApp are slanted towards positive uses such as sharing information with colleagues, promoting business products and accessing news items such that it becomes imperative for African governments to put in place policies that encourage the use of WhatsApp rather than to discourage its use. Policies that can further encourage the use of WhatsApp include reducing tariffs on smartphones so that retail prices can be lowered further. While it can be said that WhatsApp is being utilised for various positive outcomes like enhancing the quality of education in both secondary schools and universities or the improvement of media literacy among the general population, the issue of limited infrastructure of mobile phone services and the urban- rural divide that characterises many African countries inhibit many from using social media platforms. There is therefore a huge need for African governments to expedite infrastructural development so as to enable a larger segments of the population to benefit from the opportunities presented by WhatsApp and other forms of social media.

While there is an improvement in media literacy through the use of WhatsApp, there is need for African users to be educated on how to use social media effectively so has to maximise its potential. For example, the practice of using group chats as a way of influencing youths to make better life choices should become widespread and media literacy courses in academic institutions need to incorporate social media into their syllabi.

Finally, given the domino effect of social media, there is need to ensure that African citizens are educated on how to cultivate the power of sharing information and mobilising resources in a manner that enhances the value of democracy and fuels economic development by inspiring and supporting local innovations such as Ongair and Javago.

FUTURE RESEARCH DIRECTIONS

The predominance of WhatsApp and the influence that it has had on interpersonal relationships, education, politics, and other facets of life warrants a scientific inquiry into how users interact with the texts presented through WhatsApp. Building on concepts of medium veracity and the social media domino effect will enable academics to interrogate how WhatsApp is affecting our perceptions and values at a more nuanced level.

It is also important to conduct more research on the reasons people are using WhatsApp so as to develop a body of literature that goes beyond the surface motivations of cheap prices, technological determinism and peer pressure.

Whilst there has been a lot of research on certain topics such as how WhatsApp is affecting education, there has been little research on how WhatsApp is affecting citizen journalism, or how WhatsApp is affecting societal values.

As the influence of OTT companies such as WhatsApp continues to grow, it is important to look at how this is affecting the information and communication technology (ICT) sector as a whole. In other words, is the ICT industry always going to be shaped and reshaped by disruptors?

Conclusion WhatsApp has affected African societies in several ways ranging from changing how people communicate to how people are learning in secondary schools and university; and how people are building and maintaining relationships. At a cultural level, WhatsApp has resulted in complex relationship break-up issues as shown in the cases of revenge porn. It has also created an environment wherein citizen movements are able to organise and coordinate protests in a more effective manner. At a political level, WhatsApp has empowered citizen movements on one hand and attracted the ire of many governments on the other hand resulting in intermittent blocking and monitoring of WhatsApp. As citizen movements mobilise people by sharing information on several social media platforms, the domino effect of social media and the concept of medium veracity becomes more apparent as stories are passed from one social media platform to another, creating a more credible and formidable story. At an economic level, WhatsApp has enabled many businesses to communicate with their customers in an economic and efficient manner. Other businesses have managed to come up with innovative products that are thriving through WhatsApp such as Ongair. Many mobile operators have been compelled to introduce blended bundles and cheaper data costs because of WhatsApp's escalating popularity. At a technological level, WhatsApp has transformed the mobile phone industry by forcing operators and retailers to reduce the prices of smartphones. It has also made it clear that there are some legislative gaps when it comes to the provision of fair prices and competition. As an educative platform, WhatsApp has increased the levels of media literacy as people are able to create and share their own music, videos and news items. At an ideological level by affecting how people are interacting in group chats and how organisations are mobilising people, WhatsApp presents a renewed sense of connectedness or community. As the popularity of WhatsApp continues to soar, more communicative spaces will probably be created and societal values may continue to shift as people adapt to the possibilities that will be created by WhatsApp.

REFERENCES

Africapractice. (2014, April). Retrieved from http://www.africapractice.com/wp-content/uploads/2014/04/Africa-Practice-Social-Media-Landscape-Vol-1.pdf

Arimoro, A. E. (2015). Applying the law to tackle the menace of revenge porn. Creating Global. *Journal of Humanities and Social Science*, 75-80.

Aslama, M. (2009). *Public Spheres, networked publics, networked public Spheres*. New York: Donald McGannon.

Avis, W. (2015). *Mobile phone and social media interventions for youth development outcomes*. Birmingham, UK: GSDRC, University of Birmingham. Retrieved from http://reliefweb.int/sites/reliefweb.int/files/resources/HDQ1317.pdf

Boyd, D. (2014). *Its Complicated*. Yale University Press.

Bradely, J. L. (2016, March 29). *Digital Vortex. How Digital Disruption is redefining industries*. Retrieved from Cisco.com: http://www.cisco.com/c/dam/en/us/solutions/collateral/industry-solutions/digital-vortex-report.pdf

Buchanan, E. (2016). *Kenya: Anti-election protesters hit Nairobi streets despite police threats of lethal.* Retrieved from International Business Times: http://www.ibtimes.co.uk/kenya-anti-election-protesters-hit-nairobi-streets-despite-police-threats-lethal-force-1563821

Caulderwood, K. (2015, March 29). *Burundi 2015: How Twitter, WhatsApp and Facebook Influenced.* Retrieved from ibtimes: http://www.ibtimes.com/burundi-2015-how-twitter-whatsapp-facebook-influenced-protests-possible-coup-1921134

Chen, A. (2016). *Independent Work Report Fall, 2015. Understanding Pricing Effects on Mobile Data.* Retrieved from cs.princeton.edu: https://www.cs.princeton.edu/sites/default/files/uploads/ava_chen.pdf

Devezin, T. F. (2016, March 29). *Enhanced Engagement and Youth Participation in Sexual Health Programming. Effects of Smartphone Communication (WhatsApp).* Retrieved from www.africasexuality.org: http://www.africasexuality.org/wp-content/uploads/2016/02/43.pdf

Dlodlo, N. (2014). Measuring Selected M-Texting Addiction Indicators with Gender and Self. *Mediterranean Journal of Social Sciences*, 489-499.

Drago, E. (2015). The Effect of Technology on Face-to-Face Communication. *The Elon Journal of Undergraduate Research in Communications*, 13-19.

Fin24. (2016, February 10). *Watch out for these 5 WhatsApp scams.* Retrieved from fin24.com: http://www.fin24.com/Tech/Opinion/watch-out-for-these-5-whatsapp-scams-20160210

Freedom-House. (2016, June 24). *Rwanda.* Retrieved from Freedom House: https://freedomhouse.org/sites/default/files/resources/FOTN%202015_Rwanda.pdf

Fuchs, C. (2011, 8 June). *Critical Media and Communication Studies Today. A Conversation.* Retrieved from wordpress.com: https://dwmw.wordpress.com/2011/08/06/critical-media-and-communication-studies-today-a-conversation-between-dwayne-winseck-and-christian-fuchs-part-1/

Fuchs, C. (2013). *Social Media: a critical introduction*. London: Sage.

Gauntlett, D. (2011). *Making is creating: the social meaning of creativity, from DIY and knitting to Youtube and Web 2.0*. London: Polity Press.

Goldstuck, A. (2014, July 29). *Massive increase in SA smartphone purchases.* Retrieved from Mail & Guardian: http://mg.co.za/article/2014-07-29-smartphone-uptake-in-south-africa-rising-dramatically

Grootes, S. (2016, January 14). *MTN calls for regulation of messaging services.* Retrieved from Eyewitness News: http://ewn.co.za/2016/01/14/MTN-calls-for-regulation-of-messaging-services

Hall, S. (1973/1980). Encoding/decoding. In Culture, Media, Language: Working Papers in Cultural Studies (pp. 128-338). London: Hutchinson.

Harare-Bureau. (2016, March). *20 cops detained over WhatsApp Chat.* Retrieved from The Chronicle: http://www.chronicle.co.zw/20-cops-detained-over-whatsapp-chat/

Henry, J. V. (n.d.). Enhancing the Supervision of Community Health Workers with WhatsApp Messaging: Qualitative Findings From 2 Low Resource Settings in Kenya. *Science and Practice*. doi:10.9745/GHSP-D-15-00386

Hispian. (2016, June 24). *WhatsApp passes exams.* Retrieved from hisp.org: http://www.hisp.org/wp-content/uploads/2015/11/The-Hispian-18-September.pdf

Holmes, C. (2016, March 29). *Mobilising through Social Media: The Case of South Africa.* Retrieved from http://blogs.lse.ac.uk/africaatlse/2016/02/11/feesmustfall-southafrica-social-media/

ICT-Security-Monitoring-Services-Report. (2016, June 26). *ICT Security Monitoring Services Report.* Retrieved from www.ssa.gov.za: http://www.ssa.gov.za/Portals/0/SSA%20docs/CSIRT/ICT%20Security%20Monitoring%20Services%20Report%2026%20Ma%202016.pdf

IOL. (2016, June 24). *Retrieved June 24, 2016. IOL is on WhatsApp*. Retrieved from iol: http://www.iol.co.za/news/south-africa/iol-is-on-whatsapp-2033092

Jenkins, H. (2006, October 20). *Media literacy begins at home.* Retrieved from technologyreview: http://www.technologyreview.com/read_article.aspx?id=13422&ch=infotech&sc=&pg=1

Jenkins, H. (2016, March 28). *Affecting Publics and Social Media: An Interview with Zizi Papacharisi (Part One).* Retrieved from henryjenkins: http://henryjenkins.org/2015/01/affective-publics-and-social-media-an-interview-with-zizi-papacharissi-part-one.html

Jovago. (2014, June 20). *Book an Hotel using WhatsApp, Skype, Viber or Live Chat with Jovago.* Retrieved from Jumia: http://jumia.pr.co/102254-book-an-hotel-using-whatsapp-skype-viber-or-live-chat-with-jovago

Kabweza, L. S. (2016, March 29). *Revenge porn laws in Zimbabawe: Parliament considers petition.* Retrieved from Techzim: http://www.techzim.co.zw/2016/03/revenge-porn-laws-zimbabwe-parliament-petition/

Kahari, L. (2013). The effects of Cell phone use on the study habits of University of Zimbabwe First Year Faculty of Arts Students. *International Journal of Education and Research,* 1–12.

Kalyango, Y–K. (2013). *Impact of Social Media on Political Mobilisation In East and West Africa.* Retrieved from globalmediajournal: http://www.globalmediajournal.com/openaccess/impact-of-social-media-on-political-mobilization-in-east-and-west-africa.pdf

Langat, A. C. (2015). Influence of Social Media on Study Habits of Undergraduate Students in Kenyan Universities. *International Journal of Novel Research in Humanity and Social Sciences*, 42-55.

Lisulo, S. (2016, March). *MTN launches 'Prestige' to raise customer service experience.* Retrieved from Post: http://postzambia.com/news.php?id=16335

Liu, S-H. (2013). Adolescents' Interpersonal Relationships with Friends, Parents, and Teachers When Using Facebook for Interaction. *Creative Education*, 335-339. Retrieved from Creative Education: http://file.scirp.org/pdf/CE_2013051015134212.pdf

Livingstone, S. (2004). What is media literacy? *Intermedia*, 18-20.

Luescher, M. T. (2016). *Student power in twenty-first century Africa: The character and role of student organising.* Retrieved from http://scholar.harvard.edu/files/manja_klemencic/files/2016_luescher_-_klemencic_-_chapterfor_r_brooks_final1_0.pdf?m=1454077927

Mascheroni, G. V. (2015). "Girls are addicted to likes so they post semi-naked selfies": Peer mediation, normativity and the construction of identity online. *Journal of Psychosocial Research on Cyberspace*, 5.

Mavura, A. (2016, March 29). *Empowering women in Ghana and Zimbabwe through Social Media.* Retrieved from social-media-for-development: http://social-media-for-development.org/

McLuhan, M. (1964). *Understanding Media*. London: Routledge.

Mohammed, O. (2016, May 27). *How Tanzanian journalists use WhatsApp to report the news.* Retrieved from ijnet.org: http://adage.com/article/digital/bbc-teams-viber-whatsapp-distribute-content/302906/

Msonza, N. (2016, March 29). *Internet Freedom and repression in Zimbabwe.* Retrieved from boell.de: https://www.boell.de/sites/default/files/perspectives_africa_juli_2015_gamechanger.pdf

Mulupi, D. (2016, March 29). *Customer care via WhatsApp: Kenyan company finds traction abroad.* Retrieved from howwemadeitinafrica.com: http://www.howwemadeitinafrica.com/customer-care-via-whatsapp-kenyan-company-finds-traction-abroad/

MusvoZimbawe. (2016, March 29). Retrieved from musvozimbabwe.com: https://musvozimbabwe.com/2015/08/daphne-nyasha-mazara-musvo-videoake/

Ndlovu, B. (2016, March 29). *Thrilling reactions over Eritrean Law.* Retrieved from The Chronicle: http://www.chronicle.co.zw/thrilling-reactions-over-eritrean-law/

Nehanda Radio. (2013, March 11). Retrieved from nehandaradio.com: http://nehandaradio.com/2013/11/03/why-macheso-willnever-use-whatsapp-or-facebook/

Newman, N. D. (2012). Social Media in the Changing Ecology of News. *International Journal of Internet Science*, 6–22.

O'Hara, K. M. (2014). Everyday dwelling with whatsApp.*Proceedings of the 17th ACM conference on Computer supported cooperative work & social computing* (pp. 1131-1142). Baltimore, MD: ACM.

Ogunlade, O. (2015, February 12). *More than ten lucrative ways you can use WhatsApp for business.* Retrieved from Techpoint: https://techpoint.ng/2015/12/02/how-to-use-whatsapp-for-business/

Park, P. (2016, January 13). *How many people use social media in Africa.* Retrieved from cnn.com: http://edition.cnn.com/2016/01/13/africa/africa-social-media-consumption/

Phippen, A. (2016, March 29). *Sexting: An Exploration of Practises, Attitudes and Influences*. Retrieved from channel4.com: http://www.channel4.com/media/c4-news/pdf/11NSPCCresearch2.pdf

Posts, P. (2016, June 24). *Nairaland Forum*. Retrieved from nairaland.com: http://www.nairaland.com/publisher/posts

Prinsloo, C. (2015). *Digital Disruption, Changing the rules of Business for a hyper connected world*. Retrieved from gibs.co.za: https://www.gibs.co.za/aboutus/centres/Business%20Analysis%20And%20Research/Documents/digital%20disruption_GIBS_small_spreads.pdf

Rama Naidu, J. D. (2016, March 29). *When Xenophobia Rears its Ugly Head: A Challenge To Responsible and Responsive Governance*. Retrieved from mott.org: http://www.mott.org/files/pubs/WhenXenophobiaRearsItsUglyHead.pdf

Rambe, P. (2013). *Using mobile devices to leverage student access to collaboratively - generated resources: A case of WhatsApp instant messaging at a South African University*. doi:10.2991/icaicte.2013.66

Rogers, Y. S. (2015). *Interaction Design: Beyond human-computer interaction* (4th ed.). London: Wiley.

Shamabare, R. R. (2012). Are mobile phones the 21st century addiction? *African Journal of Business Management*, 573-577.

Shambare, R. (2014). The Adoption of WhatsApp: Breaking the Vicious Cycle of Technological Poverty in South Africa. *Journal of Economics and Behavioural Sciences*, 542-550.

Shapshak, T. (2016, March 29). *Why WhatsApp is South Africa's Favourite App*. Retrieved from forbes.com: http://www.forbes.com/sites/tobyshapshak/2015/09/04/why-whatsapp-is-south-africas-favourite-app/#68839f8b7939

Shriram, V. (2016, March 29). *Build Karma Points on Social Media*. Retrieved from blogs.ucl.ac.uk: http://blogs.ucl.ac.uk/global-social-media/tag/whatsapp/

Spicy-Snacks. (2016, June 24). Retrieved from Restina: http://restina.net/en/south-africa/randburg/restaurant-cafe/spicy-snacks/295326594006174

Titcomb, J. (2016, March 29). *WhatsApp reaches one billion users*. Retrieved from Telegraph: http://www.telegraph.co.uk/technology/2016/02/02/whatsapp-reaches-one-billion-users/

Trottier, D. (2015). Theorising Social Media, Politics and the State. In Social Media, Politics, and the State. Protests, Revolutions, Riots, Crime and Policing in the Age of Facebook, Twitter and Youtube (pp. 3-38). New York: Routledge.

Tuurosong, D. (2014). The Social Media Scourge among Unversity Students: A study of the University for Development Studies in, Ghana. *Journal of Asian Development Studies*, 62-74.

Vermeulen, J. (2016, March 29). *The data cost of WhatsApp voice calling in SA*. Retrieved from http://mybroadband.co.za/news/smartphones/122106-the-data-cost-of-whatsapp-voice-calling-in-sa.html

Vircoulon, T. (2016, June 24). *Burundi turns to WhatsApp as political turmoil brings media blackout*. Retrieved from The Guardian: http://www.theguardian.com/global-development/2016/jun/14/burundi-turns-to-whatsapp-as-political-turmoil-brings-media-blackout

Waldman, L. (2016, March 29). *Sexual and Reproductive Health Rights and Information and Communications Technologies: A Policy Review and Case Study from South Africa*. Retrieved from http://opendocs.ids.ac.uk/opendocs/bitstream/handle/123456789/5787/ER113_SexualandReproductiveHealth.pdf.pdf;jsessionid=5DA3C2417198E54224C0B0B8812D75BC?sequence=1

Wangui, J. (2014). *Government Goes Big on Social Media*. Nairobi: Business and Tech.

Yeboah, J. &. (2014). The Impact of WhatsApp Messenger Usage on Students Performance in Tertiary Institutions in Ghana. *Journal of Education and Practice*, 157-164.

Young, L. E. (2016, March 29). *Mobilization under threat.* Retrieved from laurenelyssayoung.com: http://www.laurenelyssayoung.com/wp-content/uploads/2016/02/Young-Party-Emotions-Experiment-Sep-2015.pdf

Zanamwe, N. R. (n.d.). Use of Social Networking Technologies in Higher Education in Zimbabwe: A learner's perspective. *International Journal of Computer and Information Technology*, 8-18.

Zimbabwean, Z. (2016, March 29). *Indecent Pictures of Undressed ZRP Officer go Viral!* Retrieved from http://iharare.co.zw/constable-chiwanza-sagas/

KEY TERMS AND DEFINITIONS

Communication: The transfer of information and meaning.
Information: Messages or content that conveys some kind of meaning.
Interpersonal Relationships: A relationship between two or more people.
Media Veracity: The extent to which a media channel is reliable or believable.
Selfies: Pictures of oneself, usually taken by a digital camera or cell phone.
Social Media: A communicative platform where users interact by generating their own content.
Social Media Domino Effect: the way information from one social medium influences other mediums.
User-Generated Content: Information that is produced by the users or interacting people.

Chapter 4
Social Media Effect on International Business Venturing in Nigeria:
A Conceptual Review

Mercy Ejovwokeoghene I. Ogbari
Covenant University, Nigeria

Moses Ayokunle Akanbi
Covenant University, Nigeria

Ugonna Kelly Ejimofor
Covenant University, Nigeria

Barnabas M. Suleiman
Covenant University, Nigeria

ABSTRACT

Entrepreneurs all over seek ways of introducing their products to international markets, but international business environments pose a lot of opportunities and threats to foreign entrants. The cultural, political and technological environment has a lot to do in entrepreneurial success in global markets. The study examined the effect of social media on international business venturing. The general objective of the study is to critically determine the extent to which social media have contributed to the international venturing of entrepreneurs in Nigeria. The study concludes that since technological products provide ample benefits for international entrepreneurs, organizations should endeavour to provide indispensable technologies that support international transactions.

INTRODUCTION

Venturing in International business, unlike domestic business, requires operating simultaneously in more than one kind of environment, coordinating one's business operations, and using the experience gained in one country for making decisions in another country.

According to Otokiti (2007),international Business refers to the carrying out business activities across national boundaries for the satisfaction of human needs and wants and the extent of a firm's involvement abroad is a function of its commitment to the pursuit of foreign markets.The demands are tough and the stakes are high. International business entrepreneurs are not only sensitive to different marketing

DOI: 10.4018/978-1-5225-1859-4.ch004

environments internationally but balancing marketing moves worldwide to seek optimum results for their businesses is the key.

In the research conducted by OECD (1997), governments support cross-border entrepreneurship and in particular exports to increase national wealth and to improve international competitiveness of the national economy. Cross-border entrepreneurship has become a more widespread phenomenon in the past decades. Traditionally, multinational enterprises (MNEs) were mainly responsible for flows of international trade and foreign direct investment (FDI), which are the prime driving forces of globalization. However, Communication is a process that involves exchange of information, thoughts, ideas and emotions. It requires a sender, a message and a receiver. Also in recent times, the social media has been a major business interactive tool due to the proliferation of internet and advancement in global technology, information seems to be disseminated faster and links between persons are easily built and this has a way of creating opportunities and risk for those using the medium; in essence the usage of social media has its own positive and negative sides in business.

Social media encompass a variety of Web-based and mobile technologies. For business professionals, social media may be used for professional networking, interfacing with partners and customers. Social media comprise a number of online and mobile resources that provide a forum for the generation, sharing, and discussion of individualized ideas and content (source?). Social media are commonly defined by specific applications and/or Web tools, most of which are widely accessible and free to use or available at minimal cost. These applications may be categorized by purpose, including such functions as professional networking (LinkedIn, Doximity), social networking (Facebook, Google_), recommending/filtering (Yelp, Delicious), media sharing (Flickr, YouTube), content production (blogs, Twitter), knowledge/ information aggregation and location-based services (Foursquare), among others.

The use of social media tools such as Facebook, Twitter, text messaging, e-mail, photo sharing, social networking, and the like does not have a single preordained outcome. Therefore, attempts to outline their effects on business venturing are often reduced to duelling anecdotes. According to Madden and Zickuhr (2011), it was discovered that as at March 2008, there were 200 million blogs worldwide, 450 million people on Facebook, 27 million tweets every 24 hours, and 1.2 billion YouTube views each day. These staggering statistics reflect the explosive growth social media has experienced over the past years. Nevertheless, the effects of social media on international business venturing for entrepreneurs seems unclear especially in developing economies as Nigeria, the international market serves as an avenue for creative entrepreneurs to earn foreign income outside their country of existence; the international business environment makes it competitive for many that want to venture in the international business due to dynamic nature of environment and culture. Social media communication factors are unpredictable and in most times they form the basis for decision for local entrepreneurs to break into new markets due to affordability. There are many ways to break into new markets which include exporting, foreign direct investment, joint venture and international partnerships, all these options seem to tie their feasibility around what is obtainable in the market to break into. For example, factors like technological change, cultural and international governmental laws are difficult to predict to get a business to be stable in a new international environment.

In retrospect the effects of social media on international business to the entrepreneurs in the developing economies cannot be overemphasized as the effects of social media on international business venturing in Nigeria has not gotten so much reference. A lot of attention considered to be important elements that account for the growth and remarkable performance of the entrepreneurial ventures in Nigeria are not directed towards the efficacy of social media. Also, the effects of social media as a major technological change factors on business entrepreneurs are not completely clear. Literatures that have to do with international marketing environment and international entrepreneurship show less on developing economy (Zahra, Ireland, and Hitt, 2000).

Overview of International Business and Social Media

International business refers to all business transactions carried out beyond international geographical frontiers or boundaries. Private companies undertake such transactions for profit; governments may or may not do the same in their transactions. These transactions include sales, investments, sourcing, manufacturing, marketing and transportation. Today, global events and competition affect almost all companies large or small because most sell output to and secure supplies from foreign countries. Many companies also compete against products and services that come from abroad. A more complex answer is that a company operating in the international business field will engage in modes of business, such as exporting and importing, that differ from those it is accustomed to on a domestic level.

Greenaway, Gullstrand, and Kneller, (2008) posit that social media, i.e., blogs, wikis, social networking sites like Facebook and LinkedIn, multimedia sharing sites like YouTube and Flickr, and social tagging sites like Digg and Yelp, represent a revolutionary shift in the way businesses interact. Social media have democratized information and empowered ordinary citizens with the ability to organize, share information, and be heard like never before in history. Social media are words of mouth on steroids and are beginning to morph from a fun and easy way to stay socially connected with friends into a dynamic and interactive way of doing business and practicing law. Businesses now interact with their customers and other stakeholders in an entirely different way.

Also, Autio (2015) defines international business as profit-related activities conducted across national boundaries. The environment for those business activities within which the international manager functions is shaped by major developments in the world. Such developments are globalization; the various regional trading blocs such as the European Union with the introduction of the Euro as its legally tradable currency; the North American Free Trade Agreement (NAFTA); the Commonwealth of Independent States (CIS); information technology; workforce diversity; the status of the emerging economies of China, India, Mexico and Brazil; and the unstable political situation in various parts of the world, such as the one in Afghanistan, the Middle East and in various parts of Africa.Furthermore, an international business is any firm that engages in international trade or investment. A firm does not have to become a multinational enterprise, investing directly in operations in other countries, to engage in international business, although multinational enterprises are international businesses. All a firm has to do is export or import products from other countries. As the world shifts toward a truly integrated global economy, more firms, both large and small, are becoming international businesses.

As their organizations increasingly engage in cross-border trade and investment, it means managers need to recognize that the task of managing an international business differs from that of managing a purely domestic business in many ways. One such fundamental difference is the fact that countries are different. As mentioned, Birkinshaw (1997) pointed that countries differ in their culture, political systems, economic systems, legal systems, and levels of economic development. Despite all the talk about the emerging "global village" and despite the trend towards globalization of markets and production, these fundamental differences still remain and are very profound and enduring. Differences between countries require that an international business vary its practices country by country. Marketing a product in Great Britain may require a different approach than marketing the same product in China; managing Italian workers might require different skills than managing Japanese workers; maintaining close relations with a particular level of government may be very important in Greece and irrelevant in Germany; the business strategy pursued in Sweden might not work in Singapore; and so on. Managers in an international business must not only be sensitive to these differences, but they must also adopt the appropriate policies and strategies for coping with them.

Another way in which international business differs from domestic business is the greater complexity of managing an international business. In addition to confronting the problems that arise from the differences between countries, a manager in an international business confronts a range of issues that the manager in a domestic business never encounters. An international business must decide where in the world to site its production activities to minimize costs and to maximize the value added. Then it must decide how best to coordinate and control its globally dispersed production activities. An international business also must decide which foreign markets to enter and which to avoid. It also must choose the appropriate mode for entering a particular foreign country. Conducting business transactions across national borders requires an understanding of the rules governing the international trading and investment system. Managers in an international business must also deal with government restrictions on international trade and investment. They must find ways to work within the limits imposed by specific governmental interventions.

Furthermore, even though many governments are nominally committed to free trade, they often intervene to regulate cross-border trade and investment. Managers within international business must develop strategies and policies for dealing with such interventions. Cross-border transactions also require that money be converted from the firm's home currency into a foreign currency and vice versa. Since currency exchange rates vary in response to changing economic conditions, an international business must develop policies for dealing with exchange rate movements. A firm that adopts a wrong policy can lose large amounts of money, while a firm that adopts the right policy can increase the profitability of its international transactions.

The Participants in International Business

According to Marios and Spyros (2007), companies of all types and sizes, and in all types of industries, enter into international business transactions. Manufacturing companies, service companies, and retail companies all search for customers outside their countries. As mentioned earlier, an international company is a business that engages directly in any form of international business activity such as exporting, importing, or international production. No matter what the nature of the international business transactions, there are different types of international companies. Small companies are becoming increasingly active in international trade and investment. Companies are having quicker turnaround times with regard to exports and are growing fast as a result of increased international business activities. Because of improved technological advancements; electronic distribution is a cheap and effective method for many small businesses. Some small businesses are reaching out to customers exclusively through the World Wide Web (WWW). Despite all these technological amenities, many small businesses that are capable of exporting have not yet begun to do so.

Acs, and Audretsch (2003) opined that there are certain myths that keep small businesses from exporting: there is no export financing available for small businesses; small businesses have no place to turn for export advice; the licensing requirements needed for exporting are not worth the effort; only large companies can export successfully; small businesses do not have the right people to assist in exporting; small businesses find it difficult to research and identify international markets; and so on. Despite all these, small businesses play a vital role in their national economies through employment, new job creation, development of new products and services, and international operations, typically exporting. Although many small businesses are affected by globalism only to the extent can function either rather independently or as parts of a tightly integrated global network. Independent operations tend to have a good understanding of local culture and are often able to adapt quickly to changing local market conditions. On the other hand, firms that operate as global networks often find it easier to respond to changing conditions by shifting production, marketing, and other activities among national units. Depending on the type of business, either structure can be appropriate. The multinational corporations' economic and political muscle makes them highly visible. Large companies generate a large number of jobs, greater investment, and significant tax revenue for the areas in which they operate. On the other hand, the downsizing of large international companies, or the closing of factories throughout the world, makes these companies visible as well. Furthermore, these companies' transactions involve large amounts of money. For example, in 1998, Daimler-Benz of Germany announced a merger with Chrysler of the United States, which was valued at $40 billion; in the same year, Exxon and Mobil, two global petroleum companies, created a merger worth $86 billion. Around the world, MNCs are already giants and growing even more gigantic. These big, traditional manufacturing enterprises, with their long planning horizons, are leading the drive toward globalization. Increasingly, too, service organizations also go global financial institutions such as Citigroup, retailers such as Wal-Mart, telecommunications companies such as MCI-WorldCom. Leadership in global trends comes from the close involvement of these companies in cross-boundary relationships.

Modes of Entry International Markets

According to Allen (2003), a firm has a number of choices when it is attempting to evaluate a mode of market entry. It should be remembered that there is no one method which is suitable to all firms under a particular set of circumstances. However, there are number of options which companies and individuals within them have to weigh up under the pertaining market situation. In choosing mode of entry, a number of specific factors will shape their choices. Speed of entry required, financial resources available, flexibility required, period of return on investment, long term objectives of the organization, degree of marketing control and degree of risk aversion.

There are various modes of entry that a multinational corporation may choose to adopt. There are a number of ways to classify the options of entry open to corporations, but no one method is accurate because of the dynamic nature of the global economy.

Export

Exporting is the marketing and direct sale of domestically-produced goods in another country. Exporting is a traditional and well-established method of reaching foreign markets. There is no need for the company to invest in a foreign country because exporting does not require that the goods be produced in the target country. There are basically two types of exporting, indirect and direct exporting.

Figure 1. Hierarchical model of choice of entry mode
Source: Allen (2003).

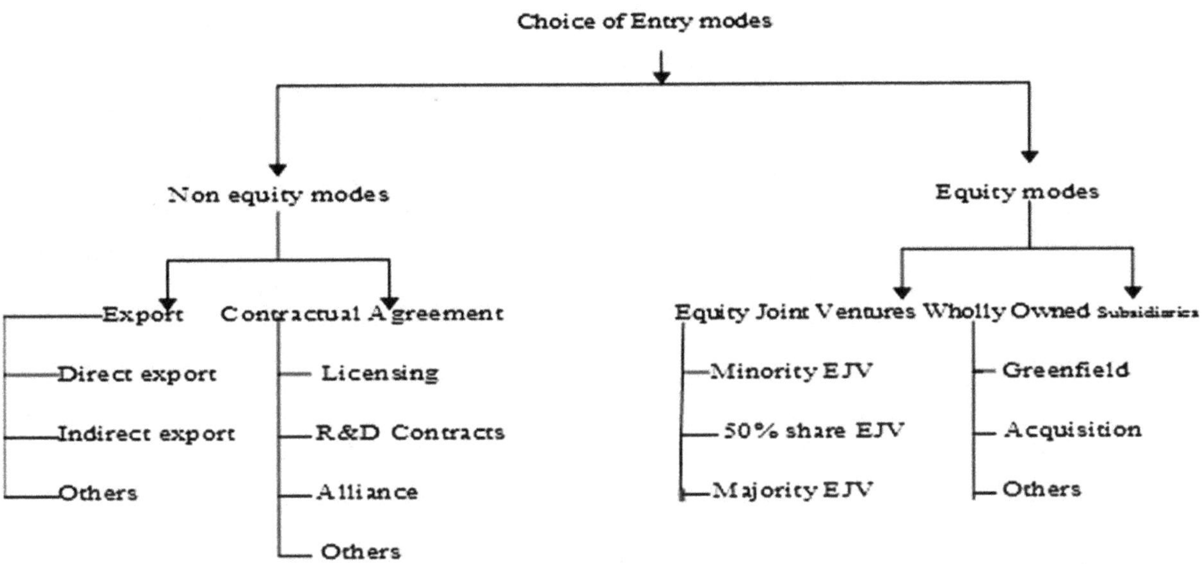

Indirect Export

This is often referred to as passive exporting or as being the result of an 'export pull effect, since people outside the company stimulate the activity. The main strategies which come under this category are:

1. Responding to unsolicited or chance orders. For some companies, this is the first introduction to international market which may stimulate them to explore the feasibility of exporting more seriously as the stage models would suggest. It does have that advantage of being new, unexpected business. It illustrates the fact that there are potential customers yet unreached in the world and reaching them would involve product adaptation costs or promotion costs.
2. Use of a range of home-based intermediaries. This can include making use of courier/express delivery services, export management companies, export houses, trading company, piggybacking, brokers and jobbers. Effectively by using the services of these intermediaries, the experience is little different than undertaking a domestic sale as there is no direct export marketing experience being gained. Courier and express delivery services will be of little use for bulk items, as they will have weight restrictions, but for low - weight high value items they are an option and it is a cost efficient way for business to reach international clients.

Export management companies are estimated to handle approximately 10% of all manufactured goods. They offer a personal service to clients who come to view them as an extension of their own business. It may even be that the buyer does not realize they are dealing indirectly with the exporter. For the exporter without the personnel or funds to run export operation, this route offers market entry at low investment. An export house will buy directly from the domestic manufacturer on behalf of an international client.

Complementary marketing as it is sometimes known, allows one firm to use the distribution channels of another firm to read an international customer. It may be that the existing exporter has excess capacity in the channel, or they may be keen to expand the range of products carried by seeking out complementary, non -competing product to those already carried. The request can also come from an existing customer for another product to be sourced. The domestic firm may then seek a third party offering a complete export package, based on its existing channels, to meet this request.

Successful complementary marketing can work well for both the piggy backer and the carrier. The former gains a ready market and the experience of an existing exporter, while the latter increases the product range and makes more efficient use of channels which would have been served anyway. All of these indirect approaches to market entry have the common advantage of being low cost, as no international sales force is required, but they do not allow the company to develop its own skills and knowledge of the exporting process. By using the expertise of others it is likely that most deals will run smoothly for the company, but crucially direct contact with the end customer is non-existent. Under these circumstances the firm must accept that its involvement in international markets is very shallow.

Direct Export

This is an active form of exporting and can involve the use of domestic and international based intermediaries. The commitment and investment required are greater than indirect exporting, but the reward can also be greater. Within the domestic market the main types are:

1. **Use of Freight Forwarders:** These companies provide advice and guidance and offer documentation and delivery service. They act on behalf of the producer, and are indispensable to an exporter who does not possess the skills in –house to handle the paperwork associated with exporting. A full – service freight forwarder will provide information on shipping, routing, schedules, charges, labelling, certification and consular requirements. By consolidating orders into larger shipments, the freight forwarder can offer a more cost effective service to the smaller exporter than they could obtain independently.
2. **Consortium Exporting:** Under this, a number of companies, normally non – competing, come together in order to combine their skills, resources, especially managerial and financial. They bid for contracts and projects as a group while remaining independent. Often these are linked to large construction project where a consortium is formed, with one lead organization driving the group forward.
3. **Export Department:** Larger companies will be able to form an in-house export department to run all exporting activities. Home-based salespeople travel abroad, perhaps in defined geographical areas, in search of orders. Those methods involving the use of international based intermediaries include:
    a. **Use of Agents:** Agents represent the exporter abroad and are normally paid on a commission basis. They will have territorial sales rights, but may work for more than one company and carry more than one product line. The exporter may remain responsible for the inventory. This is the most common means of market entry for exporters, and is a relatively inexpensive method. However, it is important that does agreement is drawn up and understood by both sides before entering into any relationship, as bad agents do exist and can be difficult to get rid of. On a positive note, they offer market knowledge and access to networks.
    b. **Use of Distributors:** Distributor operates in a very similar way to agents, but the distributor takes title (ownership) of the goods. This means that they purchase the goods from the exporter before going on to re-sell them. The advantage of this method to the exporter is that cash flow is more assured, but there is a loss of control over the price to the end – user, and no direct knowledge about who the end user is.

Licensing

A licensing agreement is a business arrangement between a licensor and a licensee whereby the later uses the former`s monopoly position and right such as Patent, Trade Mark, design or copyright having exclusive right which prevents others from exploiting the idea, design, name or logo commercially. The licensee pays a fee in exchange for the rights to use the intangible property and possibly for technical assistance.

The licensing agreement gives the following advantages to both the licensor and the licensee:

Companies have low initial investments; trade barriers are avoided; there is a great potential for utilizing location economies; the access to local data and information is possible; there is an easier ability to respond to customer's needs and wants. Whereas the disadvantages of licensing: the licensor lack the control over operations; companies face difficulties with transferring the tacit knowledge, for example negotiation of transfer price, monitoring transfer outcome, etc; there is a great chance of creating a competitor; in this case your licensee. We can conclude that licensing is appropriate if there is a well codified knowledge, strong property rights regime, and location advantage.

Franchising

Franchising is a similar entry mode to licensing. By the payment of a royalty fee, the franchisee will obtain the major business know-how via an agreement with the franchiser. The know-how also includes such intangible properties as patents, trademarks and so on. The difference from the licensing mode of entry is that the franchisee must obey certain rules given by franchiser. Franchising is most commonly used in service industries, such as McDonald's, etc.

Reasons for Entering into the Nigerian Market

- **To Broaden Markets:** Saturated home markets ...ask for market development in Nigeria (Coca Cola, McDonalds etc.).
- Multinationals seek new markets in Nigeria to fill product gap and increase return on investment.
- **To Seek Raw Materials:** Multinationals secure the necessary raw materials required to sustain primary business line (Exxon, Wal-Mart).Multinationals also seek to obtain easy access to oil exploration, mining and manufacturing in many developing nations (for example Nigeria).
- **To Seek Production Efficiencies:** By shifting to low cost regions.
- To avoid political hurdles such as import quota, regulatory measures of governments, trade barriers.

Basis of Differences between international business and a purely domestic business:

Because of different cultural, political, economic, and legal systems, and so on, countries are different. The issues at the international business level are more complex than those at the domestic level. When operating internationally, a company should consider its mission (what it will seek to do and become over the years; what is the purpose of the company's existence?); its objectives (what is the company trying to accomplish according to its mission?); and strategies (means to achieve its objectives).

According to Baumol, (2002), companies engage in international business in order to:

- **Minimize Competitive Risk:** Many companies enter international business for defensive reasons. They want to protect themselves against domestic companies that might gain advantages in foreign markets, and then use that advantage in the domestic market. Company X may fear that Company Y will generate large profits from a foreign market if left alone to serve that market. Company Y then might use those profits to improve its competitive position domestically. Thus, companies, being afraid of such activities, may enter a foreign market primarily to prevent a competitor from gaining advantages.
- **Acquisition of Resources:** Manufacturers and distributors seek out products, services, and components produced in foreign countries. They also look for foreign capital, technologies, and information they can use at home. Sometimes they do this to reduce their costs. For example, Nike relies on cheap manufacturing operations in Southeast Asian countries to make its products.
- Acquiring resources may enable a company to improve its product quality and differentiate itself from competitors, potentially increasing market share and profits. Although a company may initially use domestic resource to expand abroad, once the foreign operations are in place, the foreign earnings may then serve as resources for domestic operations; for example, McDonald's used the strong financial performance of its foreign operations to invest in more resources for domestic growth.
- **Expand Sales:** By reaching international markets, companies increase their sales faster than when they focus on a single market, that being the domestic one. These sales depend on the consumers' interest in the product and their ability to purchase the product. Ordinarily, higher sales mean higher profits and that in itself is a motive for companies to go international. Many of the world's largest companies derive over half of their sales from outside their home country, such as BASF of Germany, Electrolux of Sweden, Gillette and Coca-Cola of the United States, Michelin of France, Philips of the Netherlands, Sony of Japan, Nestle of Switzerland.
- **Diversify Sources of Sales and Supplies:** In order to minimize fluctuations in sales and profits, many companies may look for foreign markets to take advantage of the business cycle differences among countries. Sales increase in a country that is expanding economically and decrease in another that is in recession. Consequently, companies may be able to avoid the full impact of price fluctuations or shortages in any one country, by obtaining supplies of the same product or component from different countries. In addition to the aforementioned reasons for international business, there are some additional factors that have contributed to the increased international business activities in more recent years. These factors, which are sometimes interrelated, are:
    - **Increase in Global Competition:** Many companies, of any size, decide to compete internationally because new products quickly become known globally; companies can produce in different countries; and the suppliers, competitors, and customers of domestic companies have become international as well.
    - **Development and Expansion of Technology:** Even in 1970, there was no Internet as it is known today, no commercial transatlantic supersonic travel, no faxing or e-mailing, no teleconferencing or overseas direct-dial telephone.
    - The liberalization of cross-border movements.
    - The development of supporting services.

Social Media Benefits and Risks

The following are the benefits of the usage of social media.

1. **Low Costs:** Definitely, it's cheaper to use online social networking for both personal and business use because most of it is usually free. While personal use is rather simple for anyone, the business functions are underestimated by many. In a social networking site, you can scout out potential customers and target markets with just a few clicks and keystrokes, adding a boost to your usual advertisements and promotional strategies. It lets you learn about their likes and dislikes, which is tremendous. If you want to fine tune your business, then this is the way to go, whether on a budget or not.
2. **Builds Credibility:** You definitely can gain the customers' confidence if you can connect to them on both a personal and professional level. Despite having to do a bit of work, it definitely pays off as you can be tapped for an offer if someone catches wind of your products or services. As long as you don't pursue them too aggressively, you will do well here.
3. **Connections**: You are friends with people who have other friends, and so on. There is potential in such a common situation. By using a social networking site, you can do what you can and get connected with these people to form a web of connections that can give you leverage if you play your cards right. As long as you give as well as you receive, then they will most likely stick with you. These connections are definitely valuable in the long run.
4. **Opportunities to Find Jobs:** Some social sites afford people the opportunity to find jobs online by link clicks and other notable employment advertisements.
5. **It helps to Overcome Timidity**: The social media sites like facebook and twitter helpone to overcome timidity or shyness by opening up over certain developmental issues and reactions.
6. **It is a Free Tool for Advertising and Promoting one's Product:** The social media site serves a medium of advertising and publicity for companies and products, due to the amount of the large population that visit the internet for one thing or the other.

Disadvantages

1. **Lack of Anonymity:** You are putting out information about your name, location, age, gender, and many other types of information that you may not want to let others know. Most people would say be careful, but no one can be certain at any given time. As long as people can know who you exactly are, then some can find ways to do you in.
2. **Scams and Harassment:** There is a potential for failure of security in both personal and business context. While many sites apply certain measures to keep any of these cases of harassment, cyberstalking, online scams, and identity theft to an absolute minimum, you still may never know.
3. **Time Consuming:** If this is not your kind of thing, that it would just be a waste of time for you. The key to social networking is that it is supposed to be fun, whether you are just doing it for kicks or clicking around for business purposes. That should be reasonable enough for anyone, but there are those people who don't see the point. For them, it can be a disadvantage.

Now there is something to really think about. Nothing is without a blemish, but those of this type of networking shouldn't really be that much of a concern regarding your safety. As long as you go along without making big mistakes, then it is all good. You can take advantage of the Internet phenomenon that continues up to this day.

SUGESSTIONS FOR FURTHER RESEARCH

This work focused onsocial media effect on international business venturing in Nigeria, the researchers therefore suggests that further studies in this research topic be carried out on a larger population including the consideration of other international business opportunities and how political system affects businesses in Nigeria as well as assessing the impact of appropriate technology adoption on businesses in Nigeria.

CONCLUSION

International business differs from domestic business by degrees. Although laws, cultures, and economic conditions differ within countries, such differences are usually less marked than those among countries. Even though there are limitations on the movement of goods and services and the resources to produce them within countries, these limitations are usually less pronounced within than among countries. Most countries vary internally, causing companies to alter their business practices from one region to another. Take the United States, where taxes and legal requirements differ among states and municipalities. This is why many companies place their headquarters in the business friendly state of Delaware. Certain products also enjoy greater acceptance in some areas than in others, as seen in the higher per capita demand for bottled waterin California than in the Midwest. Some tastes differ regionally as well; for instance, people in parts of the Midwest prefer white-looking dressed chickens to the yellow-looking ones that people prefer elsewhere. In some areas of the United States, there are non-English-language radio and television programs. And because income levels vary, purchasing power is higher in some areas than in others. But some countries have much greater internal variation than do others. Geographic and economic barriers in some countries can inhibit people's movements from one region to another, thus limiting their personal interactions. Decentralized laws and government programs may increase regional separation. Linguistic, religious, and ethnic differences within a country usually preclude the fusing of the population into a homogeneous state, which means that business cannot be conducted in the same way throughout the country. For example, for all the reasons just given, India is a much more diverse country to do business in than Denmark. Despite all the differences among regions within countries, this diversity is small when compared to the differences among countries. To successfully conduct business abroad, companies must often adopt practices other than what they are accustomed to domestically. Differences in the legal political, economic, and cultural environment all may necessitate a company's altering every type of business activity, from production and accounting to finance and marketing.

As the communications landscape gets denser, more complex, and more participatory, the networked population is gaining greater access to information, more opportunities to engage in international business comes up. The use of socialmedia tools -- text messaging, e-mail, photo sharing, social networking, and the like does not have a single preordained outcome. Therefore, attempts to outline their effects on business performance are too often reduced to duelling anecdotes.

REFERENCES

Acs, Z., Arenius, P., Hay, M., & Minniti, M. (2005). *Global Entrepreneurship Monitor, 2004 Executive Report, Babson College and London BusinessSchool*. Retrieved March 17, 2005, from www.gemconsortium.org

Acs, Z. J., & Audretsch, D. B. (2003). *Innovation and technological change*. Springer, US.

Adams, J., & Jaffe, A. (1996). Bounding the effects of R&D. *The Rand Journal of Economics*, *27*(4), 700–721. doi:10.2307/2555878

Aldrich, H., & Zimmer, C. (1986). Entrepreneurship through Social Networks. In D. L. Sexton & R. W. Smilor (Eds.), *The Art and Science of entrepreneurship*. Cambridge, MA: Ballinger.

Allen, K. (2003). *Launching new ventures: An entrepreneurial approach* (3rd ed.). Boston: Houghton Mifflin.

Asika, N. (1991). *Research Methodology in the Behavioural Science*. Ikeja: Longman.

Autio, E. (1994). New, technology-based firms as agents of R&D and innovation: An empirical study. *Technovation*, *14*(4), 259–273. doi:10.1016/0166-4972(94)90010-8

Autio, E. (2005). Creative Tension: The Significance of Ben Oviatts and Patricia McDougalls Article Toward a Theory of International New Ventures. *Journal of International Business Studies*, *36*(1), 9–19. doi:10.1057/palgrave.jibs.8400117

Barringer, B., & Ireland, D. (2005). *Entrepreneurship: Successfully launching new ventures*. Pearson Education.

Barringer, B., & Ireland, D. (2005). *Entrepreneurship: Successfully launching new ventures*. Pearson Education.

Baumol, W. J. (2002). *The Free-Market Innovation Machine: Analysing the Growth Miracle of Capitalism*. Princeton, NJ: Princeton University Press. doi:10.1515/9781400851638

Birkinshaw, J. (1997). Entrepreneurship in multinational corporations: The characteristics of subsidiary initiatives. *Strategic Management Journal*, *18*(3), 207–229. doi:10.1002/(SICI)1097-0266(199703)18:3<207::AID-SMJ864>3.0.CO;2-Q

Brazeal, D. V., & Herbert, T. T. (1999). The genesis of entrepreneurship. Entrepreneurship Theory and Practice, 23(3), 29-45.

Buckley, P. J., & Casson, M. C. (1976). *The Future of Multi-national Enterprise*. Basingstoke, UK: Burgelman. doi:10.1007/978-1-349-02899-3

Carree, M. A., & Thurik, A. R. (2003). The Impact of Entrepreneurship on Economic Growth. In Z. J. Acs & D. B. Audretsch (Eds.), *Handbook of Entrepreneurship Research* (pp. 437–471). Boston, MA: Kluwer Academic Publishers.

Carrier, C. (1994). Intrapreneurship in large firms and SMEs: A comparative study. *International Small Business Journal*, *12*(3), 54–61. doi:10.1177/0266242694123005

Damanpour, F. (1991). Organizational innovation: A meta-analysis of effects of determinants and moderators. *Academy of Management Journal, 34*(3), 555–590. doi:10.2307/256406

Dana, L. P., Etemad, H., & Wright, R. W. (2009). The impact of globalization on SMEs. *Global Focus, 11*(4), 93–105.

Delios, A., & Henisz, W. J. (2000). Japanese firms investment strategies in emerging economies. *Academy of Management Journal, 43*(3), 305–323. doi:10.2307/1556397

Drucker, P. (1985). *Innovation and entrepreneurship: Practice and principles.* Amsterdam: Elsevier Butterworth-Heinemann.

Feenstra, R., & Hanson, G. (2005). Ownership and Control in Outsourcing to China: Estimating the Property-Rights Theory of the Firm. *Quarterly Journal of Economics, 120*(2), 729-761.

Geringer, J. M., Tallman, S., & Olsen, D. M. (2000). Product and international diversification among Japanese multinational firms. *Strategic Management Journal, 21*(1), 51–80. doi:10.1002/(SICI)1097-0266(200001)21:1<51::AID-SMJ77>3.0.CO;2-K

Grant, R. M. (1987). Multinationality and performance among British manufacturing companies. *Journal of International Business Studies, 18*(1), 79–89. doi:10.1057/palgrave.jibs.8490413

Greenaway, D., Gullstrand, J., & Kneller, R. (2008). Surviving Globalisation. *Journal of International Economics, 74*(2), 264–277. doi:10.1016/j.jinteco.2007.08.005

Hessels, J. (2007). Reactie op: Kentering in het Denken over MondialeStrategieën [Response to: A Shift in Thinking about Global Strategies]. *ESB dossier Globaliseren in Nederland, 92*, 59.

Hisrich, R., Freeman, E., Standley, A., Yankey, J. A., & Young, D. R. (2000). Entrepreneurship in the not-for-profit sector: The state of the art. *Entrepreneurship*, 321-335.

Hitt, M. A., Hoskisson, R. E., & Kim, H. (1997). International diversification: Effects on innovation and firm performance in product-diversified firms. *Academy of Management Journal, 40*(4), 767–798. doi:10.2307/256948

Jaffe, A. B., Trajtenberg, M., & Henderson, R. (1993). Geographic localization of knowledge spillovers as evidenced by patent citations. *The Quarterly Journal of Economics*, 577-598.

Madden, M., & Zickuhr, K. (2011). *Internet and American Life Project: 65% of online adults use social networking sites.* Retrieved from www.pewinternet.org/Reports/2011/Social-Networking-kSites.aspx

Marios, I., & Spyros, H. (2007). International business- A global perspective. Butterworth-Heinemann Publishing Company.

McDougall, P. P., Covin, J., Robinson, R. Jr, & Herron, L. (1994). The Effects of Industry Growth and Strategic Breadth on New Venture Performance and Strategy Content. *Strategic Management Journal, 15*(7), 537–553. doi:10.1002/smj.4250150704

McDougall, P. P., & Oviatt, B. M. (1996). New Venture Internationalization, Strategic Change, and Performance: A Follow-Up Study. *Journal of Business Venturing, 11*(1), 23–40. doi:10.1016/0883-9026(95)00081-X

McDougall, P. P., & Oviatt, B. M. (2000). International Entrepreneurship: The Intersection of Two Research Paths. *Academy of Management Journal*, *43*(5), 902–906. doi:10.2307/1556418

Menger, C. (1892). *On the origins of money*. Retrieved March 17, 2005, from http://cepa.newschool.edu/het

Moen, O., & Servais, P. (2002). Born Global or Gradual Global? Examining the Export Behavior of Small and Medium-Sized Enterprises. *Journal of International Marketing*, *10*(3), 49–72. doi:10.1509/jimk.10.3.49.19540

OECD. (1997). *Globalisation and Small and Medium Enterprises* (Vol. 1). Paris: OECD.

Otokiti, S. O. (2005). *Methodologies and Mechanics of Computational System*. New Delhi: Sultan Chan and Son.

Porter, M. E. (1990). *The Competitive Advantage of Nations*. New York: Free Press. doi:10.1007/978-1-349-11336-1

Rennie, M. W. (1993). Global Competitiveness: Born Global. *The McKinsey Quarterly*, *4*, 45–52.

Reynolds, P. D. (1997). New and Small Firms in Expanding Markets. *Small Business Economics*, *9*(1), 78–84. doi:10.1023/A:1007916217140

Tallman, S., & Li, J. T. (1996). Effects of international diversity and product diversity on the performance of multinational firms. *Academy of Management Journal*, *39*(1), 179–196. doi:10.2307/256635

Zahra, S. A., Ireland, R. D., & Hitt, M. A. (2000). International expansion by new venture firms: International diversity, mode of market entry, technological learning and performance. *Academy of Management Journal*, *43*(5), 925–950. doi:10.2307/1556420

KEY TERMS AND DEFINITIONS

Business Relationships: Business relations are connections between stakeholders in the process of businesses, such as employer-employee relationships, managers as well as outsourced business partners.

Entrepreneur: A person who sets up a business or businesses, taking on financial risks in the hope of profit.

Information Technology: Information technology (IT) is the use of any computers, storage, networking and other physical devices, infrastructure and processes to create, process, store, secure and exchange all forms of electronic data.

International Business Venturing: A business enterprise or speculation in which something is risked in the hope of profit; a commercial or other speculation.

International Business: All value-adding activities – including sourcing, manufacturing, and marketing that can be performed in international locations.

International Marketing Environment: International marketing environment is a set of controllable (internal) and uncontrollable (external) forces or factors that affect international marketing.

Multinational Enterprises: An organization that owns or controls production of goods or services in one or more countries other than their home country.

Revenue: Revenue is the amount of money that a company actually receives during a specific period, including discounts and deductions for returned merchandise. It is the "top line" or "gross income" figure from which costs are subtracted to determine net income.

Social Media: Social media comprise a variety of Web-based and mobile technological resources that provide a forum for the generation, sharing, and discussion of individualized ideas and contents.

Chapter 5
Digital Divide, Data Trash, and the Commodification of Information:
Discourses around the Digital Divide

Anusharani Sewchurran
University of KwaZulu Natal, South Africa

ABSTRACT

In an attempt to build a theoretical framework for new media within the African context, this chapter considers terminology that routinely surface around ICTs. Terms such as knowledge economy, information poverty, the digital divide, information, knowledge and data though used interchangeably or roughly to mean similar things, they are laden with value, meaning and reflect power relations within a global and local context. Routinely used terms in a field create discourses which in turn reveal sociocultural, economic practices and power relations which reflect certain hegemonies. This chapter attempts to review some of the terms in the field of ICTs in order to make visible the sociocultural and economic power relations embedded in them. The digital divide offers a key entry point into ICT discourses and the opportunities and challenges presented by new media technology within the context of Africa. The theoretical concepts of the digital divide are reviewed within the larger context of global constructions of information wealth and information poverty.

INTRODUCTION

At its heart, development – if it is to be sustainable – must be a process that allows people to be their own agents of change: to act individually and collectively using their own ingenuity and accessing ideas, practices and knowledge in the search for ways to fulfil their potential. It requires what economist Amartya Sen calls 'real freedoms', the capacity for people to participate in a diverse range of decisions that affect them, and to enjoy specific functional aspects that constitute a healthy life…Enabling greater

DOI: 10.4018/978-1-5225-1859-4.ch005

numbers of people to speak, engage and respond to one another is ultimately equipping them to take political responsibility, which is a key ingredient to establishing deep and sustainable change. (as cited by Warnock, Schoemaker & Wilson,2007, para. 4)

The heart of Sen's definition of freedom is self-determination. This theme reverberated throughout South Africa in the early 1990s. Post-1994, the transition from South Africa's apartheid past saw the hegemonic ideoscape[1] laid down by the State through the Bill of Rights and the constitution. It assertively promotes an all-embracing historical redress that seeks to remedy the legacy of Apartheid, which systemized discrimination and disempowerment across all facets of society. This reparation was intended to permeate through a series of legislation (Polity, 2004).

While these Acts in part address the communications industry, there remains dissent as regards the depth and actual 'transformation reach' of the various Acts. The Black Economic Empowerment ICT Charter (final draft, 2004) has been one response to absences in legislation (ICT empowerment charter working group, 2004, pp. 20 – 25). The Charter follows on from The Broad-Based Black Economic Empowerment Act, with the central objective being to "bridge the digital divide".

The direct causes identified for the lack of development were numerous and complex. The lack of basic and supporting infrastructure resulted in a high cost of roll-out in rural areas. This has a knock on effect on the high cost of services in rural areas, which is complicated by ineffectual governance and regulation. This has led to the unfailing tendency of corporates to focus most on commercial rather than other clients (Gillwald, 2005, pp. 469 – 472).

The challenges identified by the ICT charter are lack of equity ownership, equity management and skills development[2]. Certain areas have been identified as having critical shortages in skills. Research has been identified as problematic as it is either duplicated or inadequate and incomplete. There has also been a lack of promoting open source platforms. Given the infrastructural problems cited above, rural access was indicated as an area needing growth in addition to the urban poor. Finally, there was emphasis on promoting fairness, transparency and consistency in adjudicating matters related to the ICT sector, ownership and control (ICT empowerment charter working group, 2004, pp. 20 – 25).

If development and democracy are to be the cornerstones of progress in Africa, then New Media technology (ICTs) must be acknowledged as development catalysts if utilized in a carefully considered way. Former South African President Thabo Mbeki urged Africans, "to participate fully in the information revolution…to devise concrete ways of harnessing this new phenomenon for the advancement and enrichment of our democracies" (Mzamane, 2001, p. 15). Apart from improving quality of life, the most critical role ICTs could play is lending its potential towards democratic participation. Nicholas Garnham (2004, p. 358) rightly indicates that technology only facilitates participation but does not guarantee it. "The internet and related technologies have created new possibilities for conversation - how these potentials are actualized is not up to the technology itself, but to human decision and commitment"[3]. Ling and Donner state that, "as long as there are differences in income, skills, literacy and simple personal preference, there is likely to be some kind of 'digital divide'" (2009, p. 72). Hindman reviewing the US context confirms the lag present in the early dramatic growth of the Internet. "Disadvantaged groups—blacks, Hispanics, the poor, the elderly, the undereducated, and those in rural areas—continued to lag behind in their access to and use of the Net" (Hindman, 2009, p. 9). Hindman further indicates that recent statistics show a narrowing of the divide; however, important differences related to age, race and levels of education remain (Hindman, 2009, p. 10). Thus even in the first world contexts, ICTs mirror the social, political and economic structures out of which they evolve means business as usual despite digital affordances. It is interesting to consider how democracy is refashioned (if at all) within ICTs.

A consistent rhetoric evolving out of development discourses is the focus on democracy and the need for ordinary Africans to claim self-determination over their material, social and economic conditions. New media technology is cited as the key facilitator for democratic intercourse. This rhetoric recurs consistently among many African leaders (Mzamane, 2001, pp. 1- 4).

There must be greater clarity and less ambiguity about fostering democracy.... It is important, therefore, that as we consolidate our democracies and use them as necessary platforms for the acceleration of our development, that the progressive movement ensures that the orientation of our parties is informed by the need to empower the masses of our people, so that they themselves can participate as a conscious force in the renaissance of Africa. (Mzamane, 2001, p. 16)

The democracy rhetoric is strikingly reminiscent of Jurgen Habermas's (2004, p. 350) ideal public sphere. Indeed, the 'conscious force' that former president Mbeki refers to communes well with principles of the public sphere where public opinion can by definition only come into existence when a reasoning and critical public is presupposed.

Public sphere means a realm of our social life in which something approaching public opinion can be formed. Access is guaranteed to all citizens. A portion of the public sphere comes into being in every conversation in which private individuals assemble to form a public body.[4] They then behave neither like business or professional people transacting private affairs, nor like members of a constitutional order subject to the legal constraints of a state bureaucracy. Citizens behave as a public body when they confer in an unregulated fashion – that is, with the guarantee of freedom of assembly and association and the freedom to express and publish their opinions – about matters of general interest. (Habermas, 2004, p. 350)

Habermas's public sphere has much to offer to development initiatives, providing a template against which to measure ideal democratic participation. This is as long as the public sphere is pursued as Habermas conceived it, and issues are not just aired as a matter of publicity. Karvonen notes this difference:

At one time the process of making proceedings public was intended to subject persons or affairs to public reason, and to make political decisions subject to appeal before the court of public opinion. But often enough today the process of making public simply serves the arcane policies of special interests; in the form of 'publicity' it wins public prestige for people or affairs, thus making them worthy of acclamation in a climate of non-public opinion. (2004, p. 348)

South African leaders often exhibit a heightened sensitivity to the media's critical eye on the government's work. On many occasions the South African media and intellectuals have been reproached for being too critical of government. Habermas points out that, "although state authority is so to speak the executor of the political public sphere, it is not part of it. The state and public sphere do not overlap, as one might suppose from casual language use. Rather they confront one another as opponents" (Habermas, 2004, pp. 346 – 7).

The confrontation between the state and public opinion (criticism) is a core ingredient for self-determination and the creation of democracy. Public opinion[5] must be acknowledged as vital to nurturing democracy and must be most aggressively promoted for the good health of the state. It is in this climate of critical public debate that new media technology will realize its full potential for achieving renaissance ideals. It is indeed troubling that the South African state increasingly appears to use the very tools of democratic engagement (the new media) as tools to close society. The issue of surveillance is extremely disconcerting; however, it is beyond the purview of the chapter.

Technological Determinism and Social Shaping

Comparative political economy is concerned with regulating and developing the best model towards the end of universal communication equity. This simply translates into the creation and maintenance of equal conditions for all citizens to receive, produce, and distribute communication messages. Consequently, a host of normative assumptions circulate to form the regulation discourse and it is interesting to take a closer look at these.

As the neoliberal agenda permeates the economic spheres of society, it also brings with it supporting ideologies to govern the other spheres of life. A tendency of Neoliberalism is its role in hegemony, the maintenance of ruling class ideology, alongside which lies the ideas of technological determinism and technological neutrality. This subtle permeation will be explored as it emerges in the ICT sector. In this site neoliberalism manifests within a technologically deterministic framework which promotes the commodification of knowledge; together this triad (as it were) privileges a dominantly free-market approach despite the location of a developing context (Sklair, 2010, pp. 184 – 190).

Two major theories examine the relationship between media and society. Simply put, the one examines how media technologies determine culture and society and thus falls into the technological determinism category. The other asserts that society shapes technology, never the other way around; hence the social shaping category.

Marshall McLuhan (1986) is always associated with being a key proponent of technological determinism in the context of media. His oft quoted pithy maxim, '*The medium is the message'*, refers to the unyoking of content as a determining factor. This means that irrespective of content or consumption of media products, the technology which transmits this content changes human perception of the world and hence their own practices within it. '*The medium is the mass-age*' was McLuhan's pun on his media maxim which encapsulated the power the dominant media of an era had over society, shaping as it were the thoughts, words and deeds of individuals. For McLuhan, media determines consciousness.

Although dated, McLuhan's arguments have value as he saw media as an extension of the human body and found society quite immersed in media, likening the phenomenon to fish in water. Thus, he stated that the effects of media are subliminal yet determining even though the individual consumer may not cognize the effects (Gordon, 2002, para. 4).

In recent times, media products (and mediums) have flooded the individual's environment such that one is almost totally media-saturated. From IPODs to IPADs, smart phones and the DSTV, technology appears to move consumers to a point of never being disconnected. The user is thus interacting with geographical space (for want of a better term) via digital space which is more one dimensional.

As regards the media changing other aspects, much has been made of shortened words and phrases. Use of this rather limited stock of daring, sometimes strange abbreviations have increased along with another limited stock of emoticons[6]. This shift is shaping our youth in particular into 'iconic' individuals as opposed to literate ones which may have interesting implications for the pursuit of a robust civil society.

While McLuhan points to subtler aspects of the influence of technology on the individual, other technological determinists take the overt view of technology being an agent of social change that is free from social pressures and human agency. They prescribe a technological fix for societal ills and ascribe agency to technology itself. Progress is defined along technological lines relative to each successive stage in development.

Technological determinists would define and celebrate progress along technological lines asserting that as technology is refined and improved, so too does knowledge increase simultaneously at the same pace. While the visual impact certainly indicates the convenience of this kind of technological progress, one must be reticent at the wholesale acceptance of all other implications. The idea of knowledge doubling or increasing exponentially at the pace of technological development does not indicate any deliberation on the quality of knowledge that is increasing and increasingly available. Hence, one must necessarily enter the discourse on the distinction between knowledge, data, and information.

Even if information available was generally of a respectable quality, there is the ironic, simultaneous manifestation of information wealth with information poverty. As it is, there is simply too much information available for the capacities of the layperson to assimilate and digest. However, reiterating the conveniences of cellular telephony, McLuhan's central message of the transformative power of the medium is helpful as the cell phone has affected communication and mobility. This has yielded positive results for new potentials in development.

A key counter theorist who disagreed with some of the key thoughts of McLuhan was Raymond Williams. Williams (Fourie, 2007, p. 142) had Marxist leanings and insisted on a cultural materialist perspective and did not accept technological determinism as sound or adequate. He argued that technology could not be ascribed with 'a power of its own' without taking into account the political, economic and social milieu within which it found itself. Although Williams agreed with McLuhan on some aspects, he argued that no media could engineer change as the media was a by-product of pre-existing socio-economic and political hegemonies. Therefore, the medium will only reproduce those existing patterns of use and exchange (Fourie, 2007).

The social shaping theorists evaluate technological determinism and find it wanting as an inadequate description of technological innovation and social/cultural change. Theorists within this camp assert that technological determinism fails in establishing technology as an instrumental cause for change. They further assert that social factors do not only shape the utility value of media technologies but also shape their design and implementation. Social shaping theories thus emphasize the primacy of human agency (Lievrouw, 2002, pp. 183- 189).

It is important to summarize these two trajectories of thought on media technologies as both permeate the development media discourse in different and sometimes contradictory ways.

For the developing context the primacy of human agency applies in a very qualified way and cannot be interpreted as a universal given that is experienced equally. When addressing the development context, human agency is circumscribed tightly by the political and economic environment, perhaps even more than the consideration of media technology itself.

Following on from Moore, technology has progressed in leaps and bounds to deliver highly sophisticated devices that have become an integral part of those individuals' lives that are able to afford them. Circumventing the technological determinism - social shaping debate, one may reflect that at a larger scale, new media technologies have expanded economic connectivity and consequently increased accessibility of different national markets to one another, thus injecting growth into the global economy. Developing countries like China and India have harnessed the potential of new media technologies (NMTs) to speed up their growth, while other developing nations have used NMTs to support development efforts. Given that NMTs have such a transformative effect on the social/cultural, political and economic sphere it is critical that it is made to be universally accessible. However, alongside this tremendous technological revolution:

There is a profound contradiction in the information age. Gulfs of inequalities in the world are growing deeper as electronic and other forms of connectivity are drawing people and communications even closer to one another. (Youngs, 2007, p. 143)

Commodification

Commodification, central to the neoliberal economy, refers to the imposition of exchange relations onto social relations; while commodity fetishism refers to the distortion of use values and exchange values (Mosco, 2009, pp. 127 – 135). Within the knowledge economy, commodification and commodity fetishism manifest via the imposition of seemingly apolitical and value neutral market apparatuses. These coercive market apparatuses of repression and consent effect a blurring of boundaries between the private and the public producing an effective distortion of the ideas of knowledge, information and data. The distortion of the different categories of *information* is a necessary precursor to the manipulation and profiteering of private information.

The knowledge economy set against the backdrop of the neoliberal economic model defines information as an economic category. Talero (a member of the World Bank) quoted by Wilson states that "information is a factor of production, in a category with land, capital, and energy" (as cited by Wilson, 2002, p. 9). In addition, Hassan indicates that, "so central was access to information technology deemed to be that the United Nations categorized the issue almost as a human rights one" (2004, p. 66). The information society is hailed as a global panacea to social ills but what is remarkably silent about the information economy is that not all information is equal and not all information is valued equally. Information has emerged a commodity like any other playing in the market place to fetch the highest exchange value. Thus, information that can translate into high exchange values is regarded as valuable while information that does not translate into exchange values is regarded as insignificant.

An increasing characteristic of the knowledge economy is the way the terms knowledge, information and data are used interchangeably, meant to signify the same or similar things. In the context of the information society and (NSMI: new systems and machines of information) discourses knowledge, data and information are largely seen to be commodities signifying an economic facet; in addition, information poverty is couched in economic terms.

Information wealth and poverty adds another layer of symbolic meaning to the digital divide. Access translates into knowledge while a lack of access equates to ignorance. This is problematic as it exhibits a technologically deterministic approach – unfortunately adopted by regulatory frameworks. Set against a global neoliberal economy, the gains from ICTs are diminished due to the nexus of technological determinism, neoliberalism and Gandy's concept of the digital divide. This nexus leads to the commodification of information which impacts on citizens in problematic ways. Hence they are not prepared to deal with (protect themselves from) the increased surveillance and the commodification of personal information arising out of biometric practices (Morozov, 2011, pp. 12 – 32).

Consumer vs. Citizen

Gandy (2002, p. 448) presents the idea that the real divide in the digital age has been increasing the gulf cultivated between notions of the citizen and the consumer. This divide raises debate about the role and responsibility of government for ensuring a well-served citizenry. Gandy argues that the Reagan administration was credited with dismantling the welfare state and with ushering in the era of the *sovereign consumer* whose interests reign over that of citizens as the primary basis for evaluating social systems:

Neoliberal political philosophy has found a comfortable berth in an intellectual stream of post-modern thought that denies the possibility of an institutionalized representation of common or collective interests... Ironically, despite the obvious inequalities we observe within the marketplace, the market is praised as being more responsive than political systems to individual interests. (Gandy, 2002, p. 449)

Tracey (as cited in Gandy, 2002, p. 452) asserts that these intellectual constructs took hold after the continuous election of right-wing governments which resulted in the centrality of the *individual-as-consumer* alongside which followed the interests of *the corporation.* Hence notions of public good, public interests, community, public culture, citizenship, governance and even the nation state were "buried in the rubble of the old order" (as cited in Gandy, 2002, p. 452) – the welfare nation state. Accompanying these ideas of the centrality of the consumer, a market logic or "marketization" ensued. This referred to support for market solutions as efficient and self-regulating, paving the way for the dominant role private corporations would come to play in managing public culture (Gandy, 2002, p. 451).

This raises important questions as to the health of public culture; democracy; and civic activities on both a global and local level. Calabrese and Borchert as explained by Gandy, indicate that the construction of the individual proportionately as consumer or citizen will ultimately determine the role technology will play in *'post-industrial democracy':*

They suggest that the network environment stratified and segmented, and that a 'new class of technical and professional intelligentsia' will be engaged in activities that will be 'exclusionary, both by default and by design'. Interestingly, they suggest that the civic activities of this new class will be financed primarily by the revenue generated by the lower social strata who will be involved in the network primarily as consumers, rather than participants, or sources of argument and analysis. (as cited in Gandy, 2002, p. 456)

Simone and Fernback (2006, pp. 287 - 295) point to a deeper immobilizing conflict at the heart of the matter. They argue that the difficulty with communications policy broadly speaking lies in the attempt to converge two divergent theoretical traditions. One tradition relates to neoliberal market theory and the other tradition relates to the public sphere. As a result, Simone and Fernback argue that, "problems in developing normative guidelines and implementing consistent policy measures have reoccurred" (2006, p. 288). Notions of the public sphere, global commons and even universal access are ideas that do not in the best of times co-exist within a predominantly market model as the market model distorts social and productive relations.

Digital Divide or Data Trash?

The digital divide manifests as a lack of digital technology in some form or the other and crosses age, gender, ethnicity, class, race, nations and sometimes even education levels. It is of course most often used to emphasize the digital technology disparities between the developed and developing worlds. John Pilger in his award winning documentary *'New rulers of the world'* reflects:

This is the age when technology has advanced so much that it gives us the potential to solve the problem of poverty, disease and other social issues, ironically never before that the gap between the 'haves' and 'the have-nots' been greater. (Pilger, 2001)

It is indeed ironic and almost contradictory that while technology gallops forward at breakneck speed, simultaneously technological scarcity abounds. While Pilger points to the ever widening gulf between the 'haves' and 'the have-nots', it is also a phenomenon which is exacerbated by a troubling expansion in the number of the 'have-nots'.

Marshall (2004, p. 35) estimates that less than ten percent of the world's population of approximately six billion are internet users. Marshall also points out that with regard to technology the 'have-nots' are categorized as *data trash,* "made up of the discarded groups in any culture who are not connected to the new structures and machines of information" (Marshall, 2004, p. 35). While there is greater connectivity celebrated by mobile telephony, there still remains large tracks of citizens disconnected effectively negatively branding them as *data trash*. This kind of linguistic othering abounds in ICT literature. Hassan (2004, pp. 66-70) refers to those excluded groups as the "deleted". Klecun (2008, p. 272) describes terminology such as "sink estates" and "underwired communities" adding to the binary discourse around ICTs where those on the wrong side of the divide are cast as inefficient and miserable.

This appears a rather ruthless set of categorizations; however, given the brutality of treatment and indifference towards those occupying the wrong side of the digital divide, it is a fitting term, embedded with a most accurate connotation. However, the categorization of the 'digital divide' will be temporarily used for the purpose of clarifying a more comprehensive definition.

The digital divide describes the disadvantage of those who are unable to or choose not (for a variety of reasons) to use new media technologies, or as Marshall puts it, *the new structures and machines of information*. The divide is further refined according to a range of access or rather the lack thereof.

Lesame (2005, p. 3) describes access to technology giving greater clarity and depth to the more simplistic, existing definitions. *Mental access* refers to the lack of basic experience with new structures and machines of information (NSMI). This could be as a result of a lack of interest, a lack of perceived need or technophobia. *Material access* refers to the lack of technological devices by way of actual hardware, a lack of software and / or a lack of connectivity. It has been stated that with regard to the new media, content is no longer king. Connectivity is king (Evans, 2010, para. 1). Rural areas are particularly affected by this lack as are war-torn regions or those wracked by political instability.

Skills access refers quite obviously to the lack of requisite skills to utilize new machines and systems of information (NSMI) which is caused by inadequate education and insufficient user-friendliness. This means not having enough exposure and experience with (NSMI). A repeated critique of the attempts to address the digital divide is that provision of hardware and software is counter-productive without the concurrent investment of education (Lesame, 2005, pp. 3-4).

Usage access extends the category of *material access* and refers to the lack of sufficient usage opportunities. Linked to *usage* are the various levels of *familiarity* present in users. *Familiarity* of the full range of what is possible in the digital environment may be considered as a contributing factor towards the divide making users fundamentally unequal. An example elucidating this point is when users in more rural areas of South Africa (and some users in the older possibly technophobic category) utilize their cell phones for basic first generation telephony only, as a result of a lack of familiarity of the full range of services and applications. This is regardless of the availability of a host of 2nd and 3rd and 4th generation applications (Lesame, 2005, p. 4).

Environs access refers to the impact of space and place. The places in which people access online determine how they will utilise (NSMI). The public/private divide comes into play affecting what content would be accessed and also determines the diversity of ones use (NSMI).

Language is perhaps the final variable theorists cite as a limitation augmenting the digital divide. This refers to the predominance of English across (NSMI), which severely limits those who do not understand English, or use a minority language (Lesame, 2005, p. 5).

While the access types of *mental, material, skills, usage, familiarity, environs* and *language* provides a concise description of the digital divide, other theorists' further refine the category. Following on from the conclusions of New World Information and Communication Order (NWICO) and the MacBride commission in 1980 theorists of this trajectory reflected on the 'core-peripheral' imbalances that occur in the global information market place.

According to Hamelink (1994, p. 143), the root cause of information imbalance is the historical colonial form and pattern of trade that was imbalanced in favour of the colonisers. Within the South African context one may clearly observe the way earlier historical inequalities compound present attempts at development. Apartheid geography still lingers both in relation to the rural – urban divide as well as with the urban poor. Thus, digital inequalities are interwoven with other historical categories of inequalities such as class, gender, race, disability and even age.

While much has been made of the possible potential for the developing regions to leapfrog into the digital age, failure to address historical and structurally embedded inequalities would prove detrimental to the potentials of the digital age. Global and local attention with regard to the developing regions appear primarily to be on addressing the question of material access, and further focusing on meeting basic needs first. This effectively neglects other forms or causes of the digital divide. A compounding factor is the speed at which the digital economy develops. This adds to increasing inequalities as those who are already behind may get left further behind as the digital age gallops ahead. Conversely, new media in this infrastructure lends the capacity to increase the power and influence of already dominant groups.

Some theorists like McPhail (2013, p. 26) argue for an extension of the cultural imperialism thesis, stating that this phenomenon has morphed into an electronic colonialism. This creates and maintains all sorts of inequalities and imbalances which remain unfavourable to the developing regions. Hamelink states that, "the inadequate information capacities of most periphery countries are a serious obstacle to their own efforts to combat poverty and other deprivations" (1994, p. 148). Information imbalances limit sovereignty in that decision making is limited to the extent that decision makers lack information about their own region. This applies at the level of parliament as well as for the individual or citizen. However, the digital revolution has led to ever-increasing control and tightly guarded interests of a narrow band of global information owners.

The global information market place may be divided into the following variety of stakeholders:

- The owners of the information 'highways' or the 'common carriers'.
- The producers of information hardware and other infrastructural hardware.
- The producers of information software.
- Information producers.
- Information consumers (McPhail, 2013, p. 115).

The information highways refer to the mechanisms of actual connectivity, which are satellites; fibre-optic cabling, digital switching technology; and microwave towers. Without pre-empting the analysis which follows this chapter, these carriers are dominantly owned by MNCs from the core countries and mark the starting point of informational inequity.

Information and other infrastructural hardware refers to devices at the point of connectivity such as fixed line telephones, cell phones, televisions, various computing devices and personal digital assistants (pdas). Most hardware processors and carriers are installed in the core countries. The technology is manufactured, designed and developed and controlled by the core. If there is any participation, Youngs states that, "…the technological know-how remains under the control of the transnational companies and the periphery countries play a minor role just in assembly" (Youngs, 2007, p. 297). Green (2010, p. 67) gives the example of, "western patterns of training, behaviour and infrastructure required to keep a bank headquarters running, with 24-hour electricity supplies, western trained technical support, high-level security protocols, and communication and network infrastructures that construct this locale as a node for a network developed and controlled elsewhere".

Information software apart from the obvious one of programming, refers to the media as in the press, broadcasters, libraries and 'infopreneurs'. The critiques from NWICO-MacBride debates remain similar in this instance in that the volume and course of information flows appears largely uni-directional. Furthermore, information moves among the core countries and from the core to the periphery and rarely between the periphery countries.

The capacity to produce, distribute and access pertinent information is very unequally distributed among the world's countries and the core countries have infinitely more control. (Youngs, 2007, p. 297)

As regards information consumers, the earlier descriptors of the digital divide are applied to the individual end user within the developing world. For Africa, this situation is worsened by the high cost of access, inappropriate or weak policy regimes, inefficient telecommunication services and a lack of locally created content.

While there may be an emphasis on solving the problem of informational under-development there are different approaches to how this may be achieved; the first focuses on the interdependence of the worlds' nations thus necessitating the integration of the periphery into the global economy. The critique of this approach is that it locks peripheral regions into playing 'catch-up' according to the terms and conditions dictated by the core, thus drawing out the dependency of the periphery to the core, which could be indefinite (Mzamane, 2001, pp. 11 - 14).

An alternative approach is where the periphery would collectively take the initiative through stronger bargaining power to pressure the core for fairer trading schemes. This amounts to playing catch up by self-defined terms and conditions.

A third approach would be for peripheral countries to radically disassociate themselves from the core regions and focus on self-reliance. This seems an unlikely possibility as the global economic infrastructure is unyielding and dominantly neoliberal in nature, with far too much investment of the core regions in the peripheral regions. Hence it would be difficult to disassociate from the core-based MNCs among other things. The periphery is also still haunted by the demons of the past such as poverty, conflict and health issues which weigh down decision making processes for change.

Thus one may conclude that with regard to the digital divide there are no quick and easy solutions. A few general propositions may be for developing regions to address the issues of illiteracy by establishing community centres for public access, and encouraging the development of relevant content in the vernacular. Regulation is also important and in this context becomes a complex issue and development leaders ought to monitor this closely. This is because the digital media is global in scope and requires a more inclusive international cooperation and agreement between national legislators, and the digital media evolves so rapidly while regulatory arms (nationally and globally) move very slowly.

The million-yen question is who ought to be responsible for addressing the issue of the great divide. The options sway between formal political–economic institutions such as the rich core nations, MNCs, intergovernmental organisations, and civic lobby groups and the individual citizen/end user herself/himself.

Global Constructions of the Digital Divide

Wilson (2002, pp. 1 - 12) provides a valuable analysis of global (NSMI) discourses as she scrutinizes policy statements, unearthing a set of fundamental assumptions that appear to inform policy discourses. This refers to the technologically deterministic vision which is compounded by the construction of technology as a neutral tool for development. Wilson's survey of the literature spans 101 key documents and general but relevant websites such as UNESCO; UNDP; USAID; World Bank; IDCR – ACACIA program; ITU and the Global Knowledge program. This survey results in three salient themes:

1. The construction of the category of 'information poverty';
2. The construction of what counts as legitimate/valuable information and knowledge;
3. The development aims of these programs, in particular models of progress and catch-up to industrial country ideals (Wilson, 2002, 1-4).

Forced to Play Catch Up

Wilson indicates that most development programs are framed in a modernization theory paradigm. This results in a vision of the developing regions:

1. As stagnant needing the injection of technological progress.
2. As stuck in the past of dominantly agrarian activity while trying to meet basic needs.
3. As behind in terms of technological progress thus living in a kind of information darkness.

Wilson quotes from the United Nations Development Program and the World Development Report (1999, p. 1) which respectively reads:

Knowledge is like light. Weightless and intangible, it can travel the world, enlightening the lives of people everywhere. Yet billions of people still live in the darkness of poverty – unnecessarily, (Wilson, 2002, p. 1)

Time will be of the essence. Already the fruits of the information revolution are skewed and inequitably distributed: developing countries are largely left behind and are a little more than fringe players at best, observers at worst of the dynamics and benefits enjoyed by many industrialized countries. Indeed, a new type of poverty – information poverty – begins to afflict developing countries...Information technology has a place alongside adequate food, health care, education, and other fundamentals. By taking this place it has broadened our definition of poverty. Those people or countries who cannot or will not participate fully in the new information economy will find it all the more difficult to climb out of poverty. (Wilson, 2002, p. 3)

These two quotes indicate the way in which information poverty as a category is delineated. The idea of the digital divide is extended to wealth and poverty with time and individuals included as crucial determinants. Knowledge is equated in an unqualified way to abundance and success while the lack of knowledge is equated to poverty and darkness. The analogy in the first quote thus betrays a subtle implication of poverty being a result of ignorance - an unnecessary ignorance at that. Thus, a binary opposition emerges with regard to (NSMI) and may be tabulated as shown in Table 1.

Development Binary

Wilson's third quote was extracted from the United Nations Development Program (UNDP) and represents a further extension of the binary of the digital divide.

And let me lay my cards on the table at the start by saying that in my mind there really is no more critical question facing the developing world today than how to face up to the challenges and opportunities

Table 1. The New Systems and Machines of Information (NSMI)

Developed	Developing/underdeveloped
Rich	Poor
With NSMI	Without NSMI
Information rich	Information poor
Modern	Traditional
Knowledge	Ignorance
Light	Darkness
Proactive	Passive
Winning the race of the digital divide	Losing the race of the digital divide

offered by the Information Revolution, and particularly the phenomenon of the Internet. It is now the two edged sword that is leading the process of globalization: wounding those who don't quickly enough grasp how to use it by leaving them ever further behind, but providing unprecedented benefits for those with the courage and willingness to grasp its potential to drive change ....ICT is transforming everything it touches, from politics, to business, to culture, to education and to health. And what we have seen so far is just the tip of the iceberg. But it is an iceberg the developing world is currently at more risk of crashing into than making use of. (Wilson, 2002, p. 4)

This quote reveals assumptions of passivity amongst those considered information poor. In addition, technocratic solutions are assumed to be the magic solution to the social ills related to poverty – a sort of digital sublime (Mosco, 2004, p. 3) as it were. Furthermore, referring to information poverty devalues knowledge of people of the developing world implying a lack of knowledge altogether. Information poverty ought not to be seen on its own devoid of or excluded from general poverty. The binary extends information rich and information poor to include another dichotomy which is the equation of the traditional and the modern to the information poor and the information rich. Devaluing traditional culture occurs as a result of the particular way in which technology is cast. The concomitant effect is equating information to knowledge and classifying particular kinds of knowledge as valuable, rendering other types valueless, if they are regarded as knowledge at all.

The critique centrally is that these kind of assumptions are problematic as they inform development programs which are designed to provide as much technology as possible to developing regions so as to emerge as a reasonable facsimile of the North Atlantic region, as this is taken finally as the benchmark. The technocratic solutions do not often take into account the political, economic and cultural climate of the regions in question. A further and serious weakness is to assume a technological solution to poverty when these very problems emerge from very complex political milieu and thus the effects of the problems are not really connected to the right causes. It requires more than just access to create a climate of assimilation with technology that will have positive returns. Thus, it remains questionable how achievable these programs carrying such assumptions could be.

The tenor of the quote also implies that the fault or responsibility lies with the end user to assimilate (NSMI). Wilson states that the theme recurs across the ICT development discourses – and one may even find echoes across local leaders. Former South African President Thabo Mbeki (2001) states:

in an age of globalization, every country needs to create capacity in the area of communication and information technology or face the risk of permanent global marginalization of our countries and peoples. We cannot begin to be fully integrated into the global economy if we do not develop the necessary skills to participate in the increasingly knowledge based communication society. (as cited by Mzamane, 2001, p. 14)

There cannot be a wholesale transfer of responsibility for assimilating the information society onto the developing world as there are complex global issues at play. Trade regulations, international technological standards, and the tightening noose of MNCs and their exertion of control particularly in the developing regions, are the range of variables limiting the possibilities for universal access in the fullest sense of the term.

The information revolution offers Africa a dramatic opportunity to leapfrog into the future, breaking out of decades of stagnation or decline. Africa must seize this opportunity, quickly. If African countries cannot take advantage of the information revolution and surf this great wave of technological change, they may be crushed by it. In that case, they are likely to be even more marginalized and economically stagnant in the future. (as cited by Wilson, 2002, p. 11)

The tendency across development ICT discourses also seems to lay down an imperative of time – i.e. that it is of extreme urgency for the developing regions to participate in and *win the race* towards an information society, or risk the real possibility of marginalization, stagnation and decline. The discourses never reveal the complex political and economic quagmire that the developing peoples face in trying to achieve a win in this race. Industry is far from benign or benevolent and often beset by greed and avarice which itself adds a troubling dimension of a global scale to the political and economic quagmire mentioned above.

Despite the global political economic quagmire, policy makers still emphasize (NSMI) as the key to fast tracking developing nations out of stagnation and decline. While critics are not necessarily in opposition to the potentials of (NSMI), the key assumptions of value neutrality and technological determinism are regarded with some degree of scepticism. In addition, (NSMI) are promoted as tools for development in an unqualified way. Thus what is questionable is the assumption of a wholesale acceptance of the information society 'as it is' in the developing context.

Following on from McPhail's elucidation of electronic imperialism and in spite of the counterpoints of the east (the "Asian Tigers"), it remains clear that apart from hardware and software dominantly disseminated to the developing world, the production of knowledge, information and entertainment has assured the reproduction and maintenance of western style hegemony via established practices of knowledge production and validation. Holtendahl quoted by Wilson (2002, p. 5) states, "the criteria of what counts as knowledge continue to be defined in the academic centres of the West. The dissemination of this knowledge is based on notions of *transfer of knowledge* from the West to the South".

One issue that emerges is the constant reliance on solutions for development problems derived in the West as a result of the developing world's increasing dependence on western solutions as local, indigenous knowledges and understandings become ever increasingly under-valued, de-valued or ignored altogether. Another rather troubling issue that emerges as a result is the over reliance on knowledges and understandings defined in western academic centres. A disturbing trend is that western academic centres are not really independent free institutions of knowledge production. In some ways, institutions have become hegemonic instruments, propping up neoliberal ideology by funding academics from leading institutions in the west to endorse policy thus putting intellectual weight and adding credibility to a host of policies that together form the western economic model in which the market is deified.

This is the way in which the deck is unfairly stacked against development endeavours as alternative modalities cannot even be brought into the discussion.

The global relevance of the information society and an information economy is assumed. The global economy is taken as a given and there is no suggestion that perhaps technology or the market should slow down as has been argued by various critical development theorists and environmentalists...Rather, developing countries are impelled to work harder to catch-up. (Wilson, 2002, p. 12)

Within this political economic framework critics remain sceptical of the achievement or even pursuit of social goods like universal access, global commons, and a healthy public sphere as Habermas originally envisaged it. This is because if such things are left to market solutions, the proven track record of *the market* has been a primary concern with the "bottom line", to the exclusion of other concerns such as social goods: *the 'res publica'*.

The knowledge society set against the backdrop of the neoliberal economic model defines information as an economic category. Talero states that "information is a factor of production, in a category with land, capital, and energy" (as cited by Wilson, 2002, p. 9). The information society is hailed as a global panacea to social ills but what is remarkably silent about the information economy is that not all information is equal and not all information is valued equally. Information has emerged a commodity like any other, playing in the market place to fetch the highest exchange value. Thus, information that can translate into high exchange values is regarded as valuable while information that does not translate into exchange values is regarded as insignificant.

An increasing characteristic of the knowledge economy is the way the terms knowledge, information and data are used interchangeably, meant to signify the same or similar things. In the context of the information society and (NSMI) discourses knowledge, data and information are largely seen to be commodities signifying an economic facet; in addition, information poverty is couched in economic terms.

ICT's serve as a transmission belt to generate, access, disseminate and share knowledge, data, information, and communications and best practices. Three central features are at the heart of the knowledge revolution. Information and knowledge are instantaneously accessible, they are transportable and can be simultaneously distributed to an unlimited number of users. Indeed, they cannot be depleted. The use by one does not prevent the use or consumption by somebody else. They cannot be owned, though their delivery mechanisms can. Selling them entails sharing, not exclusive transfer. Indeed, information and knowledge represent a global public good. (as cited by Wilson, 2002, p. 14)

While the UNDP report represents an understanding of the utopian vision of the knowledge society, there is not enough acknowledgement of the very real, tangible barriers to access across a series of levels to this alleged 'global public good'. Burkett (as cited by Wilson, 2002, p. 16) alerts us to this issue when she states that,

it seems that what is needed to resolve these 'real' problems is not further information, but rather things which are much more difficult, time-consuming and probably not as glitzy as ICT development – that is, political will, recognition of personal and social responsibilities, and ultimately action on the part of governments and civil society...a call is made for a turn to social and human development to counter the increasing emergence of a "cyber-elite alliance" which is neatly camouflaged within much of the largely apolitical ICT and development discourse..

Simone and Fernback (2006, p. 288) point to a deeper immobilizing conflict at the heart of the matter. They argue that the difficulty with communications policy broadly speaking lies in the attempt to converge two divergent theoretical traditions. One tradition relates to neoliberal market theory and the other tradition relates to the public sphere. As a result, Simone and Fernback argue that "problems in developing normative guidelines and implementing consistent policy measures have reoccurred" (2006, p. 288). Notions of the public sphere, a global commons and even universal access are ideas that do not in the best of times co-exist within a predominantly market model as the market model distorts social and productive relations. With regard to the knowledge economy, commodification of information is the next frontier.

The Commodification of Information

Commodification generally speaking refers to the imposition of exchange relations onto social relations, while commodity fetishism refers to the distortion of use values and exchange values. Within the knowledge economy, commodification and commodity fetishism manifest via the imposition of seemingly apolitical and value neutral market apparatuses. These coercive market apparatuses of repression and consent effect a blurring of boundaries between the private and the public producing an effective distortion of the ideas of knowledge, information and data. The distortion of the different categories of *information* is a necessary precursor to the manipulation and profiteering of private information.

The idea of apparatuses was first introduced by Althusser (2001, pp. 79-89) who rejected the simple determinant of economics. He asserted that individuals live in a symbolic universe and meaning is made out of the process of interpellation or identification with sets of ideological representations which function to maintain dominant social relations by producing conformist subjects. Althusser identified ideological state apparatuses such as the nuclear family, education, religion, and the media whose purpose was to constrict such subjectivities. He also identified repressive state apparatuses which were the military, the police and the legal system etc. Both ideological and repressive state apparatuses colluded as it were to maintain the dominance of the ruling class, the former though consent, and the latter through (the threat of) force.

Althusser's apparatuses are extremely useful in analysing the information society. In the information society the state recedes a little in its role of creating ideological and repressive apparatuses. Technology via immensely powerful telecommunications MNCs create the technological apparatuses of new structures and machines of information (NSMI) and together with political regimes give business interests a democratic patina. Marshall (2004, p. 35) first introduced the term *new structures and machines of information* as a means of encompassing an ever widening terrain of the digital world.

The technological apparatuses of (NSMI) give rise to twin hegemonic strategies, one ideological and the other repressive. The technological apparatus of coercion creates familiarity to generate consent with new technology via a set of representations designed to create an environment of adoption of (NSMI) and to manufacture consent for a world of digitally based consumption. As Marshall aptly puts it: '*producing us as technological subjects*' (2004, p. 33). Apart from the obvious advertising for specific products, the media has provided niched areas of space by way of columns and specialized sections in the press and programming on television to construct the public as 'technological subjects':

Each Thursday the New York Times includes a section entitled 'Circuits'. 'Circuits' is devoted to new technologies generally connected to digital gadgets and new computer programs. Reviews of new PDAs, discussions of digital effects in film, or new functions in the next generation of mobile phones are regular features. The critical edge is even more lightly applied than to the entertainment sections of newspapers. For eight to ten broadsheet pages, the articles express a celebration of technology and the cult of the new. 'Circuits' represents a very long tradition with relationship to technology: new inventions express the superiority of the modern over anything traditional. It is a reading of technology in a new Darwinism: each succeeding technology improves on its previous generation and liberates humanity just that much more. The New York Times is not alone in providing a regular space for this celebration of the future in the present manifestations of technology. Every major newspaper in the world allows for these fluff pieces about the promise of technology. In the last decade, they have developed into a larger niche of the press and have become regular features on television with whole shows such as Tech TV devoted to the love of new electronic and digital gadgets. (Marshall, 2004, p. 29)

One South African version of ideological apparatuses is the 'Tech@work' show screened on a local free to air channel (n.d.). One of the telecoms giants in South Africa - Vodacom - has created 'Vodaworld', located in Johannesburg (n.d.). These are examples of how ideological apparatuses have evolved from meagre press columns and television features to a manifest, tangible show of digital life. Again, the critical edge is diminished as newer features are celebrated with no reference to the multitude of difficulties within the sector.

Althusser's repressive state apparatuses were identified as the coercive forces of the state. In this case, the repressive technological apparatuses manifest as *legalese. Legalese* refers to the legal discourses around the adoption of (NSMI). These are lengthy abstruse legal texts which upon adoption forces compliance via the threat of legal action and force from the original state repressive apparatus: the police. The prime cause for the rise of the collusive forces of these twin discourses is the economic value ascribed to the circulation of information, as information in the information society is a commodity, "a factor of production, in a category with land, capital, and energy" (as cited by Wilson, 2002, p. 9). As with commodities like land, property and energy, an entire legal discourse emerged to enforce the protection of these digital commodities, at the centre of which lies Intellectual Property Rights (IPR).

Intellectual property covers any idea or product or process that has an economic value or potential economic value. Intellectual property rights also include patent, copyright, trademark and trade secrets. In the US, IP legislation is being used to change ideas into commodities thus turning information into 'goods', and 'perishable goods' at that. Currently the extension of copyright law appears to benefit corporations more than individuals even though technically copyright now covers all creative acts, extending to e-mails, a poem and even text messages (Correa, 2000, pp. 50-61). The extension of IP law is due to the characteristics of (NSMI), one of which is digitalization. Digitalization marks the shift from analogue to binary code as a means of recording, storing and reproducing data. This renders digital texts dematerialized as they are disconnected from the physical form and may be reproduced ad infinitum, very easily and cheaply. While IP law extends to protect the rights of authors as they are cited as the key stake holders in grave danger in this new information society, it is really big business interests that IP law really works for, protecting software codes and what may arise as other valuable pieces of informational commodities.

Some theorists argue further that IPR is effectively closing the once open domain of the virtual sphere and instead assert the need for *information commons*, where information is freely exchanged without economic value or with minimal economic value ascribed to it. Others argue that IPR goes against the fundamentals of the public sphere and the fourth estate as it is aggressively utilized to hide illicit information about industry malpractice or corruption, whereas the public has a right to information. The converse practice of surveillance also arises as problematic as IPR tends to protect industry rights to privacy but not citizens.

Following on from Zimbardo's ideas of *Big Brother* (1999), the virtual space is one where there is ample potential for surveillance. Thus anonymity is a myth because of the digital trace referred to as 'cookies'. Two salient questions arise. The first asks how behaviour changes when individuals internalize the constraints of surveillance in the virtual world[7]. The second question which is of mounting concern asks who has control of and access to surveillance; and what are the implications of this control or access. The key issue is whether MNCs colluding with political regimes have acted on their power for surveillance in order to seek out and restrict, punish or eliminate dissenting or anti-hegemonic voices.

The Real Digital Divide

Gandy (2002, p. 448) presents the idea that the real divide in the digital age has been the increasing gulf cultivated between notions of the citizen and the consumer. This divide raises debate about the role and responsibility of government for ensuring a well-served citizenry. Gandy argues that the Reagan administration was credited with dismantling the welfare state and with ushering in the era of the *sovereign-consumer* whose interests reign over that of citizens as the primary basis for evaluating social systems:

Neoliberal political philosophy has found a comfortable berth in an intellectual stream of post-modern thought that denies the possibility of an institutionalized representation of common or collective interests... Ironically, despite the obvious inequalities we observe within the marketplace, the market is praised as being more responsive than political systems to individual interests. (Gandy, 2002, p. 449)

Tracey (as cited by Gandy, 2002, p. 452) asserts that these intellectual constructs took hold after the continuous election of right-wing governments which resulted in the centrality of the *individual-as-consumer* alongside which followed the interests of *the corporation.* Hence notions of public good, public interests, community, public culture, citizenship, governance and even the nation state were "buried in the rubble of the old order" (2002, p. 452) – the welfare nation state.

Accompanying these ideas of the centrality of the consumer, a market logic or "marketization" ensued. This referred to support for market solutions as efficient and self-regulating, paving the way for the dominant role private corporations would come to play in managing public culture (Gandy, 2002, p. 451).Hassan argues that the ICT revolution evolved simultaneously with neoliberalism which resulted in, "the flexibilization of labour, the free movement of capital, and an end to the state and organized labour as having a lead role in the managing of the economy" (2004, p. 68). This translates into flexible working conditions making the knowledge labour force in particular a surplus pool to be pulled in and pushed out of the work force as per capitals need. This flexibilization has made working conditions extremely uncertain even though it has resulted in immense wealth in the hands of a few. In a way, the discourse of the digital divide, where the accent is on bridging the divide, those on the wrong side of the divide are invited (as it were) into the information nirvana to inscribe their own subjugation in digital terms.

This raises important questions as to the health of public culture; democracy; and civic activities. Calabrese and Borchert as explained by Gandy, indicate that the construction of the individual proportionately as consumer or citizen will ultimately determine the role technology will play in '*post-industrial democracy*':

They suggest that the network environment stratified and segmented, and that a 'new class of technical and professional intelligentsia' will be engaged in activities that will be 'exclusionary, both by default and by design'. Interestingly, they suggest that the civic activities of this new class will be financed primarily by the revenue generated by the lower social strata who will be involved in the network primarily as consumers, rather than participants, or sources of argument and analysis. (as cited byGandy 2002, p. 456)

Although Gandy's reflections are somewhat dated, a more recent critical survey of EU (European Union) mobile narratives (Lillie, 2009, pp. 4 - 9), embedded in which lies the likeness of the ideal EU citizen of the mobile world, yields three notable findings. Policy goals articulated in "Working in the Knowledge-Based Economy" section of the EU action plan white paper for the EU Commission focuses on the need for public access as a necessary precursor to training of IT low-tech workers seen as important to stimulate 3G development. Thus public access is seen in terms of training and development of a new information workforce as opposed to the idea of information commons and universal access (ibid).

In relation to the construction of mobile users, individuals are cast in a seamless identity of work and play, "switching from information worker to media consumer" (Lillie, 2009, p. 7). Mobile users are further constructed as a fraction of the world representing only the elite who can afford the expensive gadgetry and accompanying packages:

Money rather than nationality or residency is the ticket to access in these commercial narratives...The libertarian spirit of cyberspace translates to freedom of agency and choice in the mobile universe, but the notion of "participation for all," which was key to so many cyberspace manifestos, is lost to the "pay per" market criteria... Mobile citizenship is the providence of a "pay per" nation of liberated citizen-cyborgs. (Lillie, 2009, p. 9)

In the information society consumption reigns over citizenship and individual consent is not derived so much from the state apparatuses as from contemporary technological apparatuses. The idea of technological apparatuses is distilled to include apparatuses of coercion / consent and the repressive apparatuses of legal discourses around (NSMI). Regrettably, Barber's postulation rings true that technologies can help democracy only if they are programmed or designed to do so, and otherwise remain further constrained by the politics and paradigms that inform them. "Left to the market it is likely only to reproduce the vices of politics as usual" (Barber as cited by Gandy, 2002, p. 456).

In terms of future research languagearound ICTs will continue to be of interest as it creates a discourse whichreflects and then inscribes hierarchies of power.Due to the ubiquitous nature of social media a new area that warrants attention is intellectualproperty rights as it extends to personal information and the right to privacy. Here repressive technological apparatuses manifest as *legalese. Legalese* refers to the legal discourses around the adoption of (NSMI). These are lengthy abstruse legal texts which upon adoption forces compliance via the threat of legal action and force from the original state repressive apparatus, the law.In addition much work is needed to discern the impact of (NSMI) on citizen participation in democratic discourses, structures and processes given the participatory affordances of new media.

The chapter attempted to engage with some of the routine terms surfacing when considering ICTs in the African context. While these terms appear on the surface to be synonymous and interchangeable, the chapter attempted to reflect that they are laden with value, meaning and reflect power relations within a global and local context. The key entry point was the digital divide as it impacts on the African context quite strongly indicating the opportunities and challenges presented by new media technology. The chapter goes on to consider the various forms of the divide which extends even to citizenship and consumption. Commodification of information is explored as corporates drive consumer identity as a means of managing public culture inscribed by the market economy. Finally, it is emphasized that market solutions will only replicate existing power structures and hence are not reliable in delivering normative ideals such as closing the digital divide.

REFERENCES

Althusser, L. (2001). Ideology and ideological state apparatuses. In G. D. Meenakshi & D. M. Kellner (Eds.), *Media and Cultural Studies Reader*. Malden, MA: Blackwell Publishing.

Appadurai, A. (2002). *Globalization*. Available from: http://www.polity.org.za/pol/home/

Correa, C. M. (2000). *Intellectual property rights, the WTO and developing countries: The TRIPS Agreement and Policy Options*. London: Zed Books.

Deacon, D., Pickering, M., Golding, P., & Murdock, G. (2007). *Researching communications*. New York: Hodder Arnold.

Fourie, P.J. (2007). *Media history, Media and society*. Wetton: Juta.

Gandy, O. H. Jr. (2002). The real divide: citizens versus consumers. In L. Lievrouw & S. Livingstone (Eds.), *The handbook of new media* (pp. 448–460). London: Sage.

Garnham, N. (2004). The media and the public sphere. In F. Webster (Ed.), *The information society reader* (pp. 357–366). London: Routledge.

Gillwald, A. (2005). Good intentions, poor outcomes: Telecommunications reform in South Africa. *Telecommunications Policy*, *29*(7), 469–491. doi:10.1016/j.telpol.2005.05.005

Gordon, T. (2002). *Marshall Who?* Available from http://www.marshallmcluhan.com/biography/

Green, L. (2010). *The Internet. An introduction to new media*. New York: Berg.

Habermas, J. (2004). The public sphere. In F. Webster (Ed.), *The information society reader* (pp. 350–356). London: Routledge.

Hamelink, C. J. (1994). *The politics of world communication*. London: Sage.

Hassan, R. (2004). *Media, politics and the network society*. Berkshire: Open University Press.

Hindman, M. (2009). *The myth of digital democracy*. Princeton University Press.

ICT Empowerment Charter Working Group. (2004). *The ICT Charter*. Retrieved from http://www.workinfo.com/BEE/ICTBEECharterNov04.pdf

Karvonen, E. (2004). Introduction. In F. Webster (Ed.), *The information society reader* (pp. 345–350). London: Routledge.

Klecun, E. (2008). Bringing lost sheep into the fold: Questioning the discourse of the digital divide. *Information Technology & People*, *21*(3), 267–282. doi:10.1108/09593840810896028

Lesame, Z. (2005). Bridging the digital divide in South Africa. In Z. Lesame (Ed.), *New Media: Technology and policy in developing countries*. Pretoria: Van Schaik.

Lievrouw, L. (2002). Determination and contingency in New Media development: Diffusion of Innovations and social shaping of technology perspectives. In L. Lievrouw & S. Livingstone (Eds.), *The handbook of new media* (pp. 183–200). London: Sage.

Lillie, J. (2009). A Critical Analysis of EU Mobile Narratives and the Ideal Citizen of the Mobile World. In *Proceedings of International Communication Association conference*.

Ling, R., & Donner, J. (2009). *Mobile communication*. Cambridge, UK: Polity Press.

Marshall, P. D. (2004). *New Media Cultures*. New York: Oxford University Press.

McLuhan, M. (1986). *McLuhanisms*. Available from http://www.marshallmcluhan.com/

McPhail, T. L. (2013). *Global communication: theories, stakeholders and trends*. Malden, MA: Wiley Blackwell.

Morozov, E. (2011). *The net delusion*. New York: NY Public Affairs.

Mosco, V. (2004). *The digital sublime: myth, power, and cyberspace*. Cambridge, MA: MIT Press.

Mosco, V. (2009). *The political economy of communication: rethinking and renewal*. London: Sage. doi:10.4135/9781446279946

Mzamane, M. V. (2001). *Where there is no vision the people perish: Reflections on the African Renaissance*. Available from www.unisa.edu.au/hawkeinstitute/publications/default.asp-59k-

Pilger, J. (Producer). &Lowery, A. (Director). (2001). The New Rulers of the World [Television Documentary]. London: Carlton Television.

POLITY. (2004). *The Black Economic Empowerment ICT Charter*. Available from www.ictcharter.org.za/-7k-

Simone, M., & Fernback, J. (2006). Invisible hands or public spheres? Theoretical foundations for US broadcast policy. *Communication Law and Policy*, *11*(2), 287–313. doi:10.1207/s15326926clp1102_5

Sklair, L. (2010). Transnational practices. In G. Ritzer & Z. Atalay (Eds.), *Readings in Globalization. Key concepts and major debates* (pp. 184–212). West Sussex, UK: Wiley-Blackwell.

Tech@work. (n.d.). *CNBC Africa*. Retrieved from http://www.cnbcafrica.com/tv/shows/tech-@-work/

UNDP. (n.d.). *United Nations Development Program*. Retrieved from www.undp.org/info21

Vodaworld. (n.d.). *Vodacom*. Retrieved from http://www.vodacom.com/vodacom/vodacom-world

Warnock, K., Schoemaker, E., & Wilson, M. (2007). *At the heart of change. The role of communication in sustainable development*. Retrieved from http://panos.org.uk/wp-content/files/2007/09/The-Case-for-Communication-in-Sustainable-Development.pdf

Wilson, M. (2002). Understanding the international ICT and development discourse: Assumptions and implications. *The South African Journal of Information and Communication*, *3*(1).

Youngs, G. (2007). Virtual communication, globalization and new public sphere issues. In G. Youngs (Ed.), *Global political economy in the information age: Power and inequality* (pp. 77–91). New York: Routledge.

Zimbardo, P. G. (1999). *Stanford prison experiment*. Retrieved fromhttp://www.prisonexp.org/

KEY TERMS AND DEFINITIONS

Citizen: An individual who is part of a nation state entitled to the franchise and various human rights protections and provisions as laid out but the constitution of the nation state. The term implies an individual's rights and responsibilities within a nation state.

Commodification: The imposition of exchange relations onto social relations, while commodity fetishism refers to the distortion of use values and exchange values.

Consumer: An identity construction of the individual that emerges out of market dominance. Apart from referring to an individual who consumes commodities, the term has also come to imply that an individual's only identity and purpose lies in that very consumption.

Data Trash: Was coined by P.D. Marshall (2004) to refer to the 'have-nots' of technology who are categorized as *data trash*. This kind of linguistic othering abounds in ICT literature with terms such as the "deleted", "sink estates" and "underwired communities" which together add to the binary discourse around ICTs where those on the wrong side of the divide are cast as inefficient and miserable.

Digital Divide: The uneven diffusion of new media technology which separates those with technology from those without technology resulting in economic and cultural identity formations and key differences. In recent times the digital divide also includes access or connectivity. It is of course most often used to emphasize the digital technology disparities between the developed and developing worlds.

ICT: An acronym referring to information and communication technology.

Information Poverty: The ironic phenomenon where in spite of the glut of information on the internet, people remain less informed due being unable to navigate the glut of information on the internet.

Neoliberalism: An economic model promoting laissez-faire economics which advocates a free market (or trade) with as little regulatory oversight at the national or global level. For the nation state this translates into promoting policies such as privatization, deregulation and liberalization.

Technological Determinism: Establish technology as an instrumental cause for change.

The Knowledge Economy: The digital revolution precipitating the shift from industrial capitalism to information capitalism. Where machines and industries produced standardised goods and services in industrial capitalism, ICTs, knowledge and information form the commodities that circulate in the knowledge economy.

ENDNOTES

1. Arjun Appadurai (2002) offers a useful method of analysis and description. He introduces five metaphorical 'landscapes' to define five broad areas of globalization, one of which is the 'ideoscape'. This refers to the increasing acceptance of citizenship and rights which includes political discourses of democracy.
2. This indicates a meaningful participation of black people, women, disabled people and the youth in the growth of ICTs.
3. The Internet is also said to have the capacity to connect people from diverse backgrounds but the trend seems to be fragmentation into smaller and smaller groups. The Internet can also be used for flaming conflict rather than for rational discourse (Karvonen, 2004, p. 348).
4. Habermas' concept of the public sphere is not to be equated with that of 'the public' i.e. of the individuals who assemble. His concept is directed instead at the institution, which to be sure only

assumes concrete form through the participation of people. It cannot, however, be characterized simply as a crowd.

5 This is qualified as 'critical' public opinion.

6 Emoticons refer to icon 'faces' showing a range of cartoon-like, exaggerated expressions (such as happiness, sadness, love, anger etc.).

7 One has only to review Zimbardo's work (1999) to see that surveillance has a direct behavioral change effect on individuals.

Chapter 6
Discourse and Multimodal Analysis of Netizens' Reactions to the Nigerian 2015 Presidential Elections:
A Case Study of Osun Political Parrot Facebook Site

Aretha Oluwakemi Asakitikpi
Osun State University, Nigeria

Samuel Oluwafemi Adeyeye
Osun State University, Nigeria

ABSTRACT

Words and images are channels through which identities and realities are created. In the traditional mass media the power to do this is controlled by management in accordance to prescribed rules and stakeholder desires. This concept shifts with newer media forms like Facebook which transforms the power into the hands of the netizens. This is considered in relation to postings on the Facebook pages of the Osun Political Parrot with regards the Nigerian Presidential Election. The chapter builds its analysis on the liberty netizens have through the internet and the limited monopoly the encoder has over their uploaded comments. It examines the quality of comments netizens make based on their application of intertextually derived knowledge from other media texts. Using Discourse Analysis and Multimodal Methods, examinations of word and image associations in uploaded posts and comments made on March 22-28, a week to the Nigerian 2015 Presidential Elections is done.

DOI: 10.4018/978-1-5225-1859-4.ch006

INTRODUCTION

The 2015 Nigerian Presidential Election constituted a major platform for Nigerians to decide the socio-political fate of their country through the democratic process of voting. For many, the electoral process opened up an opportunity to make their voices, fears and hopes heard not through the traditional media forms of television, radio and newspaper, but through the new age medium of the social media. Pre-election discussions through social media platforms such as Facebook provided many Nigerians an opportunity to be a part of the change they hoped for. Through personal comments, reactions and likes to uploaded posts attempts were made to convince fellow Nigerians and indeed the world of their views on the direction the much needed change should take. For many, the 2015 Nigerian Presidential Election symbolized a change which they hoped would be for good (Awobamise & Aderibigbe, 2015). Within this context, a number of discourses were created. These discourses took their roots from government statements, party campaign slogans and the personal statements of the two major parties namely: The People's Democratic Party (PDP) with the incumbent President Goodluck Jonathan as the flag bearer and the All Progressive Congress (APC) with retired Major General Muhammadu Buhari as the presidential flag bearer. These two personalities were the major themes upon which these discourses revolved.

This chapter considers the themes that arise through the online uploaded postings on the official Facebook pages of the Osun Political Parrot and the discussions they generated. The fact that this Facebook page was (and still is) operated by the then opposition party (APC) gives more insight into how texts and images are used as symbols of persuasion on the one hand (by APC dominant supporters) and on the other as symbols used to counter such persuasion (by PDP party supporters). By using Discourse Analysis and Multimodal research methods, the chapter aims at teasing out meanings and possible associated socio-political factors behind each text and image within the Nigerian political campaign platform.

The media theory of agenda setting which considers the power of the media to influence what people talk about has a different twist when it comes to the social media. The evolution of technology has put the power of creating news into the hands of citizens and not the media which is often dictated to by pressure groups and governments (Shaw, Hamm & Knott, 2010). Nevertheless, the importance of the traditional mass media forms as a major platform upon which deliberately planned discourses are created cannot be overemphasised. Through adverts by political parties, personal interviews granted to journalists by political aspirants, and televised campaign rallies; political discourses are created for social consumption and discussions (Reyes, 2015). A major deliberately constructedpolitical discourse which both parties built upon were slogans to support their campaign strategies.

For PDP, the party in power, the slogans included: 'PDP Power', '#Bring back Goodluck 2015', and 'One good term deserves another'. For the opposition party, APC, the major variables of change were significantly present in its slogan, relating them to change of leadership and ruling party. The ruling PDP had been in power since the establishment of democracy in 1999 and the need for a change in not only leadership but also party rule was emphasised. Theconcept of change was also built upon the number of accusations placed against the Goodluck Jonathan's administration ranging from corruption, insecurity in the form of Boko Haram activities of suicidal bombings, kidnappings and terror (Adegoju & Oyebode, 2015; Awobamise & Aderibigbe, 2015). The kidnapping of over 200 Chibok school girls brought Nigeria to the international spotlight with protests done nationally and internationally; physically and on the e-platform under the slogan '#BringBackOurGirls' further discredited the then Goodluck administration (Chiluwa & Ifukor, 2015).

The Nigerian political scene thus witnessed strategically created political communication discourses aimed at supporting party slogans and rebuffing opponent accusations. Such tactics seemed to have a balanced effect as they developed into major subjects that were paramount in Nigerian and foreign media outlets with some condemning and constantly attacking what was described as the lackadaisical attitude of the then Nigerian President while others supported his every actions (Awobamise & Aderibigbe, 2015).

OBJECTIVES OF THE CHAPTER

This chapter aims to achieve the following objectives:

1. To understand through a qualitative multimodal research method, the relationship between image selection and the creation of meaning within the Nigerian social media.
2. To understand through a qualitative discourse analysis, the relationship between word selection as practiced by Nigerian netizens in the creation of meaning within the context of the Nigerian social media.
3. To understand through a qualitative contextual analysis of text and images the use of Facebook as asocial media in the campaign strategies adopted by Nigerian political bodies.

THEORETICAL FRAMEWORK

Critical discourse analysis takes into consideration the processes and products of discourse and their impact on social practices (Caldas-Coulthard & Leeuwen, 2003:1). The interface between critical discourse analysis and semiotics is the multimodal analysis of communication where an understanding of how signs and symbols are combined to form patterns of meaning. This considers how meaning is created not only through the grammatical make-up of a message but also through multiple resources such as visual features (this includes colour, size, and textual qualities), punctuation, page layout, font size and type as well as associated images (Machin, 2013; Idiagbon, 2013; Wang, 2014). As Machin (2013) observes, the limitation of such studies, however, is the fact that multimodal elements do not have the concrete connotative or denotative meanings which words and grammar can bring out; yet the concept that images speak louder than words may construct and convey stronger meanings. Due to the abstract nature of semiotic images, commitments can be avoided. This factor is especially crucial when analysing images and other multimodal constructions of political discourses where a lot can be said but limited commitment derived on the part of the communicator while yet still inspiring discourses. The ability to re-contextualize meaning is also possible by adding, deleting, substituting, or evaluating meanings within the construction of a visual message.

In understanding the re-contextualization of meaning is the evaluation of how texts derived from other media sources are used within a discourse. This is referred to as intertextuality where 'voices of prior texts' are incorporated into currents texts (Ferreira & Hebrrle, 2013; Oostendorp, 2015). This reflects what Oostendorp identifies as texts moving across space in terms of meaning. Thus a meaning is derived based on a previously used or an exposure to a previously used text. Within this concept, meaning is derived not from the current text and its discourse usage but from previous exposure to the texts of similar themes. This strategy was prominently used to persuade netizens through an assumption of intertextual credibility or to rebuff such persuasions. Within the concept of re-contextualization is the creation of memes which involves the construction of names and other nouns developed within the sphere of political humour for internet use (Adegoju & Oyebode, 2015). These memes are imitations or mimics of original texts which ultimately dramatically alters the meaning and the word. From this construction new words are formed and used with the aim of dissemination through the internet space.

Major themes used by discourses were derived from various factors. A major factor used in the development of discourses was the economic situation of the country at the time of the uploading. Through the use of active sentence constructions, uploads present information and facts that suggest a possible problem while indirectly giving a solution in the person of the proposed presidential candidate. Thus, through a problem-solution presentation (Ferreira & Heberle, 2013) active yet indirect options are provided through information which is designed to help the netizens accept a proposed reality (Sani, Abdullah, Ali, & Abdullah, 2012).

RESEARCH METHOD

The method adopted for this research is a qualitative one using multimodal methods to analyse the images uploaded on the Osun State Parrot and discourse analysis to conceptualize the texts derived from the comments made by Nigerian netizens in response to the uploaded posts. The use of multimodal methods in the analysis of the uploaded images was adopted to understand the relationship between image position, colour, and text in the creation of meaning. Discourse analysis method is adopted as a complementary method to analyse the texts of the comments made by the Nigerian netizens in reaction to uploaded posts. It enables the researcher to conduct an in-depth analysis of netizens' word selection in relation to context and situation of the 2015 Nigerian Presidential election. Through an analysis of the words used by the Nigerian netizens, an understanding of meaning and context can be derived.

ANALYSIS OF DATA

The data used for this paper was retrieved from the Osun State Parrot Facebook site. The site promotes the APC party and its flag bearer General Mohammadu Buhari by uploading posts that support the party and the candidate. This section considers the three major posts uploaded during the selected time frame of March 22-28, 2015 and the responding comments by the Nigerian netizens. The posts are presented based on the dates of the uploading by the Osun Political Parrot Facebook administrators along with the associated comments of netizens to the upload. This section adopts the theory of social interactionalism to analyse the texts and the theory of information value in picture composition within the framework of Discourse Analysis and Multimodal methods to analyse the images posted on the selected Facebook pages.

Figure 1 is a posting uploaded on the Facebook page of Osun Political Parrot at 9:55 p.m. on March 22, 2015. The post will be analysed using the social interactional perspective which focuses on how people construct and perceive their social identities within their social environment (Wang, 2013). This post speaks to the Nigerian religious consciousness of evil versus good through the use of colour. Within another context, this post may not have made such a significant effect as both photographs depict both presidential candidates in a socially balanced posture of respect, authority and public control. The question posed also may be seen as innocent as it uses the colour of the outfit worn to distinguish one

Figure 1.
https://www.facebook.com/photo.php?fbid=795491110520216&set=a.359326594136672.74602.100001780046796&type=3

candidate from the other. The deliberate use of two primary colours that for many Nigerians signify opposites is an attempt by the communicator to encourage responses in favour of the colour that denotes a more positively associated spiritual meaning (the colour white) while rejecting the colour associated with negatively connoted spiritual attachments (the colour black). This construction of meaning may have an opposite connotation for example in the socio-cultural environment of the Ghanaian culture where the colour black though associated with mourning and even darkness, is more importantly associated with purification, maturity, spiritual energy (http://buzzghana.com/ghanas-kente-fascinating-stories-and-meanings-behind-the-cloths-designs-and-colours/) and this is visually represented in the symbolic representation of a black star in the Ghanaian flag.

Through a combination of colour and text, the communicator speaks to the Nigerian concept of power and spiritual dominance and even war fare. The viewers are thus encouraged through this combination to consider their answers within this ideology. It is expected that such a question, situated within the Nigerian social structure, would result in responses that would gravitate toward the colour that symbolises a more positive meaning. In choosing this colour the netizen choses the personality wearing the colour. This associated meaning ultimately ascribes unto the wearer of the colour the associated connoted meaning. Thus, in choosing the man wearing white for example, the netizen would be associating the socio-cultural attributes of that colour unto the wearer and vice versa. This is irrespective of the reality that such a personality *may not* possess such characteristics. This method of relating visual with words does not make reference to previous discourses but works on religious knowledge or assumptions in helping the social media consumer interpret or construct a meaning. The communicator draws inference not from previous discourses as consumed from other media texts such as the traditional media forms of television or radio, but from genres related to spiritual convictions that have its foundational knowledge on culture and religion. It is thus expected that the associated responses would lean heavily on such knowledge.

Picture placement is a significant variable in understanding how meaning is constructed and interpreted in relation to word and visual message portrayal. The information value of a construction of visual and verbal placement can be derived from the placement of the information provided. Visual and verbal information presented on the left represent an already given, known or acceptable information while information that is new is placed on the right. Information that is considered as ideal is presented at the top while information that is considered as real is placed at the bottom (Wang, 2013; Oostendorp, 2015). Within this framework the information which provides the option of associating the colour white with the man General Muhammudu Buhari (GMB) is considered by the communicator as ideal and a given association; while the information that relates the colour black with Goodluck Ebele Jonathan (GEJ) which comes on the left denotes the meaning of something that is new and should be taken seriously possibly as a major threat to the social and religious development of Nigeria.

The first comment to this post was made by 9:58 p.m. the same day (three minutes after the post was uploaded) and this spilled into the 23rd of March with a steady and continuous flow of comments until March 24th 2015. Within this time frame, 195 people commented on the post. Out of the 195 responses recorded for Figure 1, 44 of the responses were uploaded on March 22, 2015, (23%), 105 comments recorded (54%) on the 23rd of March and 44 comments recorded (23%) on the 24th of March 2015. The frequency and the volume of comments to this post indicate the level of interest the Nigerian public has concerning the information content as well as the associated meanings that can be derived.

This analysis will be done based on the days in which the comments were made, beginning on March 22nd, 2015. On the selected day, a total of 44 comments were recorded in relation to Figure 1 out of which 35 (80%) were in favour of GMB (Major General Muhammadu Buhari); 6 (14%) were in favour of Goodluck Ebele Jonathan (GEJ), and 1 (3%)not in favour of either personalities. The expectation of the encoder was reflected in the comments made by netizens. Fifty percent of the comments were directly related to 'white' which appeared 21 times in the comments. Some of the commentators repeated the word as a form of emphasis accompanied with exclamation marks and at times with the word 'white' written in capital letters. The colour white was more often associated in a favourable way to the personality of GMB. Three of these comments (7%) merely symbolized their response with the letter 'A' or 'A. White', giving a notion of some sort of detachment from the visual and verbal construction made yet acknowledging the importance of such a relationship. Four (9%) of the comments tried to show more of an association by putting the word 'man' which gives the response anacceptance of the implied association of colour and personality but emphasises the quality of gender through a reiteration of the word 'man'. Including the word 'man' brings to the fore discourses on gender where in Nigeria a man is a dominant gender based on the predominantly patriarchal system the country practices.

Fifteen (34%) of the comments were directly related to GMB in a positive way with emphatic statements such as 'white man with white cloth will conquer'; 'the man in white and with a pure and sacred mind'; and 'the man in white is god's choice'. Such comments relate the personality to positive cultural values ascribed to the colour white such as divine acceptance, power, and purity of mind. The comment which simply says 'white man', is in reference to the white race which is socially ascribed more wisdom and dominance in Nigeria when compared to the black race. On the flip side is the negative association of the colour black to evil and such a comment like the following captures this: 'black depicts darkness mind you'. Within this day, little reference is made to GEJ except in the form of a word in brief comments like 'GEJ' and 'Jonathan'.

A few of the comments make reference to previous text which have their foundation on popular political party slogans. Such comments stress more of the political consciousness and not the colour and word association. Ten of the comments on this day were directly related to political slogans and included comments like 'say Buhari', 'sai buhari' (a phrase coined from the political slogan 'Say Buhari) 'GMB' (the initials of the presidential candidate). There were also some comments that seemed to be more of short campaign slogans like 'buharimukeso' (translated from Hausa to mean Buhari is who we want), and 'Buhari the salvage'.

Even though there was an overwhelming support for GMB on that day, there was a singular negative and direct outburst against his personality. The comment was based on an intertextually derived discourse generated due to Buhari's boycott of a nationally organized and televised presidential debate. This boycott was highly criticized by the opposition party and this formed the basis of this comment: 'Buhari is a disgrace to Nigeria, he can't take my vote again. I have changed my mind. Now it is like he has skeletons in his cupboard'.The expression 'skeleton in his cupboard' is popularly used in the Nigerian political context to connote some hidden and disgraceful fact which if revealed could cause great damage to the personality in question.

March 23rd recorded a total of 105 (54%) comments which is more than double the total number of comments recorded under this post. The comments uploaded on this day were also the largest and with the highest frequency recorded compared to all the posts uploaded and used as data for this study. The comments started by 12:07 am and streamed in almost non-stop until 12:07 am on the 24th of March, 2015. Sixty four of the comments in response to the posting were related to the personality of GMB and (as noted in the previous day) in association with the colour white (61%). The word 'white' within this day appeared 68 times with some directly interpreted from the Yoruba language (largely spoken in the South Western region of Nigeria), Hausa (largely spoken in the Northern region) and the Nigerian Pidgin. The significance of the colour to positive spiritual connotations was more strongly reflected in some of the comments with associated words like 'light', 'peace', 'progress' and 'victory'. Within such comments a direct and positive reference is made as in the comment: 'white is for peace, I vote peace, GMB all the way'.

A distinction is made on this day between expressing preference for colour using the same colour to define the man of choice. Comments which had the colour as a central theme include: 'white is my choice', 'We pray for white (light) and white (light) shall be our portion', 'white is the best', 'I go for white', and 'my god is white'. These comments represent the hopes of the Nigerian netizens that by choosing a positive colour, something positive will happen. This is based on cultural and social associations attached to the colour. A few of the comments adapted the colour to the political slogans associated with GMB's party: 'up white because it represents change' and 'white on top 2015 till 2019'.

A few of the comments presented in their emphasis for the colour white the contrasting beliefs held about the two colours. These comments have strong religious undertones which are used in their justification for choosing GMB. Comments such as ''White is what I desire because I am a Celestial member, simply because in Celestial black means bad thing so I go for Buhari & Oshinbajo mandate'. Another explains that; 'Sai baba in white in Nigeria generally black denote mourning so I will not mourn anybody in this coming election while white denote good, happiness which implies that joy is mine in this coming election'. Within this category is a quotation from a popular Nigerian singer Dr Orlando Owoh in Yoruba which when translated reads 'I came with the day, I do not want to be associated with darkness for things concerning darkness scares me'.

More of the comments associated the colour with GMB: 'the man in white' or 'the one in white'. Some of the comments reversed this word arrangement to 'the white man' to suggest the perception of a white superior race. A major comment under this category is presented with all the words in capital letters as the netizen's way of emphasis with the words, 'VOTE FOR A WHITE MAN VOTE FOR BETTER CARE I VOTE BUHARI'. One of the comments using a Yoruba proverb insinuates that only white and good things can come out of a man like Buhari because nothing black or bad is in him in the comment 'Funfunleiyesu! Eiye o sududu, Sai buhari, Sai baba white! white!! White!!' The netzin ends the comment with the words 'white' three times with exclamation marks as a grammatical indicator of the conviction. The proverb 'Funfunleiyesu!' comes out twice within this day.

Within the selected day were comments that did not make reference to the colours but made direct personal campaigns for their candidates. A comment such as 'GOD ALMIGHTY has chosen General Buhari as the next president' gives the impression that the gods have taken a decision and based on this GMB's victory is sealed. This seems to be supported by the comment 'To God be the glory, what we should be discussing now is how Jonathan will accept defeat & handing over date'. Another netizen ambitiously counted the comments and said 'Out of 50 GMB has 42! This is great. GMB'. This comment attracts the following quick response in favour of GEJ which suggests that the individual is being fooled 'Greater indeed won tan yii ma tan arare (they are fooling you, do not fool yourself). This response represents one of the few comments that is in support of the then ruling president, Goodluck Jonathan. Even though this is an indirect statement of support, the netizen's comment implies that those in support of GMB are being fooled or deceived.

Despite this singular and lone opposing view, the comments were strongly in support of GMB. Two of these comments were represented in the form of a photograph and image depiction. The photograph is self-explanatory in the form of an uploaded newspaper page with a prominent headline reading 'I won't stay in office beyond 2015, Jonathan insists' under this are two smaller headlines which read 'president faults opposition to single tenure' and PDP holds convention January 25. Under this is a medium sized headline captioned 'N67b capital flight on bandwidth worries govt' under this is a smaller rider that reads 'to re-launch satellite in December'. The aim of this upload was obviously to provide credible sources of GEJ's untrustworthiness and bad governance. The second is an uploaded image of a night scene in a city with skyscrapers built close to each other and dotted lights in the windows. A flood light from one of the buildings makes a light image in the night sky and in the beam is the phrase 'SAI BUHARI'.

There were a few negative but indirect comments against GEJ through an associated negative meaning to the colour black. For example, the comment that was written in the Yoruba language with a direct translation says 'God will not allow me to pick black in my life. What of you?' or 'Black *bawo* (what do you mean by black?), white on point'. One of the comments is a direct, derogatory comment presented in Nigerian Pidgin with strong hate words against GEJ which reads 'Me I no go choose dis kind black in my life. Even though I like black people all over the world, no be dis one here who is black, even to his marrows and his blood! Shit.'

Within these sentimentally associated comments that relate positively the colour white to GMB as well as other comments that either promote the personality through associated campaign slogans or personal campaigns are a few comments in support of GEJ. There are just two of these comments and both are short with one saying 'Jonanijoooo' and the second simply 'GEJ 4 sure'. This second comment in support of GEJ nevertheless is quickly neutralized by the following advice in favour of GMB 'If they're deceiving you, stop deceiving yourself, Buhari is president elect, don't waste your vote'.This comment represents the third time the notion of deception will be suggested. This brings to bear the failed promises former presidential candidates presented in a bid to deceive the Nigerian populace and submit the people into oppression and poverty.

The 24th of March had comments uploaded in response to Figure 1 between 1:49 a.m. till 11:41 p.m. with 45 comments recorded. As in the previous days, a number of the comments used the colour to signify their intended choices by simply typing the word 'white'. Fourteen of these comments are a direct response to the colour with more emphasis of the colour and what it represents than the individual wearing the colour. Within this category was the phrase 'the man in white' with a larger percentage simply signifying their preference with the word 'white'. A few of the comments qualified the value of what the colour symbolizes through comments like; 'white is always pure' and 'white a replica of a DOVE'. The latter comment has the words 'DOVE' in capital as an emphasis of a quality spiritually associated with the bird which is most often white and is believed to be gentle and pure. In relation to the colour contrast quality, there were three major comments uploaded on this day. These comments include: 'Never shall I go for anything black, always WHITE is my choice', 'black colour is associated with defeat, and white success' and 'Although I'm black I go for white'. These comments depict the seriousness the Nigerian populace places on the two colours and their associated good and evil qualities. One of the comments even declared that 'Funfun ni OLORUN Buhari' (white is the colour of Buhari's God). Within this category is an upload which is highly derogatory and negatively portraying the then President Goodluck Jonathan written in Yoruba and directly translated as follows: 'You see, he is black inside, he then puts on a black outfit, his thinking in itself is black, that is Jonathan for you. Let us go out and vote enmass for Buhari'.

As 24th of March, 2015runs to an end, the comments greatly reduce in verbosity with most of the comments either stating 'Buhari' or 'GMB'. These comments represent a direct reference to Buhari outside the colour option provided in the upload. Very few of the comments go beyond these singular words and these include: 'Buhari the salvage', 'Buhari in the mighty name of JESUS' and 'Buhari is the answer'. Despite the fact that the colours are not mentioned, the associated relationship can still be discerned. Unlike the previous days where the presidential candidates were identified by name, this day witnessed netizens using the candidate's initials 'GMB' more frequently. This may be due to an uploaded photo of a voting card in the early hours of the day (1:49 a.m.) of a voting card with the logo of APC (the then opposing party) and a thumb print next to the logo. This image which was uploaded by the APC party was an awareness strategy to educate its supporters on how to vote. Under this image were the initials 'GMB'.

Despite the overwhelming preference for Buhari (GMB) three comments were in support of GEJ. These comments which include; 'B', GEJ', 'PDP' and the following comment in the form of a retort "you guys are deceiving yourselves'. Except for this, the then ruling party and its presidential candidate was not mentioned at all.

Figure 2. is a post uploaded March 25th 2015 by 8:32 am on the Facebook Page of Osun Poitical Parrot. For this posting there were 43 likes and 163 comments but 88 of these comments were recorded and this was for the 25th of March, 2015. The comments started being uploaded at 8:57a.m. and stopped at 9:49 p.m. of March 25th 2015.

Similar to Figure 1, the two major candidates are placed side by side and depicted in two different positions. The Nigerian President at the time, Goodluck Jonathan, is placed on the left with a smile and a wave of the hand. As with the previous post, he is wearing a traditional Niger Delta black outfit with a black cap to match. On the right is the then opposition party leader, General Muhammudu Buhari who is speaking into microphones, a position that suggests a public presentation or a press briefing.

Figure 2.
https://www.facebook.com/photo.php?fbid=819233158144309&set=a.149700215097610.34301.100001728985240&type=3

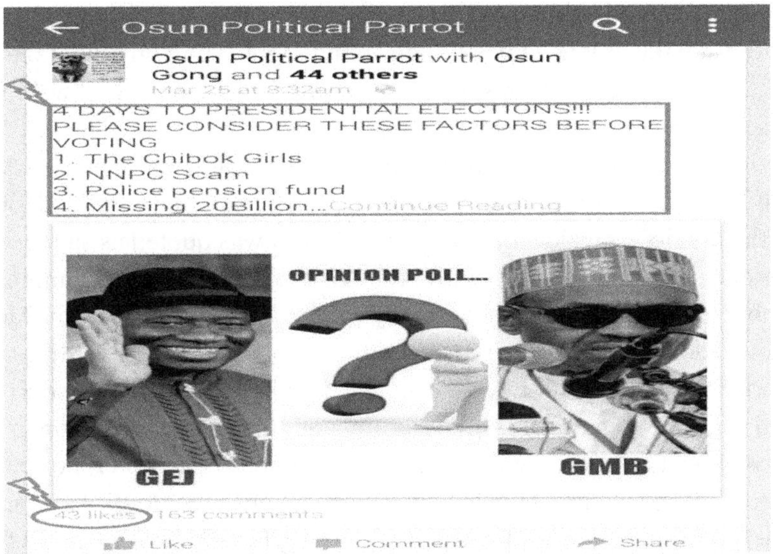

Due to the number of microphones in front of him, it is difficult to tell the exact colour the traditional outfit is, nevertheless, it is a bright one with the colour white being prominent with an olive green coloured white embroidered cap to match. Based on the information value of the picture composition, the picture on the left denotes the given (Goodluck Jonathan will be waving goodbye to Nigerians) and the picture on the right is the ideal (Muhammudu Buhari will take over the responsibility of addressing the Nigerian public). The post just as with Figure 1 does not directly suggest a candidate but rather asks a question. In this post, the question attempts to convince its readers with a prior introduction made up of an overwhelming list of information that favours one candidate over the other. The listed items aligned for readers to consider are discourses drawn from prominent public allegations made against the then ruling party and president.These allegations are intertextual in nature as there was no previous mention of the given information in previous discourses forcing the netizen to assume its source is credible. The allegations include the Chibok girls who were kidnapped by the terrorist group Boko Haram, as well as stories that suggest high levels of corrupt practices: 'The NNPC Scam', 'Police pension fund' and 'Missing 20 Billion'. The urgency for Nigerians to seriously consider these facts reflects in the first sentence that introduces the post; '4 DAYS TO PRESIDENTIAL ELECTIONS' which is presented in capital letters ending with three exclamation marks. Such a sentence construction is aimed at not only attracting the attention of the readers, but also insighting them to take the desired action even though the writer presents a surface attitude of neutrality.

The post was uploaded at 8:32 a.m. on March 25th, 2015 and the first uploaded reaction was registered by 8:57 a.m. For this post there were 163 comments though only 88 could be captured for this analysis on the 25th of March, 2015. Out of these, 55 were in favour of GMB with comments like: 'Say/Sai Baba/ Buhari', 'GMB', GMB is our man/choice', Change is inevitable', and 'GMB is better'. It is noted that the comments in response to the post does not refer to the information provided but rather base their responses on party slogans. Nevertheless, an outburst is recorded within this category which suggest a religious undertone that says; 'Sai buhari sai baba, Jonathan must go by fire by thunder'.

Unlike the previous post where few comments were in support of the then rulling government and party, the reactions to Figure 2 shows bolder support and stronger challenges against the then opposition party (APC) and its presidential candidate(GMB). The comments are derived based on intertextually derived discourses outside the uploaded post. Their foundations being on other texts from varied media sources. Sixteen of the uploaded comments were in favour of GEJ while eight were negative statements against GMB. These will be discussed below.

One of the reactions that directly challenges the post asks a question which says; ''prove this if not Buhari that's behind these'. This reaction is based on discourses that surround Buhari and his reactions when he lost the previous presidential elections. Buhari was quoted as saying that he would make Jonathan's government unbearable and soon after this statement, the Boko Haram insurgence seemed to increase in the number of daring and violent acts they carried out. This post seemed to be supported with another comment that added to the discourse the controversy surrounding GMB's academic qualifications; 'contestant bokohari with no O' Level'. The word 'bokohari' is a meme derived from the words Boko (the first word in the combination of Boko Haram) and 'hari'(the last section in Buhari). The theme behind this meme was further elaborated by the following comment which not only associates GMB with the terrorist group, Boko Haram, but also directly questions the integrity and academic capacity of GMB; 'What about the fake certificate of GMB? It's your people who took away the Chibok students and financed boko haram'. Interestingly, these direct challenges were not debunked or even defended by any of the posts in support of GMB. On the contrary, there were two major posts that strongly supported GEJ and his party. The first post which is presented in Nigerian Pidgin and directly transcribed reading 'You people had better stop all this..Jonathan is the Angel God sent to Nigeria... there is nothing Buhari can do…so just go out and vote 4 #GEJ because he is the only one that is fit to do it..even if you take Buhari's vote be sure nobody will play it okay…power *joor*'. The second post which reads; 'Majority of APC candidates left PDP for APC, same blood flow in their veins. I'd rather cast my vote for PDP than APC. We need to apply our common *sensical* knowledge here fellow Nigerians'. To this was the simple retort by the opposition, 'Stupid post'. This response represents the only post that attempts to counter comments in favour of GEJ.

Other responses that favoured GEJ include; 'GEJ TILL 2019', 'APC you are a loser, for life', 'PDP carry go', 'Ride on Goodluck', 'GEJ till 2019', 'GEJ All the way', 'PDP power'; 'Jonathan all d way. Nothing is stopping him, Not even APC'. Within this category is a major comment that directs its attack on the party and its members and not necessarily the presidential candidate. The comment; 'Osun political parrot you omitted some points. Also tell people that Aregbe of Osun state failed to pay salary for five months. And he had been receiving fed allocations without delay. 2. Remember Fakunle high school was pulled down because Tinubu wants to acquire landed property' further questions the integrity of the party members who are politically supporting GMB and even the unequal representation of facts by the Facebook site.

Figure 3 is a post uploaded on March 26 at 9:08am on the Facebook Page of Osun Political Parrot. The posting attracted 94 comments though only 74 of these postings could be captured. The posting as in Figures 1and 2 present the two major presidential candidates side by side with a distorted picture of GEJ on the left hand side wearing a black hat and white Niger Delta outfit. He is depicted a shade darker than normal with downturned lips. Under this image is the word 'JONATHAN'. Next to this is the photo of Buhari wearing a white traditional Hausa outfit with a blue cap speaking into a microphone. Underneath this is the word 'BUHARI'. Above these pictures is that of the renowned literary Nobel Peace Winner and political activist, Wole Soyinka whose white beard and hair make a contrast against his black outfit and black background. The introduction of this personality is to represent integrity based on the international acceptance Wole Soyinka has attracted over the years. The obvious distortion of GEJ's face in a picture placed on the left and bottom section of the tripod of faces indicates an already given, known or acceptable information; while the picture of GMB speaking into a microphone looking alert and focused represents information that is new. The placement of Wole Soyinka at the top of the two candidates represents information that is considered ideal and authentic. This authenticity is buttressed by the representation of detailed information that serve as a direct quotation from the Nobel Peace Winner who authoritatively gives netizens four criteria upon which voting and supporting the then ruling government would imply. These criteria imply intellectual, ethnic and corrupt blindness; fear of change; and a state of suffering from all the mentioned symptoms. This post as with Figure 2 attempts to sway support by providing information intertextually derived that highlights the negative activities and consequences of the then ruling party while presenting the opposition as the answer to the problem (Ferreira & Hebrrle, 2013).

Comments to this post started on the 26th of March at 9:11 a.m. until the 28th of March, 8:45 a.m. As with Figures 1-2, majority of the postings were in support of GMB with 31 out of the 74 postings. These included postings that reflected the party slogan of 'change' including: 'change all the way', 'change is inevitable', and 'We need change'. Some of these posts added the word 'Jonathan' to form the negative slogan against the then president 'change jonathan'. Within this category were a few comments created to give an impression of urgency by qualifying how soon such a change should take place through comments such as; 'Change in Jesus name' (this reflects concepts of instant change through prayers), 'Change Jonathan quick no waste of time'; and 'Change Jonathan shap shap'. A few showed loyalty to the opposition party with comments such as; 'APC all the way' and 'APC change federal'

Within the comments recorded under this post was the use of memes. One of such memes was recorded under supporting comments for GMB which is a combination of the words Jonathan and Haram to form 'Jonaharam'. A similar concept of word combination was observed under postings for Figure 2 which formed the word 'bokohari' (a combination of the words Boko and the last words in Buhari). Another reaction declared; 'GEJ until 2019. No way ever for the boko GEN' (the words boko Gen are derived from the word 'Boko' and 'General'). The need to associate the opposition with the dreaded terrorist group forms a discourse of interest. The activities of the group did not only affect Nigeria but became an international news item that put Nigeria in the world news as being an insecure and red zone area. Gender activists and other human rights bodies took up the slogan '#Bring Back Our Girls' in protest against the over 250 girls kidnapped from their schools by the Boko Haram. Thus associating a presidential candidate to such a group does not only suggest terror and insecurity but also stagnation and hopelessness.

It is noted that the comments of GMB supporters to this post are very brief and do not go beyond three to a maximum of four words. As observed in the comments analysed for Figure 2, there were no comments that attempted to defend GMB against comments that challenged his personality and party. The only exception recorded for Figure 3 was a response to the post 'Gej till 2019' to which the GMB supporter sharply retorted; 'you are illiterate'. To this was the quick response 'You are actually the one suffering from the above syndromes'. This online conversation was followed by two more comments in support of GEJ: 'GEJ till 2019' and 'No GEJ no election'. The latter post which seems more like a threat was not challenged nor referred to by any of the GMB supporters. Indeed the GEJ supporters were more verbose and insistent in their comments. For this post the following comments were recorded in favour of GEJ: 'LET THEM SAY GEJ FOR VICTORY INSHA ALLAH TAHALA', 'I say no to Buhari, sai

Figure 3.
https://www.facebook.com/photo.php?fbid=819661724768119&set=a.149700215097610.34301.100001728985240&type=3

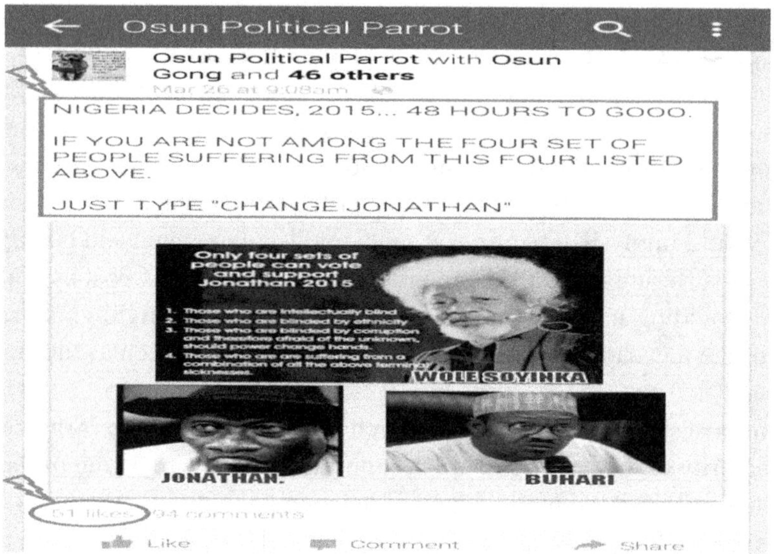

Jonathan', and 'GOODLUCK is working'. Some were comments against the party such as; 'APC will soon cry that they have rigged'; while others showed support for the ruling government's party; 'PDP is the way to NIGERIA' and 'PDP is the winner'.

Using the Nobel Winner as a symbol of credibility in Figure 3 was not regarded nor referred to responses to the post while the items listed as representing his thoughts were also not mentioned. The only referral in relation to the personality and his thoughts was a negative statement uploaded in the Nigerian Pidgin Language and transcribed as follows: 'Wole Soyinka please go and sit down. It is the same old age syndrome that is disturbing Buhari that is obviously disturbing you also'. This brings to bear a major intertextual discourse regarding the age of Buhari which for the opposition was a factor that proved his inability to rule. Public and media messages promoted allegations of his ill health and frailty due to old age and were used to support the argument that he may die due to the pressure of governance just as a former president (Musa Yar'Adua) did. The possibility that GMB could become physically incapable to rule and the confusion a possible demise of a president while in office could cause was a scenario many Nigerians did not want to witness again.

DISCUSSION

The three uploads analysed in this chapter aimed to create a reality by presenting to netizens a problem-solution scenario through the use of images and active text elements in their sentence and phrase constructions. The APC-biased uploads drew their information from previous media discourses which negatively portrayed the then ruling government and their party (PDP and Goodluck Jonathan) in a bid to convince the public of the superiority of APC's candidate and party (General Muhammadu Buhari and APC). The first upload which attracted the most comments played on the religious sentiments of the Nigerian public which resulted in an overwhelming response in support of GMB. The next two posts were not as successful as it presented information derived from other media texts for which the supporters of the ruling government at the time was able to attack and counter. It was noted that for the last two posts, the netizens in support of GMB could not defend their candidate using intertextual discourses. Thus though the support for GMB was overwhelming, the comments in favour of the then ruing government presented evidence that showed more insight from a better informed audience.

The analysis of the data derived suggests that the three uploads through an associated use of text and image addressed more of the emotional than the informative in deriving support from their supporters. This was prominently displayed through comments of APC supporters that had strong hate words against the then Nigerian President. The use of emotion which borders on ethnic, religious and social factors are campaign strategies that have dominated the Nigerian politcal sphere for some time (Idiagbon, 2013). Before the advent of social media platforms, this strategy was efficient as the media consuming audience were limited to traditional media forms such as the televsion, radio and newspaper which was most often dominated and operated by ruling politcial parties. Using politcal and financial might, other political voices were suffocated and hardly ever heard. The social media platform is able to break this monopoly, giving the Nigerian netizens various media channels through which they can source for information, confirm information and even upload information. This was evident in the counter comments that a number of PDP supporters used to defend their candidates and parties.

The social media platform has indeed opened the doors for Nigerian netizens to consume and be a part of the information production world of the mass media. The frequency and almost immediate response to the posts uploaded shows that Nigerians are very active participants of the social media media world. The data indicates that the Nigerian netizens are able to emphasize their points through capitalization, punctuation marks, the creation of memes and even images. Their adoption and use of various indigenous lingua franca (Yoruba, Hausa and Pidgin) is a prominent feature in the conceptualization of ideologies that normally may have been challenging using the English Language. This suggests that the Nigerian netizens are not only conscious of the power of the social media, but are social media literate enough to be able to communicate and manipulate the platform, thus making them not only consumers but also producers of information.

FUTURE RESEARCH

For future research a more diverse data should be considered which will take into consideration various political parties and their construction of meaning through text and images. For the Social media communication analyst, such a research can help in the understanding of how the social media can be used in persuading a complex nation like Nigeria. With the increase of political consciousness in Nigeria, the politcal analysts will find such a research ideal in helping to predict the possible politcal direction future campaigns may take while capturing the political evolution of the Nigerian society.

CONCLUSION

This chapter attempted to analyze three major posts as uploaded on the official Facebook Page of the Osun Political Parrot a week before the 2015 Nigerian Presidential Elections. Though the postings were heavily biased in favour of the then opposition party (APC), netizens in support of the then ruling party were able to defend and support their party and its candidates through information derived intertextually from other media sources. The overwhelming support witnessed through the responses of netizens in support of the then opposition party bordered more on emotional and political sentiments and not on intertextually derived information which meant more of quantity than quality in their responses. Nevertheless, the general active participation of the Nigerian netizens to the uploaded postings through their reactions, responses and uploaded images indicates that the boundaries of the traditional mass media have been broken leaving Nigerians free to not only be consumers but also producers of mass media messages. Through their online participation, the Nigerian netizens have been able to challenge traditional agendas of political parties to assert their voices and reactions through the new technology provided on the social media platform (Shaw et al., 2010).

REFERENCES

Adegoju, A., & Oyebode, O. (2015). Humour as discursive practice in Nigerias 2015 presidential election online campaign discourse. *Discourse Studies*, *17*(6), 643–662. doi:10.1177/1461445615602378

Awobamise, A., & Aderibigbe, A. (2015). An analysis of Western and Nigerian media depictions of President Jonathan's presidential bid. Global Media Journal (African Edition), 9(1), 63-79.

BuzzGhana. (2015, October 14). *Ghana's Kente: Fascinating Stories and Meanings behind the Cloth's Designs and Colours.* Retrieved from http://buzzghana.com/ghanas-kente-fascinating-stories-and-meanings-behind-the-cloths-designs-and-colours/

Caldas-Coulthard, C., & Leeuwen, T. (2003). Critical social semiotics. *Social Semiotics*, *13*(1), 3–4. doi:10.1080/1035033032000133481

Chiluwa, I., & Ifukor, P. (2015). War against our children: Stance and evaluation in #BringBackOurGirls campaign discourse on Twitter and Facebook. *Discourse & Society*, *26*(3), 267–296. doi:10.1177/0957926514564735

Ferreira, S., & Hebrrle, V. (2013). Text linguistics and critical discourse analysis: A multimodal analysis of amagazine advertisement. *Ilha do Desterro*, *64*, 263–267.

Idiagbon, A. (2013). Meaning conception and perception in political campaign adverts – A multimodal discourse approach. *English Language and Literature Studies*, *3*(1), 84–93.

Machin, D. (2013). Introduction: What is multimodal critical discourse studies? *Critical Discourse Studies*, *10*(4), 347–355. doi:10.1080/17405904.2013.813770

Oostendorp, M. (2015). Multimodal construction of theidentity of politicians. *Critical Discourse Studies*, *12*(1), 39–56. doi:10.1080/17405904.2014.962066

Reyes, A. (2015). Building intimacy through linguistic choices, text structure and voices in political discourse. *Language & Communication*, *43*, 58–71. doi:10.1016/j.langcom.2015.05.002

Sani, I., Abdullah, M., Ali, A., & Abdullah, F. (2012). Linguistic analysis on the construction of satire in Nigerian political cartoons: The example of newspaper cartoons. *Journal of Media and Communication Studies*, *4*(3), 52–59. doi:10.5897/JMCS11.129

Shaw, D., Hamm, B., & Knott, D. (2010). Technological change, agenda challenge and social melding: Mass media studies and the four ages of place, class, mass and space. *Journalism Studies*, *1*(1), 57–79. doi:10.1080/146167000361177

Wang, J. (2014). Criticising images: Critical discourse analysis of visual semiosis in picture news. *Critical Arts (South-North Cultural Media Studies)*, *28*(2), 264-286. doi: 10.1080/02560046.2014.906344

KEY TERMS AND DEFINITIONS

Critical Discourse Analysis: A method which takes into consideration the processes and products of discourse and their impact on social practices.

Intertextuality: A study of analysing how 'voices of prior texts' are incorporated into currents texts where meaning is derived not from the current text and its discourse usage but from previous exposure to the texts of similar themes.

Memes: Imitations or mimics of original texts which ultimately dramatically alters the meaning and the word.

Multimodal Analysis of Communication: A process of understanding how a combination of signs and symbols can form patterns of meaning.

Netizen: A technical term used to describe the users (citizens) of the social media network.

Semiotic Images: A technical word used in describing how images are used to create meaning.

The Re-Contextualization of Meaning: An evaluation of how texts derived from other media sources are used within a discourse.

Chapter 7
Nature and Outcome of Nigeria's #NoToSocialMediaBill Twitter Protest against the Frivolous Petitions Bill 2015

Oyewole Adekunle Oladapo
University of Ibadan, Nigeria

Babatunde Raphael Ojebuyi
University of Ibadan, Nigeria

ABSTRACT

Focusing on the #NoToSocialMediaBill Twitter protest against the introduction of The Frivolous Petitions (Prohibition, etc.) Bill 2015, in Nigeria, this chapter examined the composition of the loudest voices that framed the protest, explored the level of engagement of those voices with the content of the bill, and analysed the nature and outcome of the protesters' reactions to the bill. To achieve these objectives, the study adopted a mixed method sequential design consisting of quantitative and qualitative content analyses with framing as the theoretical framework. Findings show that hashtag and retweet features of Twitter together functioned effectively as collective framing tools in the #NoToSocialMediaBill Protest. However, the protesters adopted fear-induced collective frames that ignored the contents of the bill. Though the frames drew global media and human rights activists' attention in support of the protesters' goal of resisting an attempt to restrict freedom of speech, they did not achieve the same with the recall of the initiator of the bill.

INTRODUCTION

Central to democratic development in any modern state is an environment that is conducive for a free press, and ultimately, freedom of speech. Since its restoration of democratic governance in 1999, Nigeria has consistently recorded low press freedom ratings compared to other equally democratic African countries (Freedom House, 2015). Apart from frequent cases of clamp down on media houses and journalists, there have been other subtle means of manoeuvre and censorship in media-government relations (Human

DOI: 10.4018/978-1-5225-1859-4.ch007

Rights Watch, 2015). Besides, Nigerian press system has exhibited essentially partisan nature and lack of ideal professionalism as a result of ownership influence (Ojebode, 2013), political affiliation and state control. The aforementioned complications limit the ability of the Nigerian media, and by extension the Nigerian citizens, to defend the nation's democracy despite the fundamental assumption that survival of democracy is unsure without a free press that provides a good platform for public debates. Inherent in this assumption is the fact that democracy is susceptible to abuse from the people to whom state power is entrusted. As the conventional media seems to have failed to provide the space for the citizens to contribute to the development of the Nigerian democracy, social media has created an alternative platform through which Nigerians, especially the youths, can participate in the democratic debates. Apart from a number of initiatives and agitations to ensure a free Nigerian press, ordinary Nigerians have also risen in defence of democracy using the free platforms provided by the Web 2.0 technologies (Vesnic-Alujevic, 2012), and by extension, the social media (Campbell & Kwak, 2012; Chatora, 2012; Bosch, 2013; Price, 2013; Liu, 2014) to post socio-political commentary and participate in politics (Vesnic-Alujevic, 2012).

The dynamism, pervasiveness and spontaneity of public reactions to the introduction of the Prohibition of Frivolous Petitions and Other Related Matters Bill—officially titled *The Frivolous Petition (Prohibition, etc.) Bill 2015*—in the Senate, Nigeria's upper legislative arm, provided a case of how citizens could actively use the social media to defend the democratic principle of freedom of expression. The bill was sponsored by Senator Bala Ibn Na'Allah of APC, Kebbi South (Alade, 2015) and was passed into Committee Stage after Second Reading on December 2, 2015 (*Today,* December 6, 2015). Many Nigerians reacting through Twitter and other social media platforms considered the attempt by the Nigerian Senate to pass the bill into law as threatening to free speech and alternative platform for participating in mainstream governance and politics which social media offer. Citizen campaign against the bill started in the form of Twitter hashtag activism #NoToSocialMediaBill and culminated into #MarchOnNASS with a parallel offline march to the Senate in Abuja, Nigeria's capital city.

The timing of this bill coincided with global movement for inclusion of youths in ensuring accountability in governance. ActionAid (2015) indicates that youths are an excluded population in global development. It notes also that Millennium Development Goals (MDGs) did not make any specific provision for youth inclusion in development, the costly oversight which the post-2015 development agenda, Sustainable Development Goals, seeks to address. As a result, *ActionAid* sees addressing this problem of unjust distribution of societal power and the structures that sustain it as critical to sustaining democratic development. One important way it identifies is through "young men and women playing the role of democracy watchdogs– through lobbying, campaigning and collaborative partnerships". The kind of direct involvement in governance, which ActionAid proposes, remains an aspiration in Nigeria where governance is still the preserve of the old politicians who have continued to recycle themselves in power. In the current democratic dispensation, Nigeria slides backward in youth representation in governance as known old faces were nominated, elected and appointed into major political offices.

Recently, Nigerians have constituted a virile community of civil activists in social media and Twitter has emerged as the hub of civic mobilisation in the country (Chiluwa & Adegoke, 2013). While they are structurally excluded from mainstream political power, Nigerians, mainly youths, utilise Twitter and other social media to contest and negotiate the actions and decisions of government in the light of the implications of those actions and decisions on the citizens and democratic development. Traditionally, the role of defending democracy is entrusted to the press. That is why constitutional protection of the freedom of the press is characteristic of democratic countries. Over the years though, constitutional provision for freedom of the press does not automatically translate into a free press. Democracy thus becomes threatened in a country where there is a sharp discrepancy between constitutional guarantee of the right and the operational reality of the press. This has been the experience of Nigeria in the past few years with social media having been filling the gap created by a tactically gagged mainstream media as the case was in Nigerian Twitter and other social media users' response to the Prohibition of Frivolous Petitions and Other Related Matters Bill.

After the era of hyperbolic reference to the influence of social media use in citizen activism birthing coinages such as "Twitter Revolution" and "Facebook Revolution" especially during the Arab Spring (Olorunnisola & Martin, 2013), most studies on citizen political activism on social media have focused on elections and the activities of the executive arm of government at the expense of other issues equally critical to democratic consolidation (Hawthorne, Houston & McKinney, 2013; Lee *et al.*, ., 2013; Dyagilev, and Yom-Tov, 2014; D'heer & Verdegem, 2014; Kasmania, Sabrana, &Ramle, 2014). Studies that focus on use of social media to mobilise citizens against legislative process are hard to come by. To fill the gap created by this neglect, this study focuses on Twitter protest that developed among Nigerian Twitter users in response to a perceived anti-free speech law proposed by the Nigerian Senate.

The introduction of the Bill was a major event interpreted as contradictory to the mood of change that ushered in the Buhari-led Federal Government of Nigeria on May 29, 2015. ActionAid (2015) acknowledges ensuring the right to information, association, and freedom of speech and opinion for all, including the marginalised youth, as critical to attaining change, the promise of which earned the Nigerian ruling party the All Progressive Congress (APC) massive victory over the People's Democratic Party (PDP) during the 2015 general elections. Nigerian social media users, especially Twitter community, responded to the perceived attempt to gag freedom of speech specifically provided by social media platform. In the process, using the hashtag #NoToSocialMediaBill, these Twitter users portrayed the bill as anti-social media. However, as powerful and compelling as these Twitter responses might appear, any reference to them as informing the final decision of the Senate to withdraw the bill may lack scholarly legitimacy and substance.

While mainstream media have already attributed the withdrawal of the bill to pressure from Nigerians as they did during the Arab Spring (Olorunnisola & Martin, 2013), the report of Senate Committee on Judiciary, Human Rights and Legal Matters presents a different reason: "the passage of the bill would conflict with some provisions of extant acts, penal code, criminal code and the cybercrime act which have sufficient provisions to address the issues that the Frivolous Petitions Prohibition Bill sought to address (Vanguard May 18 2016)." The committee, therefore, recommended the withdrawal of the bill because it infringes on the provisions of the extant laws. Therefore, there are no empirical proofs to back any claim to the potency of this social media activism as the basis for the withdrawal of the bill. In order to empirically situate the outcome attributed in the media to the #NoToSocialMediaBill Twitter protest, this study asks three questions:

1. What is the constitution of the loudest voices that framed the #NoToSocialMediaBill protest?
2. What is the level of engagement of those voices with the content of the bill?
3. What nature of reactions to the bill is contained in those voices?

Empirical analysis of these sub-themes will reveal the extent to which the collective intelligence of the people, as demonstrated in the framing of the bill by Twitter users, had the tendency to influence the legislature's handling of the Bill. Findings of this study would also provide justifications for the assertion that in democracy it matters that the many are smarter than the few (Landemore, 2012). Ultimately, findings of this study have implications for recognition of the role of social media as the platform for citizens' participation in governance and the overall democratic project.

THEORETICAL FRAMEWORK

Framing is a theory popularised in media studies by Robert Entman. As Entman describes it, "[t]o frame is to select some aspects of a perceived reality and make them more salient in a communicating text, in such a way as to promote a particular problem definition, causal interpretation, moral evaluation, and/or treatment recommendation for the item described" (Entman as cited in Scheufele, 1999:107). deVreese (2005:53) notes that "[b]y virtue of emphasizing some elements of a topic above others, a frame provides a way to understand an event or issue". The framing process thus confers on the framer the power to define how an issue is to be interpreted by others. This position is captured in Reese's (2007) description of framing cited in Hossain (2015:523) as "an organizing principle that connects media content to broader issues related to political and social power". Despite the many controversies that trail the theory, scholars agree that framing is about giving salience to perceived reality.

As a theory, Scheufele situates framing in the context of media effects. Most framing studies have, however, concentrated on the top-down, transmission model of communication of the mass media age which places the audience at the receiving end of media discourse. As a result, attention has been called the need to apply framing to the growing user-generated contents on interactive digital media. "This bottom up, or emergent, conceptualization is important to consider because it changes the focus of framing studies from traditional media and hierarchical gatekeepers to ordinary, everyday citizens that easily and regularly produce (online) media, and who ultimately become ''active contributors, creators, commentators, sorters, and archivers of digital news content'' (Nisbet, 2010 as cited by Groshek & Al-Rawi, 2013: 564). Scholars have responded to this call by adopting framing in studying collective actions as well as social media activism.

Benford & Snow (2000) conduct a review of literature on collective action frames and social movements and conclude that framing is critical to the attainment of social movement objectives. Scholars studying collective actions on social media generally have also adopted framing. For instance, Groshek & Al-Rawi (2013) studied critical framing of the 2012 U.S. presidential campaign in Facebook and Twitter. Specifically in the area of social media activism, scholars have also found framing a heuristic approach. For instance, Zhang & Kramarae (2014) use framing to study SlutWalk social media activism in China. Chen & Liao (2014) also employ framing in studying the "Sunflower Movement" in Taiwan which protested against the Legislative Yuan's misplaced priority on services trade agreement at the expense of an oversight mechanism for the Cross-Strait agreement. This study also adopts framing as theoretical framework in line with growing attention being given to the theory in the study of social media activism.

LITERATURE REVIEW

Social Media Activism

In recent times, social media have become prominent in discussions on different aspects of civic and political interactions in democratic countries across the world. Influence of social media looms so large that some scholars have concluded that social media have changed the world as they have altered the make-up of the human society (Lewis, Gray & Meierhenrich, 2014; Zhang & Leung, 2015). A group of scholars have also argued that it is rather the world that has changed social media, replicating in various social media platforms certain inequalities that are entrenched in traditional human societies (Miller, Costa, Haynes, McDonald, Nicolescu, Sinanan, Spyer, Venkatraman, & Wang, 2016). On both sides of the argument, however, there are evidences in support of reciprocal influence between social media and the human society. That is why social media have become popular platforms for furthering different kinds of agenda prominent among which is citizen activism.

Social media activism in general becomes prominent as a result of the hegemonic nature of the mainstream media. Media critics have focused attention on anti-democratic manifestations in mainstream media that make them serve the interests of the elite at the expense of those of the ordinary citizens: their inequalities of access, representation and political/ideological power as well as their economic and structural integration with globalizing capitalism and consumer culture (Carroll and Hackett, 2006:83). Despite the fact that some scholars have critiqued social media for exhibiting many democratic deficits (Sandoval & Fuchs, 2010; Allmer, 2014; Gehl, 2015), other scholars have found that social media are driven by a set of logics that are different from those that drive the commercialised mainstream media and that the commercial status of social media companies does not neutralise the democratic value of the platforms they provide (Miller, Costa, Venkatraman, & Wang, 2016). The latter position justifies the now common practice of initiating on social media by ordinary citizens collective mobilisation efforts against diverse forms of formalised oppression. Social media activism has been on the rise lately because the mainstream media have remained inaccessible to the excluded class who also need their voice to be heard in matters of public importance. Arab and Asia provide good examples of social media activism where youths have used the social media platforms to oppose the established hegemony (Al-Kandari & Hasan 2012; Chatora 2012; Bosch 2013). It is noted, however, that the success of social media activism still depends on its connection with traditional media and civil society (Iskander, 2011).

Among the myriad social media platforms that exist, Twitter is being used extensively for state-centric activism because a growing number of politicians and state actors across the world are adopting the platform (Lee, Ryu, Mon, & Park, 2013; Barbera´ & Zeitzoff, 2014). Twitter thus enables ordinary citizens to become a part of the political discussion equation, allowing for a form of direct networking with political elite without the traditional go-between such as the mass media, multiplying and diversifying the voices that are heard in the public space (D'heer & Verdegem, 2014). These attributes make Twitter a suitable site for activism in defence of social media freedom in general. Twitter activism, therefore, is a form of citizen-championed digital activism which draws upon the affordances of the microblog to negotiate issues of domination, oppression, marginalisation, and societal power imbalance. Among different forms of activism that take place on Twitter, this study focuses on only one, Twitter protest. It is a protest because it mobilised a strong objection to an action or decision of constituted authority, and a hashtag campaign because it trended on the hashtag feature of Twitter.

Nigeria also is not behind in the use of Twitter for activism. For instance, the *#OccupyNigeria Movement* of 2012 was started on Twitter. The #BringBackOurGirls hashtag used to demand the rescue of over 200 Nigerian schoolgirls abducted by the Boko Haram terror group also made the Twitter moment of the year 2015. In #NoToSocialMediaBill Twitter hashtag campaign started by @Gbemisola on 2 December 2015, which is the focus of the present study, the same Twitter was used extensively to contest a bill proposed in the country's Senate but perceived by Twitter users as anti-free speech. The study, therefore, explores how the framing of the Twitter hashtag protest furthers or constraints the attainment of the protest objectives.

The State of Free Speech in Nigeria

Freedom of speech forms the bedrock of media and journalism functionality and citizens' participation in the affairs of a state. Nigeria, however, has been notorious for violations of the fundamental human right of freedom of speech since her first experience of military rule in 1966. Military eras in Nigeria were characterised by proscription of the constitution, clampdown on media houses and seizure of press publications, brutalisation, incarceration and murder of journalists and human rights activists, and promulgation of obnoxious laws against free speech (George, Shadare & Owoyemi. 2012; Elebute, 2015). Scholars have also noted that the expectation that the state of freedom of speech in Nigeria would improve as a result of restoration of democratic rule in 1999 did not materialise as journalists continued to experience state-sponsored intimidation, brutalisation and arrest (Abone & Kur, 2014). The situation places limitation on the ability of the press to perform, without fear of state suppression the watchdog, function of the fourth estate of the realm: "monitor those in authority, expose their limitations and provide checks and balances to the three other estates, namely the legislature, the executive and the judiciary" (Akinwale, 2010: 63).

The continuous restriction of freedom of speech and of the press was a major reason behind advocacy for Freedom of Information Act in Nigeria as the key promoters of the Freedom of Information Bill were journalists and civil society groups (Abone and Kur, 2014). The Freedom of Information Bill was first presented to the National Assembly by two journalists, Tony Anyanwu and NdukaIrabor, and re-presented by another journalist, AbikeDabiri (Ojebode, 2011). The Act in Section 2(1) specifically provides that: "Subject to the provisions of this Act but notwithstanding anything contained in any other act, law or regulation, every citizen of the Federal Republic of Nigeria, has a legally enforceable right to, and shall, on application be given access to any record under the control of a government or public institution." Although journalists were found to be knowledgeable about the provisions of the bill (Abone & Kur, 2014), the knowledge has not translated into any significant difference in the nature of journalistic reporting in Nigeria (Amadi, 2014). In other words, Freedom of Information Act makes "infinitesimal difference in the reportorial prowess of journalists who work in an environment where the dread of job security is acute" (Amadi, 2014: 14). Contextual factors have rendered the provisions of the Act difficult to apply for journalists.

Besides Nigerian government's intolerance of free speech, the 1999 Constitution of the Federal Republic of Nigeria is also found to constitute an obstacle to journalists' enjoyment of freedom of speech. Harry (2015:32, 34) observes that "there are several constitutional landmines on the path of press freedom in Nigeria" as journalists are nowhere in the Constitution "isolated for power the way they are isolated for functions". Tony Iredia, a veteran broadcaster, maintains that the Section 39(2) of the 1999 Constitution provides protection only for media owners and not for journalists (Harry, 2015:34). This complicated situation provokes the conclusion that even though the Nigerian Press is free in principle, it remains bound in chains in everyday practice (Harry, 2015: 34). In this emasculating context is citizen activism is growing on social media platforms in Nigeria.

The Trigger of #NoToSocialMediaBill Twitter Protest: The Frivolous Petitions (Prohibition, etc.) Bill 2015

The #NoToSocialMediaBill Twitter Protest developed in response to the Nigerian Senate consideration of a bill for a law which was perceived to be anti-social media and anti-free speech by Twitter users.

As contained in the explanatory memorandum to the bill, the *Frivolous Petitions (Prohibition, etc.) Bill* "seeks to prohibit frivolous petitions intended to report the conduct of any person of an investigation, inquiry or inquest without a duly sworn affidavit". Section 1 of the bill provides that any such petition must be accompanied by "a dully sworn affidavit in the High Court of a State or the Federal High Court confirming the content to be true and correct and in accordance with the Oaths Act".

Section 3(1) of the bill proposes anybody that violates the above stated provision "shall be deemed to have committed an offence and upon conviction shall be liable to an imprisonment for six months without an option of fine." Section 3(2) adds that anybody who uses such petition or causes it to be used has "committed an offence and upon conviction shall be liable to an imprisonment for a term of 2 years or a fine of N200, 000.00 or both." The bill specifically names paper, radio or any medium of whatever description in Section 3(3) and adds in Section 3(4) media such as "text message, tweets, WhatsApp or... any social media post" as affected media platforms. It identifies as defining frivolity in the context of the bill abusiveness, falsity and malice. Also, the bill in its Section 2 renders frivolous any petition that is submitted without an affidavit. The bill offers protection against frivolity targeted at "any person and/or group of persons, an institution of Government or such other bodies established by law". The bill in its Section 2 renders frivolous any petition that is submitted without an affidavit (see Attoh, 2016; *Policy and Legal Advocacy Centre*, 2016).

The foregoing review of the bill reveals that summarisation of the threat poses by the bill as anti-social media is at best a distraction from the far-reaching implications it has for free speech. Irrespective of medium, the bill criminalises any allegation or petition that is published without a sworn affidavit. This has grievous implications for journalism, especially investigative journalism. The frivolity clause which defines the bill is also missing in christening it anti-social media bill. What implications could this form of atomization that characterised the #NoToSocialMediaBill activism have for public engagement with the bill through the Twitter platform? Is it possible that an activism that was built on an atomized position produce sufficient intellectual counterarguments to expose the threats the bill poses to free speech irrespective of medium?

METHODOLOGY

Methods and Data

A mixed method sequential design consisting of quantitative and qualitative content analyses was adopted for the study with specific focus on Twitter as a platform for citizen activism. The quantitative content analysis was conducted first to establish a quantification system for the tweets. To get the primary data, the study relied on the hashtag #NoToSocialMediaBill and used a tweet as the unit of analysis, drawing purposively the most retweeted tweets for analysis. The tweets that attracted 20 or more retweets were isolated and purposively selected for final analysis. Using these criteria, the researchers analysed a total of 280 tweets. Retweets have been found to be tool of amplification as each retweet potentially exposes to a tweet all the followers of the person that retweeted it. Hashtag that trends on Twitter also occupies a unique space where users can follow the story as it unfolds. Retweeted hashtag thus combines the potentialities of retweet and hashtag, thus resulting in super visibility of tweets. The retweet feature of Twitter is thus conceived in this paper as a framing tool. The retweet function captures the essence of framing as it makes the view expressed in a tweet more salient, leveraging on the networks of each user doing the retweet.

The qualitative content analysis aspect of the study was based on the findings of the quantitative content analysis. Its focus was to explore the significations embedded in the tweets beyond quantification. Specifically, it was used to examine the quality of opinion leadership the most retweeted tweets provided for the #NoToSocialMediaBill activism. It thus provided the basis for determining moving the analysis from the level of number to meaning to establish the quality of arguments employed to counter the perceived hegemonic move of the political elite against free speech in the Nigerian social media space.

Three reasons informed researchers' choice of Twitter:

1. It is popularly being used in Nigeria, as in different parts of the world, for state-centric activism;
2. Campaign against the Frivolous Petition Bill started on Twitter before it filtered into other media; and
3. A growing number of politicians and state actors across the world are adopting the platform, giving citizens direct access to them (Lee, Ryu, Mon, & Park, 2013; Barbera´ & Zeitzoff, 2014).

Data Analysis

Quantitative content analysis was used to disaggregate the categories of voices that framed the #NoToSocialMediaBill Twitter protest. To conduct the quantitative analysis, the researchers developed a coding sheet with a five-item categorisation matrix—*Media; Non-media Groups; Political Parties; Notable Politicians; and Individuals* (non-politicians). These categories provided the bases for identifying the composition of the loudest voices in the Twitter protest. Besides, deductive qualitative content analysis was used to explore the contents of the tweets for dominant themes in the campaign. Catanzaro (1981) and Marshall and Rossman (1995) as cited by Elo and Kyngas (2007) explain that deductive content analysis is employed when the primary aim of the study is to retest existing data in a new context, or when the objective is to test concepts, models, hypotheses, categories, or themes. Through inductive theorizing (Clarke, 2010), the researchers used the study objectives as the categorisation matrix to guide the sorting and coding of data. The researchers read the corpus of the tweets to understand how the Twitter activists engaged with the content of the Frivolous Petitions Bill, their reactions to the bill, and the reasons they advanced against the passage of the bill. After deep immersion in the data through reading and re-reading of the corpus of the tweets, four main themes and other seven sub-themes emerged. These themes guided our presentation and discussion of findings. To ensure credulity and transferability (Bryman & Teevan, 2005 cited by Clarke, 2010), the researchers included several examples, illustrations and excerpts from the tweets to substantiate and exemplify each of the themes that emerged as presented in the following sections.

PRESENTATION OF FINDINGS

As noted earlier, retweet serves as an amplification of the original tweet. In this section, guided by the core objectives of this study, we present the composition of the voices that dominated the #NoToSocialMediaBill Twitter Protest, the level of engagement of those tweets with the contents of the bill, the justifications provided in the tweets for branding the bill anti-social media, and the outcome of the Twitter protest as regards the call for non-passage of the bill and recall of its sponsor.

Composition of the Loudest Voices that Framed the #NoToSocialMediaBill Protest: A People-Centred Leadership Propelled by Individual Twitter Users

The loudest voices belong to individual twitter users who share the common concerns about the implications of the bill for freedom of expression. Media and political figures' participation in the Twitter protest only provided the necessary popularity crutches that such activism ordinarily attracts. Of all the tweets that attracted 20 and more retweets, tweets of individual Twitter users constituted 78.6%. Tweets of different media organisations, local and international, online and traditional, combined with those of advocacy groups, politicians and political parties amounted to only 21.4% of the total tweets analysed. Table 1 presents the distribution of the tweets according to their sources.

The leading role played in the #NoToSocialMediaBill by individual Twitter users indicates that the individuals perceived the bill as a direct threat to their public voice. As reflected in their engagement with the content of the Bill, individual Twitter users have come to regard social media as an alternative voice which empowers them to take active part in governance and other aspects of public life. They took to Twitter to protest this perceived threat to their social media voice and resounded their message of rejection of the bill through retweeting views they considered as representative of their concerns as participants in the democratic project. The two retweets from political parties were from the opposition Peoples Democratic Party (PDP). Despite that most of the reactions to the proposed bill were directed at the ruling party–All Peoples Congress (APC)—PDP could not exploit the situation to its advantage. The Twitter agitators refused to identify with the partisan alignments of the political parties as they collectively agreed that what was at stake was too sacrosanct for partisanship. A Twitter user expressed this position thus: "Can the senators see that there is no division on #NoToSocialMediaBill? We have a consensus. So, they should rethink it" (@Omojuwa 2 Dec. 2015). Not even the sentiment expressed in most of the tweets that social media were used to defeat Goodluck Jonathan's government by APC politicians could sway the campaign in favour of PDP.

Some politicians, however, identified with the position of ordinary citizens on the bill and condemned the bill as dictatorial and anti-people. Notable was the position occupied in the #NoToSocialMediaBill activism by three politicians who posted the four tweets recorded. First to emerge was an account named after former President Olusegun Obasanjo (@SegunObasanjo) with two tweets. The first of the tweets was an acerbic attack on the national relevance of the Senate: "If social media bill is the next agenda of the Nigerian Senate, the 8th assembly should be added to Nigeria's problem (486 retweets)." The second tweet is a challenge to the ruling APC government: "This current govt cannot pretend not to have benefitted from the freedom of social media during the last election (396 retweets)." The third tweet is from

Table 1. Distribution of tweets according to nature of source

Tweet Source	No. of Tweets	Percentages
Media	42	15%
Non-media groups	12	4.30%
Political parties	2	0.70%
Notable Politicians	4	1.4%
Individuals (non-politicians)	220	78.6%
Total	280	100%

an account named former Military President Ibrahim B. Babangida (@General_IBB) who reinforced Obasanjo's position that "Social Media played a vital role in the last general election", adding in the same tweet that: "I believe this is not a Military Administration (117 retweets)." The third voice is that of a PDP serving senator, GodswillAkpabio (@SenAkpabio). His own position is succinctly expressed thus: "I believe that the generality of social media users should not suffer for the reckless acts of a few (74 retweets)." That the media and other groups also attracted 19.3% of the most retweeted views in the protest mirrors their alignment with the pro free speech concerns of the protesters. Much more, solidarity of media and civil society groups has been identified as critical to the success of citizen activism on social media (Iskander, 2011). Overall, individual users provided the leadership that is needed to make the #NoToSocialMediaBill people-led. Individual Twitter users were clear about the goal of the campaign: it was a movement against a proposed anti-people law. The occupation of the leadership position in the #NoToSocialMediaBill by individual twitter users indicates that those individuals see the campaign as a fight for their own socio-political relevance in Nigeria.

Tweets' Engagement with the Bill: Emotional Response to Perception of Threat at the Expense of Draconian Content

This section examines the extent to which #NoToSocialMediaBill protest reflects engagement with the contents of the bill in question. In other words, it examined how informed the sentiments expressed in the bills are. The #NoToSocialMediaBill started trending on the 2 Dec. 2015 when the bill passed second reading in the Senate. However, the hashtag started trending when the bill had not been made public. Concerned by the growth of the activism without the bill being made public, a prominent figure in Nigerian Twitter Community (@ogundamisi) warned: "May I also say that the MOB mentality is dangerous even on social media", asking "Why can't anyone publish the draconian bill? (22 retweets)" When another leading voice in the Nigerian Twitter Community (@Omojuwa) also tweeted: that day: "Our guys at NASS have said they will make the entirety of the bill available tomorrow. We wait (80 retweets)," He equally confirmed that the protest against the bill was intensifying without the bill being public yet. When the bill finally emerged, only two tweets indicated that the bill does not pose any direct threat to freedom of expression on social media other than it does in other media. After having read the bill, @ogundamisi clarifies his position on the activism thus: "I am more than happy to [?] #NoToSocialMediaBill but I have read the attached & still can't find any mention of SM (46 retweets)." Also, @Omojuwa reviews his position when the bill was made public: "If this is the bill in question, I see no attempt to gag Social Media here except there is another bill (47 retweets)." The activism continued despite the realisation that the bill is not anti-social media that it was called.

While the entire loud frames in #NoToSocialMediaBill protest are silent about the frivolity aspect of the bill, which is the primary focus of the bill, they express concerns over the severity of the punishments the bill proposes for violating its provisions. Only one tweet was found among the super tweets to directly question the plausibility of having to swear on an affidavit before writing petitions. @remirex notes that: "To publish a petition you will need to [swear on] an affidavit... Who issues affidavit, govt. I just want to make common sense" (60 retweets). The Twitter user finds an inescapable trap in having to seek from the government approval of petition written against the same government. The fear is that government can frustrate the process so much that meaningful activism becomes impossible. Despite low engagement with the intricacies of the bill, there is a high degree of awareness that mere publishing what the bill defines as frivolous attracts jail term and or fine. For instance, @ebuka_akara posted:

"Two year jail term for tweeting 140 characters. That's nearly a week per character (385 retweets)." @ Oke4chukwu also retorted: "Our senators are funny. They propose 2 years jail 4 tweeting 140 characters. That's nearly a week per character." Chai (143 retweets).

The foregoing analysis suggests that the entire #NoToSocialMediaBill protest was built on perception of threat that ignored the contents that produced the threat. As a result, the leadership that drove the Twitter protest ignored intellectual response to the content of the bill. The atomization of the bill as anti-social media thus results in missed opportunity. For instance, realisation that the bill was not just anti-social media but anti-free speech presents the agitators with an opportunity to reframe the entire protest as anti-free speech, irrespective of platform. The opportunity was, however, missed and the dominant frame remained to the end of the protest. The bill even became surnamed in the mainstream media as anti-social media bill. The opportunity to extend the frontiers of the protest to address the poor state of freedom of speech in the country was sacrificed for the emotional appeal. As a result of this missed opportunity, the best that the protest could achieve is to maintain the freedom of speech status quo.

Twitter Community's Reactions to the Frivolous Petitions Prohibition Bill: Many Reasons to Kill the Bill

Here, we examined how the Twitter users justified their position on the bill as anti social media. Seven key sub-themes emerged from exhaustive textual analysis of the corpus of the selected tweets. The themes summarised the arguments that the Twitter community presented against the bill as reasons it should not be passed into law. The sub-themes are presented as follows:

Gagging the Voice You Used to Secure Legitimacy

The most prominent of the arguments is that since APC used social media extensively to oust Goodluck Jonathan and PDP from the presidency and emerged the country's ruling party, it does not have the moral right to sensor it. This position is captured in the tweet of @claraobiageria: "The APC led government are now trying to kill the cobra they nurtured. Am sorry it has grown" (109 retweets). One @ebuka_akara also articulates this position thus: "The same social media that played a huge role in their victory at the polls? They want to bite the hand that fed them" (26 retweets). These tweets as other tweets also do acknowledge the centrality of the role played by members of the then opposition APC in building a virile social media sphere for public criticism. The reasoning is that if APC members now serving at different levels of government as members of the ruling party together were once a part of the social media "culture of frivolity" as the bill described it, they have lost the moral right to legislate against it.

Fighting the Masses, Ignoring Socio-Economic Woes

Next in the order of magnitude is the position that since the APC led federal government cannot solve the myriad socio-economic problems plaguing the country, it does not have the moral right to initiate a bill that censors social media and people's right to democratic free speech. @1_am_pixelhub summarises the concerns of the country thus: "Fuel scarcity, missing soldiers, Boko Haram, missing girls, rape, now we have #NoToSocialMediaBill. This government tho" (109 retweets). The Twitter Community sees as more imperative ameliorating the existential sufferings of Nigerians than a bill on frivolous petitions. They see the bill as a trivialisation of the growing difficulties that ordinary citizens experience in the country.

A Tactical Violation of Extant Laws

Another argument mustered against the bill is violation of existing laws. A group of tweets affirmed that the existing defamation law sufficiently addresses the object of the Frivolous Petitions Prohibition Bill. The bill is thus seen as a mere duplication of the existing law, and as a result, needless. One @ responficient11 presents this position thus: "It is largely unnecessary, the govt can sue for defamation if it feels anyone has told lies against it" (30 retweets). Another Twitter user, @dr_kaduru, addresses a more fundamental issue alongside: "If someone has made a libellous comment, sue the person. If you think the courts don't work, strengthen the judiciary", indicating that the process of implementing the existing laws needs more attention than amassing new laws.

A Gross Infraction of Access to Information and Freedom of Speech

A number of tweets also maintain that the provisions of the bill violate the tenets of the Freedom of Information Act, and by extension, the freedom of speech. An example of such tweets is @a_onipede: "Hw did we just transit from a govt that passed FOI to 1 that wants to gag d social media? This scam is real bro! @Osekita" (25 retweets). None of the tweets, however, highlights the intricacies of such violations. In place of the bills, Twitter users call for better bills that address the concerns of Nigerians: "Yeyedey smell! Years counting, they could not pass the PIB but want to trample on our freedom of expression" (@MickeySunny, 28 retweets); "What abt bills to improve our sorry Naira Value, education system, transport sector, trade and investment?" (@mryungkelz, 25 retweets). In place of the Frivolous Petitions Prohibition Bill, Twitter users express the need for bills that address societal problems.

Providing Shields for the Political Elite

Many Twitter users perceive the intent of the bill as an attempt to shield politicians and government officials from public oversight and criticism. A Twitter user, @ImohUmoren, succinctly expresses how social media have provided a virile platform for oversight on Nigerian politicians thus: "For the first time in our history we are aware of the sheer level of corruption and incompetence of these politicians" (32 retweets). One @Chukwukajerri imagines what it would be like if the bill become successfully passed into law: "If freedom of speech is take away, then dumb and silent we may be led, like sheep to the slaughter. Let us all say #NoToSocialMediaBill" (53 retweets). @owuyi also notes that: "The social media (Twitter especially) have been able to increase political awareness and orientation. They fear that. (111 retweets). Popular in the #NoToSocialMediaBill activism is the argument that since the bill has failed the test of pure intention, it does not merit being passed into law.

An Assault on Democracy

It also resonates through the #NoToSocialMediaBill activism that the bill furthers the dictatorial end of the ruling party. The bill is thus considered an assault on democracy. A tweet by @antonio_zimo states: "I'm just wondering if we are about to become North Korea" (79 retweets). Another tweet by @d_problemsolver rings out this point to the government: "Someone shdpls remind them that DEMOCRACY is about Freedom of SPEECH, Freedom of TWEETS. No be Military REGIME ooo" (132 retweets). Yet another tweet by @JackieUko concludes that: "A government that wants to take away our FREEDOM OF SPEECH is not democratic. That is all shades of DICTATORSHIP" (30 retweets). A comparison of the present administration with the immediate past also rings through the tweets as in the tweet of @ owuyi: "GEJ, his wife and everyone in his administration were criticised and abused. But freedom of expression was never threatened" (329 retweets). Popular position in the tweets is that armed with social media, ordinary citizens can check the excesses and dictatorial tendencies of the government.

A Bill Sponsored and Hawked by Inept, Dictatorial Politicians

A large amount of the tweets use the #NoToSocialMediaBill activism as an occasion for public shaming of politicians and government officials directly or remotely related to the bill or the general deteriorating condition of the country's situation. The harshest attack is directed at Senator Na'Allah who was the brain behind the bill. Responding to a picture of poorly dressed Senator Na'Allah, @aligthebaptist tweets: "Is this what we paid N21.5M wardrobe allowance for? Fellow Nigerians please see your senator" (153 retweets). Senator Dino Melaye is caricatured as a wife beater. President Muhammadu Buhari's dictatorship past is also exhumed. Senator Ben Bruce's double standard is also upbraided: "Ben Bruce knew of this bill but never mentioned it in any of his tweets but yet claims to be an advocate of freedom" (@ Dili_1810, 53 retweets). The senators and APC government with its change agenda were all dismissed as a fraud. In all, there is a clear boundary work done in #NoToSocialMediaBill activism as irrespective of political interest or party affiliations, the Twitter community stand distance away from politicians and their tendencies. The Twitter activists sent a clear, consensus message that that condemns the bill.

The Outcome of the Twitter Protest: A Mix of Failure and Success

In addition to demanding the withdrawal of the Frivolous Petitions Bill, the #NoToSocialMediaBill protest also initiates a move to recall the sponsor of the bill, Senator Na'Allah. As @MrAyeDee expresses the justification for the move: "Best way to say #NoToSocialMediaBill is by helping recall its architect Se. Na'Allah" (171 retweets). Adding a tone of urgency to the recall move, @AndyMadaki also tweets: "You don't have to wait till 2019.... You can recall your senators or reps. They should be scared of us, not us scared of them" (43 retweets). The recall move is specifically directed at natives of Kebbi State who have the constitutional right to recall the senator: "If you are from Kebbi, do participate in this recall move. Please, keep retweeting 4 people of Kebbi" (@AnthonyEhilebo, 31 retweets). Despite the mobilisation against the senator, the recall did not happen. The bill also progressed to public reading stage despite the Twitter protest against it only to die a natural death five months later at the committee stage where its eventual withdrawal was recommended to the Senate on May 17, 2016.

In the case of Arab Spring and similar activisms that have utilised social media extensively, attributed results were almost always immediate. However, in the case of the #NoToSocialMediaBill protest, there was a time lag of over five months between the protest and the attributed result. This makes it difficult to directly attribute the withdrawal of the bill to the Twitter protest. However, that the Senate committee that recommended the withdrawal of the bill indicates that the bill is "capable of unduly abridging the fundamental rights of the citizen to privacy and free speech" aligns with the concerns expressed in the protest (Ojiabor, 2016).

DISCUSSION

The combination of hashtag and the retweet feature of Twitter is found to be an effective affordance for collective framing on which activists can rely for projecting and amplifying their core concerns to the public. The protesters utilised these features of Twitter to resound their positions. Consistent with popular position in literature, the #NoToSocialMediaBill protest provides sufficient proof that users have confidence in social media as tools for socio-political activism and consider access to them as a form of democratic empowerment against the excesses of the political elite (Chiluwaand Adegoke, 2013; Lewis, Gray & Meierhenrich, 2014; Zhang and Leung, 2015). The Twitter protest reflects the tension between two prominent positions on social media-society relationship: an attempt to use social media to change socio-political situations on the part of user and a counter-attempt by the political elite to change the social media landscape through legislation, suggesting that the two positions are not mutually exclusive (Lewis, Gray & Meierhenrich, 2014; Zhang & Leung, 2015; Miller, Costa, Haynes, McDonald, Nicolescu, Sinanan, Spyer, Venkatraman, & Wang, 2016).

That the ordinary citizen constitutes the nucleus of the protest and provides a virile leadership for it indicate that unregulated social media landscape to a great extent is free from the anti-democratic trappings identified by Carroll and Hackett (2006) which make it difficult for mainstream media to defend the cause of ordinary citizens. Social media thus provide a voice for articulating a position that diverges from the interest of the elite whose influence looms over the mainstream media. This finding confirms that ordinary Nigerians are actively utilising social media as an alternative route to public participation in governance in the absence of traditional participation opportunities. ActionAid (2015) identifies this form of participation most especially by the youth as an important means of attaining meaningful change. The findings support also the position of Leung and Lee (2014) that the Internet environment is influential in furthering counterhegemonic goals especially during heightened political activities when political interests are activated irrespective of political orientation. Twitter users' lines of political divide became blurred in the #NoToSocialMediaBill activism, making it possible for the participants to pursue a single agenda against the position of the elite political class.

However, the pattern of Twitter use in the #NoToSocialMediaBill protest dismisses any basis for technological determinism. While citizens' voices were loud enough to attract the attention of local and international media and human rights organisations to perceived threat to free speech in the Nigeria social media landscape, those voices were not potent enough to mobilise support for the initiated recall of Senator Na'Allah who sponsored the bill in question. In addition, the recall aspect of the protest did not attract much attention like the central objective of the protest did. This questions the potency of mobilising citizens on social media against certain political structures that are protected by stringent regulations. That a legislator could only be recalled by the direct votes of the people he or she represents in the legislature reduces the chances of popular, nationwide social media protest in achieving that.

In addition, while the main problem the initiator of the failed bill intended to solve was that of frivolity practiced in public spaces like social media, the agitators did not see frivolity as a problem. In fact, frivolity, to the agitators, is synonymous with freedom of speech. To take away their right to be frivolous is to deny them the essence of democratic free speech. As a result of their confusion of right to frivolity in speech with right to freedom of speech, it becomes unclear what exactly the agitators defended in the protest. This again demonstrates the limit of public knowledge of constitutional provisions for freedom of speech in a country like Nigeria where there exist in extant laws many trappings for the exercise of freedom of speech. In a context where social media are being used to bridge the participation gap created by the exclusion of youths from mainstream governance, citizens will have to come to terms with the demands of public speech to be accorded due respect and recognition.

Another critical weakness of the #NoToSocialMediaBill protest is its poor engagement with the content of the bill. While the activism focuses on the immediate threat to freedom of speech on social media, it ignores the complicated realities that freedom of speech persistently faces in Nigeria. This shows that framing is both an enabler and a constrainer of cause. While framing the concerns as anti-social media drew wide attention, it constrained stakeholders from seeing the implications of the bill beyond social media. Scholars have noted that while there is constitutional guarantee of free speech in Nigeria, the press remains in shackles (Harry, 2015). Had the social media protesters contextualised their agitations within the existing realities of freedom of speech in Nigeria, they would have staged a better and more effective campaign in defence of democratic free speech in Nigeria. Although the protest, in the long run, achieved its goal which is the withdrawal of the bill perceived as anti-social media, it only maintained the status quo of freedom of speech in Nigeria.

FUTURE RESEARCH DIRECTIONS

That the agitators were more concerned about freedom of speech than they were about the problem of frivolity raised in the failed bill is a concern. The boundary problem persists whether freedom of speech includes freedom to be frivolous when speaking in a public space like social media. It constitutes a serious challenge to policy making as any attempt to control the practice of public frivolity in speech might end up being an attack on free speech. Options for resolving this problem is one area that future studies may explore. There is also the need to study the role of frivolity in the potency of social media activism against the elite. Also, since this study concentrates only on the loud voices that framed the protest, a lot of intellectual contributions might be missed by ignoring tweets that did not attract the attention of the crowd. Focusing future studies in this direction will contribute a lot to knowledge about the nature of tweets that do not attract popular attention in Twitter protest.

CONCLUSION

In order to empirically situate the outcome attributed in the mainstream media to the #NoToSocialMediaBill Twitter protest, this study examined the constitution of the loudest voices that framed the protest which was organised against the introduction of the Frivolous Petitions Bill, explored the level of engagement of those voices with the content of the bill, and analysed the nature of their reactions to the bill. Findings reveal that while emotion-driven collective frame is potent in achieving the withdrawal of the bill, it is ineffectual in achieving the recall of the initiator of the bill. While an emotionally charged amplified noise can be effective in achieving goals that resonate with the global ideals of democracy, a potent voice driven by knowledge of context is necessary for achieving goals that are rooted in the intricacies of local laws and politics. For social media activism to further democratic development cause in a qualitative way, this and other concerns must be attended to.

REFERENCES

Abone, G. O., & Kur, J. T. (2014). Perceptual influence of freedom of information Act on Journalism Practice in Nigeria. *Arabian Journal of Business and Management Review*, *3*(7), 23–35.

ActionAid. (2015). *Democracy watchdogs: Youth-driven participatory monitoring and accountability in the Post-2015 development agenda*. Author.

Akinwale, A. A. (2010). Repression of press freedom in Nigerian democratic dispensations. *Africa Development, 35*(3), 47–70.

Al-Kandari, A., & Hasan, M. (2012). The impact of the internet on political attitudes in Kuwait and Egypt. *Telematics and Informatics, 29*, 245-253. 10.1016/j.tele.2012.02.005

Alade, A. (2015, December 6). Social media censorship bill: We'll take your input, Senate assures Nigerians. *The Vanguard.* Retrieved from http://www.vanguardngr.com/2015/12/social-media-censorship-bill-well-take-your-input-senate-assures-nigerians/

Allmer, T. (2014). (Dis)like Facebook? Dialectical and critical perspectives on social media. *Javnost-The Public*, *21*(2), 39–56. doi:10.1080/13183222.2014.11009144

Amadi, F. A. (2014). Reportorial activism and Nigerian journalism after the enactment of Freedom of Information Act. *New Media and Mass Communication, 21*, 10–15.

Attoh, N. R. (2016). The Frivolous Petitions Bill and permissible restrictions on freedom. *Social Science and Research Network*. Retrieved March 28, 2016 from http://papers.ssrn.com/sol3/papers.cfm?abstract_id=2713169

Barbera, P., & Zeitzoff, T. (2014). *The Empirical Determinants of Social Media Adoption by World Leaders and its Political Consequences*. A Working Paper.

Benford, R. D., & Snow, D. A. (2000). Framing processes and social movements: An overview and assessment. *Annual Review of Sociology*, *26*(1), 611–639. doi:10.1146/annurev.soc.26.1.611

Bosch, T. (2013). Youth, Facebook and politics in South Africa. *Journal of African Media Studies*, *5*(2), 119–130. doi:10.1386/jams.5.2.119_1

Campbell, S. C., & Kwak, N. (2012). Political involvement in mobilized society: The Interactive relationships among mobile communication, network characteristics, and political participation. *Journal of Communication*, *61*(6), 1005–1024. doi:10.1111/j.1460-2466.2011.01601.x

Carroll, W. K., & Hackett, R. A. (2006). Democratic media activism through the lens of social movement theory. *Media Culture & Society*, *28*(1), 83–104. doi:10.1177/0163443706059289

Chatora, A. (2012). Encouraging political participation in Africa: The potential of social media platforms. Pretoria: Institute for Security Studies. Retrieved from http://www.issafrica.org/publications/situation-reports/encouraging-political-participation-in-africa-the-potential-of-social-media-platforms

Chen, B., & Liao, D. (2014). *Social Media, Social Movements and the Challenge of Democratic Governability*. Paper presented at Taiwan Democracy Project, Stanford University.

Chiluwa, I., & Adegoke, A. (2013). Twittering the Boko Haram uprising in Nigeria: Investigating pragmatic acts in the social media. *Africa Today*, *59*(3), 82–102. doi:10.2979/africatoday.59.3.83

Chong, D., & Druckman, J. N. (2007). Framing theory. *Annual Review of Political Science*, *10*(1), 103–126. doi:10.1146/annurev.polisci.10.072805.103054

Clarke, J. N. (2010). The domestication of health care: Health advice to Canadian mothers 19932008 in todays parent. *Family Relations*, *59*(2), 170–179. doi:10.1111/j.1741-3729.2010.00593.x

deVreese, C. H. (2005). News framing: Theory and typology. Information Design Journal +. *Document Design (Amsterdam)*, *13*(1), 51–62.

Dheer, E., & Verdegem, P. (2014). Conversations about the elections on Twitter: Towards a structural understanding of Twitters relation with the political and the media field. *European Journal of Communication*, *29*(6), 720–734. doi:10.1177/0267323114544866

Dyagilev, K., & Yom-Tov, E. (2014). Linguistic factors associated with propagation of political opinions in Twitter. *Social Science Computer Review*, *32*(2), 195–204. doi:10.1177/0894439313506850

Elebute, A. (2015). Issues on the gagging of Nigerian press with obnoxious laws. *International Journal of Arts and Humanities, 4*(1), 207-223. DOI: 10.4314/ijah.v4i1.15

Elo, S., & Kyngas, H. (2008). The qualitative content analysis process. *Journal of Advanced Nursing*, *62*(1), 107–115. doi:10.1111/j.1365-2648.2007.04569.x PMID:18352969

Gehl, R., W. (2015). The Case for Alternative Social Media. *Social Media + Society,* 1 –12. DOI: 10.1177/2056305115604338

George, O., Shadare, S., & Owoyemi, O. (2012). Military interventions in the Nigerian politics: 'A timed bomb' waiting to explode? The avowal of a new management elites. *International Journal of Business. Human Technology*, *2*(5), 191–198.

Groshek, J., & Al-Rawi, A. (2013). Public sentiment and critical framing in social media content during the 2012 U.S. presidential campaign. *Social Science Computer Review*, *31*(5), 563–576. doi:10.1177/0894439313490401

Harry, I. (2015). Towards a global democratization of free speech: A critical appraisal of constitutional landmines on the trajectory of press freedom in Nigeria. *International Journal of International Relations. Media and Mass Communication Studies*, *1*(2), 25–36.

Hawthorne, J., Houston, J. B., & McKinney, M. S. (2013). Live-tweeting a presidential primary debate: Exploring new political conversations. *Social Science Computer Review*, *31*(5), 552–562. doi:10.1177/0894439313490643

Hossain, M. D. (2015). Manufacturing consent: Framing the liberation war of Bangladesh in the US and UK media. *Journalism*, *16*(4), 521–535. doi:10.1177/1464884914524516

Human Rights Watch. (2015). *World Report 2015: Nigeria*. Retrieved from https://www.hrw.org/world-report/2015/country-chapters/nigeria

Idris, S. (2015, December 6). Senate vows to go ahead with anti-social media bill despite criticism. *Today*. Retrieved from https://www.today.ng/news/national/48193/senate-vows-to-go-ahead-with-anti-social-media-bill-despite-criticism

Iskander, E. (2011). Connecting the national and the virtual: Can Facebook activism remain relevant after Egypt's January 25 uprising? *International Journal of Communication*, *5*, 13–15.

Kasmania, M. F., Sabrana, R., & Ramle, N. (2014). Can Twitter be an effective platform for political discourse in Malaysia? A Study of #PRU13. *Procedia: Social and Behavioral Sciences*, *155*, 348–355. doi:10.1016/j.sbspro.2014.10.304

Landemore, H. E. (2012). Why the many are smarter than the few and why it matters. *Journal of Public Deliberation*, *8*(1), 1–12.

Lee, J., Ryu, H., Mon, L., & Park, S. J. (2013). Citizens' use of Twitter in political information sharing in South Korea. *iConference 2013 Proceedings*, (pp. 351-365). doi:10.9776/13210

Lewis, K., Gray, K., & Meierhenrich, J. (2014). The structure of online activism. *Sociological Science, 1*, 1-9. DOI: 10.15195/v1.a1

Miller, D., Costa, E., Haynes, N., McDonald, T., Nicolescu, R., Sinanan, J., & Wang, X. (2016). *How the World Changed Social Media*. London: UCL Press. doi:10.14324/111.9781910634493

Ojebode, A. (2011). Nigeria's Freedom of Information Act: Provisions, strengths, challenges. *African Communication Research*, *4*(2), 267–284.

Ojebode, A. (2013). Ethical dilemma revisited: PBO newspapers and the professional elbowroom of the Nigerian journalist. *Journal of African Media Studies*, *5*(3), 295–312. doi:10.1386/jams.5.3.295_1

Ojiabor, O. (2016, May 17). Finally Senate throws out anti-social media bill. *The Nation*. Retrieved from: http://thenationonlineng.net/finally-senate-throws-anti-social-media-bill/

Olorunnisola, A. A., & Martin, B. L. (2013). Influences of media on social movements: Problematising hyperbolic inferences about impacts. *Telematics and Informatics*, *30*(3), 275–288. doi:10.1016/j.tele.2012.02.005

Policy and Legal Advocacy Centre. (2016). *What you need to know about the Frivolous Petition Bill 2015*. Retrieved from: http://www.placng.org/new/upload/FrivolousPetitions.pdf

Price, E. (2013). Social media and democracy. *Australian Journal of Political Science, 48*(4), 519-527. Retrieved from:10.1080/10361146.2013.846296

Sandoval, M., & Fuchs, C. (2010). Towards a critical theory of alternative media. *Telematics and Informatics*, *27*(2), 141–150. doi:10.1016/j.tele.2009.06.011

Udofa, I. J. (2011). Right to freedom of expression and the law of defamation in Nigeria. *International Journal of Advanced Legal Studies and Governance*, *2*(1), 75–84.

Umoru, H., & Erunke, J. (2016, May 18). Senate throws out anti-social media bill. *Vanguard Newspaper*. Retrieved from http://www.vanguardngr.com/2016/05/senate-throws-anti-social-media-bill-2/http://www.vanguardngr.com/2016/05/senate-throws-anti-social-media-bill-2/

Vesnic-Alujevic, L. (2012). Political participation and web 2.0 in Europe: A case study of Facebook. *Public Relations Review*, *389*(3), 466–470. doi:10.1016/j.pubrev.2012.01.010

Yassin-Kassab, R., El Fattah, A. A., Soueif, A., Barghouti, M., Lalami, L., Shehadeh, R., . . . Haddad, J. (2016, January 23). 'I was terribly wrong' - writers look back at the Arab spring five years on'. *The Guardian*. Retrieved from https://www.theguardian.com/books/2016/jan/23/arab-spring-five-years-on-writers-look-back

Zhang, W., & Kramarae, C. (2014). SlutWalk on connected screens: Multiple framings of a social media discussion. *Journal of Pragmatics*, *73*, 66–81. doi:10.1016/j.pragma.2014.07.008

Zhang, Y., & Leung, L. (2015). A review of social networking service (SNS) research in communication journals from 2006 to 2011. *New Media & Society*, *17*(7), 1007–1024. doi:10.1177/1461444813520477

KEY TERMS AND DEFINITIONS

Collective Framing: The act of many people interpreting an issue or event from a similar perspective and presenting it to be seen from the same.

Hashtag Campaign: A topical discussion on Twitter with the configuration of a keyword with the hash sign (#) started to achieve a set goal.

Loud Frames: Predominant perspectives from which an issue or event is interpreted and projected.

Public Frivolity in Speech: Publishing or causing to be published in a public space information that is capable of damaging the reputation of another without considering the seriousness of doing so.

Social Media Activism: The practice of mobilizing human and material resources on social media platforms in support of collective actions for or against certain issue involving domination, oppression, marginalisation, and societal power imbalance.

Social Media Protest: A form of social media activism in which concerned individuals or groups express a strong objection to an action or decision of constituted authority.

Twitter Activism: A form of citizen-championed social media activism which draws upon the affordances of the microblog to negotiate issues involving domination, oppression, marginalisation, and societal power imbalance.

Section 2
Gender, Children, and the Media

The role of the media in promoting gender and children issues in Africa.

Chapter 8
Portrayal of Women in Nollywood Films and the Role of Women in National Development

Suleimanu Usaini
Covenant University, Nigeria

Ngozi M. Chilaka
Covenant University, Nigeria

Nelson Okorie
Covenant University, Nigeria

ABSTRACT

This study investigates how women are portrayed in Nollywood films, as well as the interpretation of their representations. It aims at understanding how the images of women are reflected in films, with a focus on investigating the influence of such portrayals on their role in national development. The methods adopted were Quantitative Content Analysis (five Nollywood films were content analysed) and Focus Group Discussion (three sessions of FGD were organised). Data collected and analysed show that over two-thirds of major female characters analysed were portrayed as dependent, 80% were depicted in such situations of physical, sexual, and emotional abuse, while only 30% of the major female characters were portrayed as career professionals and intellectuals. It was observed from the analyses that portrayals and representations of women have negative influences on their contributions towards national development. The study concludes, therefore, that positive portrayal of women in Nollywood films should be encouraged. This can only be made possible through changing the narrative style of the film scripts. This is a call for more female script writers and directors to be involved in charting the narratives that will adequately give women a voice, new roles, and the right representation in Nollywood films.

DOI: 10.4018/978-1-5225-1859-4.ch008

INTRODUCTION

The media has become inevitable in this present society. From newspapers to magazines, radio to television, the Internet to films, these various media outlets influence their audience in one way or another. The media inform, entertain, and educate us. Fischer (2010, p.1) says that "the impact of the media on our daily lives is tremendous, and most of the time we are not aware of this huge influential factor". The media can "move our emotions, challenge our intellects, and insult our intelligence. They can also influence people's beliefs, values, attitudes, behaviours, perceptions and opinions on various issues without them even realising it" (Baran, 2010). Furthermore, the mass media, beyond the three core functions of providing information, education and entertainment, play a significant role in shaping the perceptions and worldviews of the audience through the messages they communicate and the interpretation they also give to the messages. It, therefore, comes as no surprise that, due to its especially powerful and pervasive nature, the media contributes to shaping people's perception and opinion on issues having to do with social relationships and gender relations in the society.

Among the mass media of communication, television is seen as the most influential. By combining pictures and sound, TV can communicate messages which are impossible to convey as effectively by radio and or print. Television is the central cultural arm of society, as a culture's primary story teller. It is the chief creator of entertainment and information for heterogeneous mass publics (Baran, 2009).

Jonah (2011, p.1) points out that "Film is one of the channels of mass communication which is transmitted via the television which has grown over the years. It started as cinema and later grew into home video which many Nigerians patronize today". Sambe (2008, p.142) defines film as series of motionless images projected into a screen so fast as to create in the mind of anyone watching the screen an impression of a motion. Nigerian films industry otherwise known as Nollywood films has come of age.Nigerian films daily flood the global channels with series of films which dwell on various aspects of life (Jonah, 2011).

Nigerian films are deeply rooted in the Nigerian cultural traditions and social texts that focus on Nigerian community life (Onuzulike, 2007). Daramola (2008) supports this that the Nigerian film industry holds up a mirror to the society and projects what is seen in the society.

As the twentieth century ended, research evidences showed that Nigerian films were filled with negative and stereotypic images of women. This is done to the disadvantage of women as these images promote continuous domination by men and subordination and subjugation of women (Okunna, 2002). "Given the great influence that Nollywood has over African culture, such effect is reinforced by a massive consumption of Nigerian films by Africans living in Africa and off the shores of Africa" (Onuzulike, 2007, p. 237).

According to Amobi (2010, p.1), the treatment of women in films has occupied the realm of discourse for several decades now. Feminist scholars, critics and women's movements have relentlessly challenged the stereotypical representations of women in Hollywood films (the American film industry), criticising their sexist depictions as whores, jilted mistresses, emotional cripples, sex-starved spinsters and psychotics. The Nigerian film industry, Nollywood, is no different as this entrant to the world of cinema has also been criticised for its portrayal of women as sex objects, weak, cold-hearted, materialistic, vengeful, vicious, diabolical, and scheming.

This study investigates why women's roles are mostly captured in domestic spaces which shows them as a marginalised group whose functions in some of the films cannot go beyond those of being mothers, housewives, and other related roles. In contrast, men are depicted mostly as leaders both in military and civilian spheres of rulership and as successful businessmen.

The portrayal of women in these films demands great scrutiny to identify the image of women which can have adverse effects on women viewers and the extent to which it can with a view to reordering the trend. This study, therefore, investigates the portrayal of women in Nigerian films and how their representations influence perceptions of the roles they play in the society as the backbone of the family.

LITERATURE REVIEW

The Nigerian film industry is fast growing and its global reach is quite phenomenal. It is said that Nollywood has become the third largest film industry in the world after Hollywood and Bollywood (Eke, 2012). Aladese (2013) agrees that Nollywood is one of the foremost filmmaking industries in the world and has earned a reputation for being one of the most prolific filmmaking industries in the world. Despite the success story of Nollywood, a lot of critics are concerned about the themes reflected in the films. Some researchers have argued that Nollywood films portray negative themes that damage the way other countries perceive Nigeria.

Akinfeleye and Amobi (2011, p. 6) opine that Nollywood films are internationally dismissed for their low production quality while scholars posit that they defy all theoretical explanations to phenomenon. Osofisan (2006) adds that most Nollywood films portray cultural and ideological deficiencies, promoting alien customs and inferiority complex among the people. He criticises the financiers and script writers for their interest in filmmaking as business to make quick money, and consequently ignoring the aesthetic or ontological dimensions of film production.

Akpabio (2007, p. 95) conducted a study that revealed that, "A majority of respondents (53.6%) expressed the view that there is glorification of negative themes and storylines in Nollywood films. Respondents who expressed disagreement accounted for 22.5% while close to a fifth of the respondents (18.7%) preferred to remain neutral on this score. Based on this finding, the study concluded that there is indeed glorification of negative themes and storylines in some home video productions". Akpabio (2003) enlists some themes portrayed in Nollywood films such as: evils of polygamy, extra marital affairs, different forms of evil rituals, cultism, occultism and witchcraft, conflict between western and indigenous cultures, religious divergence and complexities of the Nigerian society, just to mention a few.

Ebewo (2007) has a different view as he sees nothing wrong with a film dealing with any of these themes. He opines that critics frown at the fact that they recur, film after film. Aladese (2013, p. 20) argues that while some believe the themes that recur in Nigerian films are negative, others believe that it is these very themes that allow viewers to be aware of the society a part of which they are. She also claims that some researchers are of the opinion that Nigerian films do not portray the country in a negative light by showcasing negative themes but reflect the culture of the society they emanate from. The argument thus is that the films are reflections of the society. They capture our cultural make-ups and differences. The contents of the films are indigenous and they address issues that relate to the day-to-day lives with which the audience can relate.

Therefore, what is the role of women in or their contribution to national development? Women make useful contributions to the development of society. Obi (2001) opines that one of the ways they do this is through their ability to keep neat homes and environments, produce smart and healthy children and also cook neatly. She further expatiated on this that women can contribute to the development efforts of their country, improve their families' health, diet, productive ability, socio-cultural status and also be able to discharge responsibilities as mothers, wives and members of the society effectively. According to Sani (2001, p. 139), women more often than not derived their social status from performing dual roles of wife and mother. As a mother, the woman is the primary custodian of the cherished values of the society. This she does by devoting her valuable time and energy to teaching and inculcating societal values into the growing babies, caring for the well-being of all other members of the family, which earned her a high social status. Haralambos & Holborn (2008, p. 83), citing Goldthorpe (1983), do not believe that a male where present should automatically be considered head of a household. They believe that the head should be defined as "the family member who has the greatest commitment to and continuity in the labour market".

Therefore, husbands and wives should be allocated to classes as individuals rather than as part of a family unit. Igwesi (2012, p. 218) further posits that developing countries such as Nigeria need not only understand and emphasise the importance of gender equity in national development but also to ensure that at all levels in the society females are given the opportunity to harness their in-built potentials and are able to contribute immensely to national development. National development in this context refers to the plans, policies and actions of a nation to improve the lives of its citizens. This presupposes that national development is not just the responsibility of the government or political representatives of the people, but the responsibility of all and sundry. This development can be social, economic, political, educational, quality health care, reducing poverty, providing job opportunities, etc. Therefore, the goal of every national development agenda is to improve the lives of the citizens within the context of a growing economy, and an emphasis on the good of the community at large.

Igwesi (2012) recalls that there are many women who have contributed immensely to the socio-economic, political and human development of the society. Some are politicians, academics, entrepreneurs, leaders of different organisations, career professionals and bread winners. Examples are Professor Alele Williams, former Vice-Chancellor of University of Benin; Dr. Ngozi Okonjo-Iweala, Nigeria's former Minister of Finance and former Vice President, World Bank, who contributed to the economic growth and development of Nigeria, and late Professor Dora Akunyili, former Director General of National Agency for Food Drug Administration and Control (NAFDAC) and former Minister of Information also contributed to the fight against fake drugs and rebranding campaign of Nigeria's image in the international community. This is just to mention few Nigerian women who have made their marks in the country as Ministers, members of the National Assembly, Deputy Governors, etc, and have contributed useful ideas to the growth, development and sustainability of their immediate communities and the country at large.

According to Sani (2001, p. 4), historical records show that there were situations where women not only contributed to the socio-economic development of their communities, but were also involved in the territorial struggles. Examples of such women were Queen Amina of Zazzau, Queen Kambasa of Bonny, Chief (Mrs.) Olufunmilayo Ransome-Kuti, Nana Asma'u, and Queen Emotan of Benin Kingdom. Their positions were not merely passive, supportive roles, but powerful, constructive, sometimes self-sacrificing roles. Some of these historical women, especially the legendary figures, sacrificed themselves in the service of their communities.

Onwubiko (2012, p. 69) states that, "in Nigeria today, the women folk have come a long way; in business, politics, education, sports and other professions. Women have made an indelible mark in their effort to conquer the limitations of the past which have sought to place them permanently in the kitchen and bedroom". Sani (2001, p. 221) argues that in order to further strengthen the effort of women in national and international development, policy makers, international organisations and non-governmental organisations, should continue to do the following:

- Promote women's access to economic resources.
- Encourage women to acquire both formal and informal education.
- Increase women's access to appropriate, affordable and qualitative health care and related services.
- Eliminate all forms of violence against women.
- Encourage the media to always portray positive images of women.
- Increase women's participation in decision-making.
- Establish national mechanism towards the emancipation of women.
- Integrate women into environmental management.

Empowering Nigerian women towards national development should be a matter of national top priority which demands the attention and genuine commitment of every responsible member of the society. The Nigerian nation owes the women folk the responsibility of removing those artificial and institutional barriers based on religion, culture or traditional considerations which have incapacitated the ability of Nigerian women to participate effectively and freely in national affairs particularly at the political and economic levels (Onwubiko, 2012).

Theoretical Framework

The theory adopted for this study is the Cultural Norms Theory. Folarin (2005, p. 96-97) in his view states that, through selective presentation and emphasis on certain themes, the mass media created the impression among their audience that such themes were part of the culture or clearly defined cultural norms of society. As a result, some members of the society tend to pattern their behaviour along the line of such media presentations.

This theory of mass communication suggests that the mass media selectively presents, and emphasises certain contemporary ideas or values. According to this theory, the mass media influences norms by reinforcing or changing them. For example, the cultural norm theorists argue that television programmes presenting an active lifestyle for women as promiscuous and dependent on their male counterparts can change the attitudes of viewers in that direction (Ballack, 2009). According to Okenwa (2000, p. 22), the cultural norms theory postulates that the mass media through selective presentations and emphasis on certain themes create impressions among the audience that common cultural norms concerning the emphasized topics are structured or defined in specific ways. Furthermore, the media can potentially influence behaviour by reinforcing existing norms; creating new norms; and modifying existing norms.

Okenwa (2000, p.22), states that, "the media can be adequately utilised for the transmission of culture both internally and externally". Oyero (2011) opines that cultural norms theory offers a broad range of interesting ideas about how media can affect culture and provide many different views concerning the long-term consequences of the cultural changes affected by media. The theory directly addresses questions about the way media might produce profound changes in social life and argued that media might have the power to intrude into and alter how we make sense of ourselves and our social world. This theory posits that through the presentations and depictions of women, Nigerian films have created an impression among their audience that such portrayals are part of the culture of the society and as a result they begin to mould their behaviour in line with the film's depictions. Therefore, it might likely influence the perceptions of the audience towards the roles of women in their active participation towards national political, socio-economic and human development.

Research Questions

1. What specific roles are assigned to women in the selected Nollywood films?
2. How is the ability of women to contribute to national development portrayed in the selected Nollywood films?
3. What behavioural traits are female characters assigned in the selected Nollywood films?
4. To what extent do these representations of women in the selected Nollywood films showcase their active participation in the process of national development?

Methodology

The researchers adopted a quantitative and qualitative approach to the study. Quantitative content analysis was used to collect data concerning how women were portrayed in the selected Nollywood films. The researchers got access to the list of some Nollywood films approved for distribution and viewership in the year 2012 by the National Film & Video Censors Board (NFVCB). A total of Eighty (80) films produced in English were got. The sample for the study was drawn from the list of films. The list formed the sampling frame for the sampling process, and the sampling method used to select the five films examined was systematic sampling with a random start, adopting an interval of five. The researchers decided to study only five out of the 80 films (over five per cent of the total number). The five selected films were: *Mr. & Mrs.*, *Damage*, *Ties that Bind*, *Fazebook Babes*, and *Midnight Whispers*. The conclusions reached in this study were based mainly on the five selected and analysed films.

Also the researchers went further to do a Focus Group Discussion (FGD) on the perceived influence these films have on women's participations in national development. Three (3) sessions of FGD involving a representative sample of viewers of Nollywood films from different socio-economic groups were conducted. The participants ranged according to age, sex, and educational level, and were drawn from undergraduate and postgraduate students of University of Lagos, Akoka, and some selected residents of Akoka community.

The participants in the first group (18-25 years) consisted of participants who were in their third and fourth years of their first degree at the University of Lagos. The second group (26-34 years) were postgraduate students studying at Masters Level on part-time basis also at the University of Lagos. Participants in the third group (35-45 years) were predominantly less educated people including artisans, traders, and shop owners who were residents of Akoka at the time this study was conducted.

Results

For results, see Tables 1 and 2 and Figures 1 and 2.

Table 1. Distribution of major female characters in the films

Film Title	Major Female Characters
Mr & Mrs.	5
Damage	2
Ties that Bind	3
Fazebook Babes	3
Midnight Whispers	2
Total	n=15

Table 2. Frequency of behavioural traits attributed to females

Feminine Traits	Percentage
Caring/ Loving	67%
Childlike	40%
Crying/Whining	53%
Dependent	73%
Fearful	33%
Happy	53%
Scared/ Desperate	47%
Total	n=15

Figure 1. Percentage of stereotypical and non-stereotypical female professional role

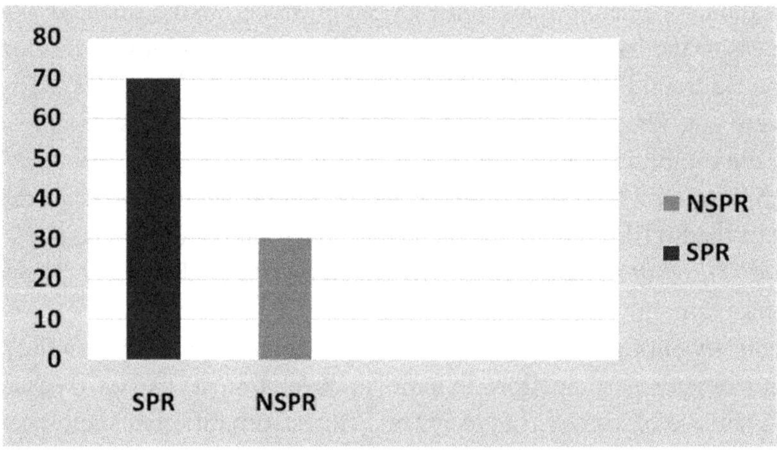

Figure 2. Percentage of stereotypical and non-stereotypical female behavioural traits

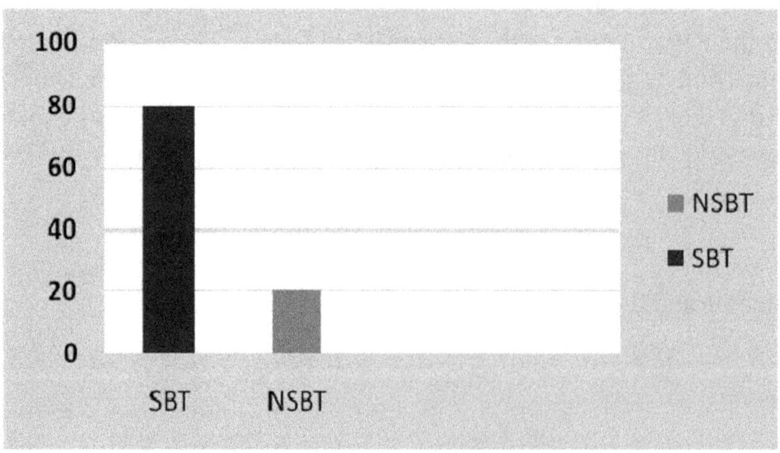

DISCUSSION

Nollywood films are deeply rooted in Nigerian cultural traditions and social contexts that focus on Nigerian community life (Onuzulike, 2007). "Given the great influence that Nollywood has over African culture, such effect is reinforced by a massive consumption of Nigerian films by Africans living in Africa and off the shores of Africa" (Onuzulike, 2007, p. 237). As the twentieth century ended, research evidences showed that some Nollywood films were filled with negative and stereotypic images of women. This is done to the disadvantage of women as these images promote continued domination by men and subordination and subjugation of women (Okunna, 2002).

The researchers, therefore, sought to analyse and evaluate the portrayal of women in some selected Nollywood films, the interpretations given to their roles in the films, and audience perception of such roles in relation to women's contribution to national socio-economic, political and human development. For the five (5) Nollywood films selected and analysed, the unit of analysis used was every major female character in each film. A total of 5 major characters were analysed. See Table 1 which presents the number of female characters in each film analysed.

What specific roles are assigned to women in the selected Nollywood films?

From the data analysed, only 30% of the female characters were portrayed in high ranking positions, career professionals and intellectuals while 70% of the female characters in the films were not portrayed as career professionals and high ranking positions (see Figure 1). Rather, they were portrayed as being dependent and subordinates. It was also observed that assigning high ranking positions in film mostly to men leaves female characters as the inferior gender, lacking the ability to intellectualise. It is no wonder that scholars such as Signorielli (2001) agree that independence and intelligence are highly recognised traditional values of boys.

Thompson and Zerbinos (1995) posit that if women were portrayed in high occupations or intellectual roles such as lawyers and doctors they were often sexually objectified. This can be seen in the film, *Mr. & Mrs.* as Linda, a dedicated banker is seen wearing extremely tight fitting clothes that expose sensitive parts of her body and portrays her as sexually attractive.

Consequently, in previous studies males were more likely portrayed in more leadership positions than the females (Baker and Raney, 2007), and this can be seen in the analysed Nollywood films. For example, in *Mr. & Mrs.,* Ken Abbah plays the part of a rich and influential son of a minister, who heads the family business and is married to a lawyer whom he insists that she becomes a full-time house wife. *Damage* has Taiwo a successful businessman who loves to be in control of his wife.

Also, Figure 4.2 clearly reveals that 70% of the female characters in the films were depicted in stereotypical domestic roles (taking care of the family and other home management chores) and contexts. About 25% of female characters were not portrayed performing domestic activities. Based on this statistical presentation it is clear that female characters in Nigerian home videos are assigned to more of domestic roles than professional or high ranking positions.

Durkin and Nugent (1998) opine that in the popular media, women are more often shown in the domestic setting, as housewives than their male counterparts. This can be clearly observed in films such as *Mr. & Mrs, Ties that Bind, Damage,* and *Midnight Whispers,* in which female characters were constantly depicted in domestic roles such as cooking, washing, taking care of the children and the home.

How is the ability of women to contribute to national development portrayed in the selected Nollywood films?

Oyinade, Daramola, and Lamidi (2013, p. 100) state that, "Nigerian women had contributed enormously to what is known as the family unit as they specialised in maintaining their households, rearing children and emotionally supporting their respective husbands." Responses from both the male and female members of the Focus Group Discussion conform to the ideal of mass culture, agreeing that Nollywood, as a cultural industry, has some filmmakers who produce films that portray women perpetually in negative lights, using stereotypic roles, and the narratives do not represent the ideals that women are active participants in the growth and development of their immediate communities that translate into national development.

Chisom Agugua who was a participant in the first group is of the opinion that women take up exceptional roles in the society. She goes further to mention women such as Dora Akunyili and Ngozi Okonjo-Iweala who have or are still making marks in relation to how they have used their position to transform Nigeria politically. Osaze Ezodaghe also from the first group was of the opinion that socially, women nurture the children even in schools; women are mostly in charge of training the children both in school and at home and this helps the children to learn more from the women while growing up. The way they are trained determines the way they grow up to become and how much they can impact in their society.

Blessing Ume says that it was observed that women were more than men statistically and this makes them form a larger percentage of workforces in the country. She also says in her own words, "You educate a man, you educate a man; you educate a woman you educate a generation". Joy Eze (second group) says that because men are hardly around, women have more responsibilities in taking care of the children. It is this nurturing and upbringing that determines what they turn out to be in the society. She also observed that school teachers are more of women because they have an exceptional ability and skill in raising children. Chima Ohameze, one of the participants in the third group agrees to this fact by saying that his wife gives him strength and supports both emotionally and physically. He goes further to say that he does not know what he would have done with his children if his wife was not there. In his words, 'I no sabi cook, I no sabi change napkin, I no dey even get chance carry my pikin go school...na my wife dey do all those ones, my own na to hustle for money'. Shola Akinwunmi (Third group) states that, "A child is relevant to the society if he or she is properly trained and mothers have an upper hand in this aspect of child development".

Contrary to the views of the FGD participants so far, the films selected for the study were few of the many Nollywood films that do not give women a good image in the society as mothers and homemakers. The films have also been criticised for their portrayal of women as sex objects, weak, cold-hearted, materialistic, vengeful, vicious, diabolical, and scheming. This has in no way presented them as agents of development in all spheres of the society.

Angela Olusanya (First group) says that Nollywood has not succeeded in showcasing women as contributors to national development because there are just few or no films where a woman is seen taking up an important role that contributes to the society politically, culturally, or socially. Rather, they are underrated and subjugated by the opposite sex. Osaze says that the reason why women cannot be assigned important roles in a film, especially political role. In his own words, "you cannot be watching a film and a woman plays the role of a president; this is because when a president is mentioned no one thinks of a woman first. Everyone assumes it is the role of a man so it is only realistic a man plays that part in the film".

Seun Balogun (Second group) was of the opinion that the portrayal of women in Nollywood films is in contrast with what they actually contribute to national development and these representations have in no way showcased them as agents of social change. Blessing Ume gives examples of Nollywood actresses Mercy Johnson and Ini Edo who act vulgar and obscene roles; she says that such actresses might not be taken seriously as contributors to national development or agents of social change as the roles they play in films are negatively influencing youths rather than empowering them to be useful in the society. She also mentions that though the blame is not solely the actors' but also film producers'. But being realistic, the audience do not think of the film producer first but the actors who play these roles. It is the role they play that people get to watch and it is what people see that people believe. If they portray women in a negative light, the impression lasts in the minds of the audience. Rather than see women as agents of social change which they really are, they are rather influenced by the unfavourable narratives concerning women portrayed in the films.

It can be concluded from the analyses of the selected films and the responses of the participants of the Focus Group Discussion (FGD), that women arenot adequately presented in the selected Nollywood films as being actively involved in national development through the roles they play in the films.

What behavioural traits are female characters assigned in Nollywood films?

In order to analyse the behavioural traits female characters are assigned in Nigerian home videos, it was important to identify the behavioural traits that are stereotypically assigned to women. The behavioural traits for this study were derived from Fischer (2010) who gathered these measures from Thompson and Zerbinos (1995) and the Bem Sex Role Inventory (Bem, 1974). The behavioural traits coded as stereotypically feminine are: caring/loving, childlike, crying/whining, dependent, fearful, happy, and scared/desperate.

From Figure 2, almost all the female characters with a total of 80% were recorded as having exhibited the identified stereotypical behavioural traits (SBT) in film while less than 20% of them did not exhibit such behavioural traits (NSBT). It was also realised after analysing the female characters in each film that 11 out of 15 female characters possessed at least four (4) of these behavioural traits and so were described as having stereotypical feminine behavioural traits. By virtue of these results, it is clear that, except for some insignificant others, most female characters in Nigerian home videos are assigned specific behavioural traits. The data presented in Table 4.3 elucidate the results further.

To what extent do these representations of women in the selected Nollywood films showcase their active participation in the process of national development?

Drawing from the tables, a large percentage of the women play subservient roles. They are mostly portrayed as victims of abuse, always dependent on the opposite sex. From the analyses of these films, it was observed that it was to a low extent that the representations of women in the selected Nollywood films showcased their active participation in the process of achieving national development.

Women were not portrayed as diverse personalities and with active roles in the changing world. It even reinforces the stereotypical image of women as sex objects, consumers and slaves (Ali and Khan, 2012). Helen John, a participant in the first group says that this is where the media is important to changing people's orientation. When home videos start to give women important roles to play in a film, people's perspective begin to change and they begin to see women in a different light. Amarachi Arinze agrees by saying that the media should project what is ideal rather than carrying on with what seems to be the status quo. Seun Balogun (Second group) is of the opinion that these portrayals have made women see themselves as the weaker vessel and this has affected their role in national development as they have become reluctant and docile towards decision making even in their respective homes and the society at large. Rather they have become over dependent on the men like the films portray them. Joy Eze was of the opinion that women are beginning to lose their identity because if they understand that they are agents of social change they will be selective in the roles they play in the film industry. Titus Johnson says that some films portray women as agents of social change, but Mabel Okafor negates that by saying that Nollywood films do more of promoting the patriarchal society by showcasing the man's domineering attitude over a woman, and the women are left with no choice but submission. She adds that instead of showcasing women as home builders which they really are, they showcase women as society spoilers. Chima Ohameze agreed by saying that the kind of films women act does not portray them as homemakers and bearers of future leaders. He also compared the kind of films women act abroad to that of Nigeria saying that women are highly respected abroad than in Nigeria".

Future research endeavour in this area should investigate the role of the filmmakers in the portrayal of women in Nollywood films. There are few women who are actively involved in film production in Nigeria. Some of them are: Emem Isong, Stephanie Okereke Idahosa, Uche Jumbo, Monalisa Chinda, Ini Edo, and Funke Akindele. It is important to carry out a research into what strategies they are adopt to correct the negative portrayal and to ensure that women get favourable depictions in the Nollywood films, taking into cognizance societal and cultural expectations of the female folks.

CONCLUSION

Based on the findings of this study, the researchers conclude that women were not well represented in the selected Nollywood films and this has created doubt in the minds of the audience if women are actually capable of contributing to national development. It was clear that the audience are influenced by the films they are exposed to. When the films continuously position women as dependent, sex symbols, the weak gender incapable of carrying out certain tasks, these images begin to stay in the minds of the audience as they start to see women in that light. Nwabudike (2012, p.82) agrees with this when she states that "repeated exposure to certain stereotypes may cause people, especially people of impressionable age, such as youths, to believe that what they view in these films are actually realistic".

Furthermore, because of the images these films have created in the minds of the audience, they do overlook the efforts women make as agents of social change in their communities, and women are given little or no chance to contribute to national development. Though these portrayals sometimes depict reality but the media should rather change this trend and project the ideal because these negative gender stereotypes are not necessarily created by the Nollywood films; they are projected and reinforced by them.

The Nollywood industry is a veritable aspect of the Nigerian media and so, it is certain that Nigerian films form a crux of the Nigerian media (Aladese, 2013). This is why their negative representations of females are being given adequate attention as such representations are also affecting their roles as change agents.

The role of a woman starts from the family, and the family is an agent of socialisation. A woman's role in the family cannot be overemphasised. Women occupy important positions in the home and therefore play a crucial role in home management and training of the children. Fischer (2010) notes that socialisation, specifically gender socialisation, begins with the family as the primary agent and later other agents of socialisation become more influential, e.g. peers, school, and media. It is no wonder then that Nollywood films serve as channels through which gender stereotypes are projected and maintained. As was stated earlier, these Nollywood films are not necessarily responsible for creating these gender stereotypes but they play a major role in projecting and sustaining them. It can be deduced that Nigerian films seem to be reinforcing some of the negative images resident in the minds of the audience concerning the roles of women in promoting national development.

From the data analysed and discussed in this study, there is need to change the narratives of the inadequate representation of women in Nollywood films. Stereotypes can be positive or negative, and they are inevitable as part of film narratives. Therefore, positive stereotypes of women in Nollywood films should be encouraged. This can only be made possible through changing the narrative style of the film scripts. There seems to be a lot of dynamics when it comes to film production, especially in Nollywood, and it begins with the narrative the script writer wants to tell. The script serves as the map for the film production which the director and the actors/actresses follow. The script, therefore, determines the roles the female talents are to play. If the storyline does not provide a narrative that will aptly capture and represent women as valuable assets and active participants in nation building, there is no way the female talents can play contrary roles in the films. This, therefore, calls for more female script writers and directors to be involved in charting the narratives that will adequately give women a voice, new roles, and the right representation in Nollywood films.

NOTE

According to the Zonal Director, National Film and Video Censors Borad (NFVCB), South-West Zone, Ikoyi, Lagos, Mr. Edward Edion, the eighty films approved for viewership in the year 2012 is not the sum total of all films in circulation in the Nollywood market, becuase not all the films go through the NFVCB for censorship and approval due to sharp practices in the industry. Therefore, even the board does not have a complete record of all films or home videos released into the market.

REFERENCES

Akinfeleye, R., & Amobi, I. (2011). Nollywood video films as a medium for reconstructing the African cultural identity. *Communication Review*, *5*(1), 1–23.

Akpabio, E. (2003). Themes and content of Nigerian home video films. *Unilag Sociological Review*, *4*(October), 138–139.

Akpabio, E. (2007). Attitude of audience members to Nollywood films. *Nordic Journal of African Studies*, *16*(1), 90–100.

Aladese, T. (2013). *Gender stereotyping in Nigerian home videos* (Unpublished undergraduate thesis). Department of Mass Communication, College of Development Studies, Covenant University, Ota, Ogun State, Nigeria.

Ali, G., & Khan, L. (2012). A language and construction of gender: A feminist critique of Sms Discourse. *British Journal of Arts and Social Sciences*, *4*(2), 342–360.

Amobi, I. (2010). *Audience interpretation of the representation of women in Nigerian Nollywood films: a study of women from different social contexts in Nigeria*. Retrieved from http://profteri.wordpress.com/2010/11/12/online-journalism-syllabus/

Baker, K., & Raney, A. A. (2007). Equally super?: Gender-role stereotyping characters in childrens animated programs. *Mass Communication & Society*, *10*(1), 25–41. doi:10.1080/15205430709337003

Ballack, X. (2009). *Cultural norms theory*. Retrieved from http://ballack.xomba.com/cultural_norm_theory_mass_communication

Baran, S. J. (2009). *Introduction to mass communication: Media literacy and culture* (5th ed.). New York: McGraw Hill.

Baran, S. J. (2010). *Introduction to mass communication: Media literacy and culture* (6th ed.). New York: McGraw-Hill.

Bem, S. L. (1974). The measurement of psychological androgyny. *Journal of Consulting and Clinical Psychology*, *42*(2), 155–162. doi:10.1037/h0036215 PMID:4823550

Daramola, Y. (2008). Mass media and society in Nigeria: Selected functional perspectives. In R. Akinfeleye (Ed.), *Mass Media and Society: A Multi-perspective approach* (pp. 31–45). Lagos: Department of Mass Communication, University of Lagos.

Durkin, K., & Nugent, B. (1998). Kindergarten childrens gender-role expectations for television actors. *Sex Roles*, *38*(5), 387–402. doi:10.1023/A:1018705805012

Ebewo, P. (2007). The emerging video film industry in Nigeria: Challenges and prospects. *Journal of Film and Video, 59*(3), 46-57.

Eke, M. (2012). *Nollywood, women, and cultural identity*. Paper presented at the International Conference on Nollywood, Women, and Cultural Identity. Retrieved from http://www.cfplist.com/default.aspx

Fischer, S. (2010). Powerful or pretty: A content analysis of gender images in children's animated films. Graduate faculty. Auburn University. Retrieved from http://etd.auburn.edu/etd/bitstream/handle/10415/2065/ThesisSabrinaFischer.pdf?sequence=2

Folarin, B. (2005). *Theories of mass communication*. Ibadan: Bakinfol Publications.

Haralambos, M., & Holborn, M. (2008). *Sociology: Themes & Perspectives*. London: Collins Educational.

Igwesi, B. N. (2012). Enhancing women participation in national development through a change in the gender system of Nigeria. *Asian Social Science*, *8*(1), 217–223.

Jonah, A. A. (2011). *Film as a vehicle for cultural promotion and unity in Nigeria*. Department of mass communication, The Federal Polytechnic, Bida, Nigeria. Retrieved from www.imim- ng.org/downloads/Alice%20Jonahacticle.doc

Nwabudike, N. (2012). *Stereotyping in Nigerian movies and social interaction among students of Bells University of Technology* (Unpublished undergraduate thesis). Department of Mass Communication, College of Development Studies, Covenant University, Ota, Ogun State, Nigeria.

Obi, C. (2001). Women's political participation through economic empowerment. Lagos: Human Development Initiatives.

Okenwa, S. A. (2000). *The mass media: Uses and regulations*. Enugu: Bismark Publications.

Okunna, C. S. (2002). *Gender and communication in Nigeria: Is this the twenty-first century?* Department of Mass Communication, Nnamdi Azikiwe University, Akwa, Nigeria. Retrieved from www.portalcommunicacion.com/bcn2002/n_eng/programme/prog_ind?papers/o/pdf/o005se04_okunn.pdf

Onuzulike, U. (2007). Nollywood: The influence of the Nigerian movie industry on African culture. *The Journal of Human Communication, 10*(3), 231-242.

Onwubiko, C. P. (2012). Empowerment of Nigerian women towards national development. *Journal of Resourcefulness and Distinction*, *2*(1), 68–78.

Osofisan, F. (2006). *From Nollywood to Nollyweight? Or reflectons on the possiblities of literature and the burgeoning film industry in Nigeria*. Retrieved from http://www.africultures.com/index.asp

Oyero, O. S. (2011). *Theories of mass communication. Unpublished lecture notes, Department of Mass Communication*. Ota, Nigeria: College of Development Studies, Covenant University.

Oyinade, R., Daramola, I., & Lamidi, I. (2013). Media, gender, and conflict: The problem of eradicating stereotyping of women in Nigeria. *Kuwait Chapter of Arabian Journal of Business and Management Review*, *2*(12), 100–114. doi:10.12816/0001275

Sambe, J. A. (2008). *Introduction to mass communication in Nigeria*. Ibadan: Spectrum Books Limited.

Sani, H. (2001). *Women and national development: The way forward*. Ibadan: Spectrum Books Limited.

Signorielli, N. (2001). Television's gender role images and contribution to stereotyping: Past, present and future. In D. Singer & J. Singer (Eds.), *Handbook of children and the media* (pp. 341–358). Thousand Oaks, CA: Sage.

Thompson, T. L., & Zerbinos, E. (1995). Gender roles in animated cartoons: Has the picture changed in 20 years. *Sex Roles*, *32*(9/10), 651–673. doi:10.1007/BF01544217

KEY TERMS AND DEFINITIONS

Behaviour Trait: An action and attitudinal composition commonly observed in a person or group of people. It captures and defines the person's or group's identity and peculiarities.

Culture: The sum total way of life a people in a community, and it includes their values, norms, mores and artefacts. It is their identity and it defines who they are.

Feminism: A movement in all ramifications (social, political and economic) that has at the base of its ideology, advocacy for women's rights with the belief in equality of all sexes.

National Development: A concept that explores how the socio-economic living standards of the citizens in a nation or country can improve greatly through valuable and positive changes orchestrated by the government in partnership with the people.

Nollywood: The name used to refer to the Nigerian film industry after the likes of Hollywood in United States of America, and Bollywood of India.

Portrayal: The act of depicting or representing a person, a place or thing in a work of art and literature, for example a film. In this context, it refers to the depiction or representation of women as a group in Nollywood films.

Stereotype: A widely held or accepted belief, idea, or image about a person or particularly a group of people.

Theme: The central idea or message that recurs or permeates a film as well as other works of arts.

Chapter 9
Education, Gender, and Child-Rights:
Salient Issues in SDGS Years in ADO-ODO/OTA Local Government Area of Ogun State, Nigeria

Taiwo Abioye
Covenant University, Nigeria

Esther Ajiboye
Covenant University, Nigeria

Kehinde Oyesomi
Covenant University, Nigeria

Segun Omidiora
Covenant University, Nigeria

Olusola Oyero
Covenant University, Nigeria

ABSTRACT

Promoting and improving quality education is one of the core missions of the United Nations at ensuring sustainable future; hence, the slogan: Change towards a better quality of life starts with education. This paper examined the place of education, gender and child rights within the current status of Sustainable Development Goals (SDGs) in Ado-Odo/Ota local government of Ogun State, Nigeria. Questionnaire and interviews were used as instruments of data collection. School children between ages 7 and 18 in both private and public schools formed the study population. A sample size, 1000 respondents, was drawn from the population out of which 976 responded effectively to the questions. The findings revealed that education and child rights remain in a precarious state in the local government. There was a limited awareness about child rights among children in primary schools and secondary schools; teaching materials and instructors were grossly inadequate in many of the schools sampled and basic needs such as water and electricity were unavailable. It was also observed that the number of enrolled male children in schools is 24% higher than the females. These challenges should be put into consideration when formulating policies for education in developing countries. There is therefore the need to prioritize education, especially female education, as well as child rights in general in the local government through adequate funding, investment in teachers and creation of awareness about the rights of the child.

DOI: 10.4018/978-1-5225-1859-4.ch009

INTRODUCTION

One of the core missions of the United Nations at ensuring sustainable future is improving the quality of education, hence the slogan: 'Change towards a better quality of life starts with education'. Quality education entails provision of school building and amenities, as well as textbooks, qualified and trained teachers, etc., which would ensure the full development of the human personality and the sense of its dignity. As contained in the Millennium Development Goals (MDGs) and the new Sustainable Development Goals (SDGs), education is one of such rights to ensure future sustainability and the goals can only be realised when the rights of children are targeted and fulfilled (Oyero, 2010).

Global organizations such as The World Bank, UNDP and UNICEF recognize the relevance of education in the drive for sustainable national development around the world. Thus, Education for All (EFA) as an international initiative was first launched in 1990 to bring the benefits of education to "every citizen in every society", with six main goals:

- To expand and improve comprehensive early childhood care and education, especially for the most vulnerable and disadvantaged children.
- To ensure that by 2015 all children, particularly girls, those in difficult circumstances, and those belonging to ethnic minorities, have access to and complete, free, and compulsory primary education of good quality.
- To ensure that the learning needs of all young people and adults are met through equitable access to appropriate learning and life-skills programmes.
- To achieve a 50% improvement in adult literacy by 2015, especially for women, and equitable access to basic and continuing education for all adults.
- To eliminate gender disparities in primary and secondary education by 2005, and achieve gender equality in education by 2015, with a focus on ensuring girls' full and equal access to and achievement in basic education of good quality.
- To improve all aspects of the quality of education and ensure the excellence of all so that recognized and measurable learning outcomes are achieved by all, especially in literacy, numeracy and essential life skills.

In 2000, 189 countries and their partners adopted the two EFA goals that align with Sustainable Development Goals (SDGs) 5, which refers to gender equality and empowerment of all women and girls and goal 10, which refers to the reduction of inequality within and among countries.

The idea of women in development is an issue to be prioritised as women play a significant role in fulfilment of the rights of the child because many of child rights would not be fulfilled without the women. As primary care-givers to children, women need to be appropriately positioned to perform their responsibility. Positioning women then begins from the cradle, such as receiving necessary education, healthcare and empowerment to function. Unfortunately, women in developing countries are generally marginalised. Many young women face serious threat to their empowerment through child marriage and denial of educational opportunities. It thus follows that development must focus on women too if something tangible would be achieved. Thus, there exist wide-ranging and deep linkages among gender, child rights, education and the Sustainable Development Goals.

Statement of Research Problem

The dawn of year 2015, the year meant for the achievement of MDGs, lends credence to the statement-'the future is now'. While setting goals is important to expedite efforts and monitor progress, care needs to be taken to ensure that real efforts beyond the rhetoric of goals are geared towards the achievement of the stated goals. As regards the MDGs, we do not really know whether the primary targets and beneficiaries of these goals are aware of any effort being made to improve their situations. Besides, it is important to ascertain the level of progress made thus far on MDGs in order to chart a clearer path for future agenda. While there have been global and national reports on the MDGs progress, there is need to understand with specific indicators in a small scale setting, the reality of the new Sustainable Development Goals (SDGs). It is in this direction that this study was undertaken to study the place of child rights, education and gender in development agenda after the 2015 target.

Objectives of the Study

The following objectives were set to be achieved in the study:

1. To find out the extent to which pupils/students are aware of the SDGs and Child Right Acts.
2. To compare the learning conditions (quality of education, conduciveness of classroom, availability of teachers and teaching materials) of public schools with private schools.
3. To assess if there are disparities in the boy-girl school enrolment in the local government (LG).
4. To investigate the factors responsible for the non-enrolment of school-age children.
5. To evaluate the feasibility of achieving child rights, education and gender balance in post- 2015 development agenda.

Research Questions

1. What is the awareness level of SDGs and Child Rights among pupils/students in Ado-Odo Local Government Area (LGA)?
2. What are the learning conditions (quality of education, conduciveness of classroom, availability of teachers and teaching materials) of public schools compared to their private counterparts?
3. Are there disparities in the boy-girl enrolment in the LG?
4. What are the factors responsible for the non-enrolment of school-age children in the LG?
5. What is the feasibility of achieving child rights, education and gender balance in post- 2015 development agenda?

MILLENIUM DEVELOPMENT GOALS, SUSTAINABLE DEVELOPMENT GOALS, EDUCATION, AND CHILD RIGHTS IN AFRICA

We cannot talk about SDGs without looking at the MDGs. The African continent is the contextual framework for examining the challenges confronting the attainment of MDGs by 2015 with particular reference to education. But Africa has 52 independent countries and each country is different in many respects, yet they all fall within the category of developing or underdeveloped countries. None of them has attained 100% literacy rate; life expectancy is below the world average; quality of life is low with many citizens living below poverty line which is less than one US dollar a day. These are some of the major indicators used by the World Bank to rate a country's standard of living. By way of case study, Nigeria will be used to represent the African scenario in this study. The choice is informed by the fact that it has the largest population with diverse peoples, linguistic groups and cultures at different stages of development. Moreover, many of Nigeria's social, religious, economic, political and developmental problems and challenges are replicated in the other African countries. Comparisons and contrasts will be made when necessary or possible.

According to Oyesomi, Oyero and Okorie (2014), child right issue is a perennial issue that has been discussed and debated among media experts, educationists and social scientists. Child Rights Acts 2003 contained the rights of children. One of such rights is right of a Child to free, compulsory and Universal Primary Education (UPE): Universal primary education is a must for every child. Therefore, every child has the right to be educated at least at the primary level. This means that every parent or guardian should ensure that his or her child irrespective of the gender is educated.

At the international level, it is rare to find a voice outside of academia that will question the value of the MDG-education. To question the goal of universal primary education comes too close to questioning the right of all children to receive a basic education, as enshrined in the Convention on the Rights of the Child. As the human-rights approach, championed by international non-governmental organisations (NGOs) as well UN agencies, is competing ever more fiercely with human capital theory as the dominant paradigm for understanding development, there is little opportunity to consider whether the cost of expansion in enrolments is sustainable in terms of its impact on quality. However, within the many countries where class-sizes have surged in advance of the implementation of reforms intended to improve quality (Riddell, 2003), this question has become a matter of public debate. EdQual researchers based in Tanzania, for example, have come across this debate in the popular media. Statistics published in the Global Monitoring Reports, however, do suggest that whilst national and international resources are channelled towards primary education, many countries are managing to raise retention rates and reduce repetition rates simultaneous to enrolment expansion.

In Kenya, significant milestone in the promotion of the rights of the child was the enactment of Children's Act by Kenya's Parliament in 2002. Concerted efforts towards domestication of the UNCRC (1990) were made when the government directed the Law Reform Commission (1984) to look into children's issues and ways of implementing the convention (Nwanko & Okwemba, 2002). The bill was enacted into law in 2002 and it provides for the legal instruments for the development of policies as well as legal framework for education and promotion of children's rights (ibid.). It recognises that children have a right to education, parental care, religious education, healthcare, and protection from abuse, protection from harmful cultural practices, protection from sexual exploitation and drugs (RoK, 2001).

MDGs are a tool through which people can hold governments and policy makers to account regarding the eradication of poverty. Analysts on the progress of MDGs argue that whatever happens post 2015 must include education and creation of rights awareness and targets to all people. Access to free education, health and protection by all is one such key target enshrined in the rights of the child. It was for this reason that the study was undertaken, particularly to determine whether children were aware of children's rights since child rights and contemporary global development agenda are inseparable (Commonwealth Secretariat, 2006).

With the new developmental agenda, Sustainable Development Goals (SDGs), the SGDs has been proposed to attend to the shortcomings of the MDGs. The Sustainable Development Goals (SDGs) are a proposed set of targets relating to future international development. The Sustainable Development Goals offer major improvements on the Millennium Development Goals (MDGs). The SDG framework addresses key systemic barriers to sustainable development such as inequality, unsustainable consumption patterns, weak institutional capacity, and environmental degradation that the MDGs neglected.

THE PLACE OF EDUCATION, GENDER, AND CHILD RIGHTS TOWARDS ATTAINMENT OF THE SDGs

Oyero and Oyesomi (2014) is of the opinion that when a child is securely attached to the family through active and co-viewing mediation, it will help the child process what is right and wrong in what he/she is viewing, and learn to cope realistically with social pressures and expectations. Education is therefore the key to the development of the society especially the child. The right to education is inseparable from the child's right. Education constitutes the core of human development. It is a vital tool for transformation and key to the sustainable development of a nation. Education is the most crucial instrument for empowering young people with knowledge and skills, which in turn provides them access to productive employment. The Millennium Development Goals (MDGs) adopted in September 2000 at the United Nations Millennium Declaration has two of the eight goals devoted to education. They are goal 2 (to achieve universal primary education) and goal 3 (to promote gender equality and empower women). Over the years, Nigeria has expressed a commitment to education, in the belief that overcoming illiteracy and ignorance will form a basis for accelerated national development. However, regardless of the controversy surrounding education in Nigeria; it is very crucial to the development of the community and nation. There are inequalities in access to education. Millions of poor people and their children are excluded from the processes and outcomes of education (Otive, 2006).

According to Odekunle and Okuwa (2012), the global debate on the place of education and training in social, economic, political and technological development is important. It has been empirically proven and universally acknowledged that unless the citizens of a nation are well educated and appropriately trained, the achievements of rapid economic and social development of such a nation cannot be guaranteed. Education has long been seen as the bedrock of national development and the primary level of education is the foundation required for this development. Therefore, basic education in Nigeria represents the most facilitating and perhaps the most contributory factor to the education system. It is the most fascinating because therein lies the greatest foundational activities in the educational systems.

When children are helped to be knowledgeable and self-aware about their rights, they grow up in the society with the ability to protect and respect their rights and those of others. Knowledge of child rights is critical in the overall development of all children. The use of communication media is critical in the achievement of this awareness among children. Kumar (2006) notes that creating awareness about the Convention, its application and monitoring is a necessity which calls for a joint effort of all the parties concerned with children. He argues that communication media can be a powerful tool for advocacy and awareness creation provided it is used in a planned manner. More importantly, the media serves as a major vehicle for sustainable development in a society through the choice and use of appropriate language that would effectively create and sustain awareness among the masses. Besides, the awareness roles of the mass media across multiple platforms are performed through language (Owolabi & Nnaji, 2013). Hence, if awareness on child rights as well the need for education is to be increased, attention should be paid to the discursive forms and practices that the mass media employs.

With reference to goals 2 and 3 of the MDGs, gender as well as child rights cannot be dissociated from the attainment of universal primary education. The Nigeria Millennium Development Goals Report 2010 estimated that there is an average potential that children everywhere, boys and girls alike, will be able to complete a full course of primary schooling by 2015. The goal to eliminate gender disparity in primary and secondary education preferably by 2005 and all levels of education no later than 2015 was also reported to have an average potential to be actualised. The report further notes that the availability of a supportive environment for both of these MDGs could be rated as 'good/fair' (p.12).

To achieve the targeted 100% gender equality by 2015, the MDG Report 2010 noted that it would be helpful 'to identify those social, economic, traditional, cultural and religious factors that militate against gender equality', and consequently, these should be addressed. Furthermore, appropriate programmes such as 'the training and supply of female teachers to schools in rural areas' should be developed to boost girl-child enrolment (p.24).

Based on data collected from Child Labour Survey of the Federal Office of Statistics (FOS), in conjunction with International Labour Organization (ILO), Okpukpara and Chukwuone (2001) opine that there are econometric forces that influence school enrolment. As such, the child, parent, household and community in both rural and urban areas of Southern and Northern Nigeria play measurable roles in the increasing or decreasing rates of child enrolment in and attendance of schools. This is closely related to Olaniyan's (2011) findings which note that socio-economic factors that range from household incomes to parents' education are major determinants of child enrolment. In some other situations, a child's schooling is influenced by the child's gender, age and educational level of the household head (Okpukpara and Chukwuone, 2001). In fact, Olaniyan (2011) reveals that it is more likely for an educated mother to enrol her female child in school, the same way a father's education positively influences the male-child's enrolment. Other factors that influence a child's enrolment in school, as provided by Okpukpara and Chukwuone (2001) include age of the child, relationship of the child to the household head, household composition, regional differences, location and access to school" (for example, Southern Nigeria has more children in school than Northern Nigeria, and there are more children from urban non-poor homes than those in rural and poor homes). Both studies suggest that there is more gender equity (in relation to child enrolment) in urban areas than in rural areas.

In addition to all these, Kainuwa & Yusuf (2013) posit that female children are more likely to drop out from school because of patriarchal belief patterns, some of these for instance suggest that females are care-givers and are needed more around the home to perform house chores. There is also the issue of early marriage of very young girls in the Northern region of Nigerian. A significant fraction of the respondents (41.1%) also ascribe high drop-out rates of girls to absence of proper security. Alika and Egbochuku (2009) add that the learning environment and learning experiences provided for female education are also variables that increase drop-out rates.

The United States Embassy in 2012, reports that the Northeast and Northwest zones of Nigeria have the highest record of non-school attendance. For instance, it observes that 72% of primary age children never attended school in Borno state; whereas, the Southern zones record less than 3%. Factors responsible for primary school drop-outs largely include: monetary cost and insufficient interest (Nigeria Education

Fact Sheet, 2012). With the new developmental agenda, Sustainable Development Goals (SDGs), it is hoped that the SGDs would correct the shortcomings of the MDGs.

METHODS

This study was based on a survey research design. Questionnaire and interview guides were used as instruments of data collection. The population for this study were school-age children between 7 and 18 in both private and public schools in Ota local government area of Ogun State. Primary four to six pupils (Pry 4-6) were sampled in primary schools, and Junior Secondary School 1 to Senior Secondary School 3 (JSS 1-SS3) were sampled in secondary schools. A sample size of 1000 respondents was drawn from the population. A total of 976 respondents were able to effectively respond to the questionnaire.

Demographic Data

As stated earlier, nine hundred and seventy six (976) students participated in this study, out of which 62% were male and 38% female. These students were sampled from two schools; public and private, 52% attend public while 48% attend private schools. The class distribution revealed that 45% were in primary school, 30% in junior secondary school while 25% were in senior secondary school. Their ages range from 7years to 18 years; 45% were between 7-10 years of age, 33% between 11-14 years of age while 22% were between 15-18years years of age. Only 44% reported payment of school fees while 56% paid other levies (PTA, Excursion) excluding school fees.

Results

Table 1 shows that there is a significant difference in the level of awareness of Child Right Acts and SDGs among male and female children at X^2=182, 2 degree of freedom and $p<0.05$.

In determining the level of awareness of Child Right Acts and SDGs among children, the findings revealed that 25% could be categorised as having high level of awareness of Child Right Acts and SDGs, while majority (42%) had average level of awareness, 33% have low level of awareness. This implies that many of the students sampled have average level of awareness of Child Rights Acts and SDGs.

From Table 2, it is clear that there were provisions of learning facilities to students in private schools than to their public school counterparts. This implies that in public schools, provision of basic facilities for learning was inadequate.

There is a significant relationship between available facilities in learning environment and type of school at X^2=207, 4 degree of freedom and $p<0.05$.

Table 3 shows that there were teachers for all subjects. It was however pointed out that some teachers teach more than one subject because it was noted in some schools that some teachers were teaching up to four subjects.

Table 4 shows that there is a significant relationship between sex and type of school at X^2=10.4, 1 degree of freedom and $p<0.05$. It is evident from the findings that there were more male students than their female counterparts.

Table 1. Level of awareness of Child Right Acts and MDGs among children

	High	Average	Low	Total
Female	56 (5.7%)	98 (10.1%)	218 (22.3%)	372 (38.1%)
Male	188 (19.3%)	314 (32.1%)	102 (10.5%)	604 (61.9%)
Total	244 (25%)	412 (42.2%)	320 (32.8%)	976 (100%)

Table 2. Learning conditions in private and public schools

	Public	Private	Total	Ratings
Adequate Chairs and desks	194 (19.8%)	388 (39.8%)	582 (59.6%)	1st
Marker boards	20 (2%)	389 (39.9%)	409 (41.9%)	2nd
Electricity	56 (5.7%)	314 (32.2%)	370 (37.9%)	3rd
Computer	13 (1.3%)	333 (34.1%)	346 (35.4%)	5th
Decent walls, roofing, windows etc.	47 (4.8%)	301 (30.8%)	358 (36.6%)	4th
Total no of students in public and private schools	505 51.7%	471 48.3%	976 100%	

Table 3. Availability of teachers in all subjects

Type of School	Yes	No	Total
Public	505 (51.7%)	-	505 51.7%
Private	471 (48.3%)	-	471 48.3%
Total	976 (100%)	--	100%

Table 4. Gender (dis)parity in schools

	Private	Public	Total
Male	267 (27.4%)	337 (34.5%)	604 (61.9%)
Female	204 (20.9%)	168 (17.2%)	372 (38.1%)
Total	471 (48.3%)	505 (51.7%)	976 (100%)

DISCUSSION OF FINDINGS

In defining who a child is; 43% described a child as anybody under 18 years, 51% described a child as anybody who lives with parents, while 6% described a child as anybody who is not working. The greater number of students (73%) believed that children are entitled to basic needs of life (food, shelter, education, clothing, etc.) However, participants' responses differed on who should provide these basic things of life; 82% were of the opinion that shelter should be provided by parents against 18% that were of the opinion that shelter should be provided by government, 96% were of the opinion that food should be provided by parents against 4% that were of the opinion that it should be provided by government and 64% were of the opinion that education should be provided by parents against 18% that were of the opinion that this should be provided by government. Furthermore, 18% were of the opinion that education should be provided by religious bodies. Medical facilities was seen as the responsibility of the government (77% respondents), 18% were of the opinion that it should be provided by the parents while 5% believed that it should be provided for by the religious bodies. The security of lives and property was seen as the responsibility of the government (77% respondents), while 23% were of the opinion that it should be provided for by parents. However, only 64% participants were of the opinion that they had access to these basic needs of life, while 36% declined.

Participants were asked to describe availability of teaching materials; 44% had access to textbooks, 62% described their classrooms as conducive, 100% had teachers for subjects, while 44% perceived their teachers as having love for their jobs. Only 26% of the respondents aspired to be teachers, while 74% did not want to be teachers because they were not comfortable with the ways teachers are treated.

Findings are discussed in relation to proposed research questions below:

RQ1: Whatis the awareness level of pupils/students on SDGs and Child Rights?

If majority of the children report that they have fairly average knowledge about SDGs and Child Rights Acts, the implication of this is that they do not know their entitlements in the society, and when children do not know their basic rights, they could pass for 'uneducated' citizens. It should be noted that education is not limited to classroom education; social awareness is a core aspect of education. This suggests that children who were thought to be educated are not getting the best of education. More so, this could be an index that the sensitisation efforts of the Local Government are not as widespread as expected. Kiprotich and Ong'ondo (2013) are of the opinion that when children are helped to be knowledgeable and self-aware about their rights, they grow up in the society with the ability to protect and respect their rights and those of others. Knowledge of child rights is critical in the overall development of all children.

RQ2: What are the learning conditions (quality of education, conduciveness of classroom, availability of teachers and teaching materials) of public schools compared to their private counterparts?

Except for the availability of chairs and desks which is below 30% in public schools, all other required facilities and infrastructure required for learning in the public schools sampled are inadequate and few. The environment in which a child learns is a great determinant of the value placed on that child. If children are exposed to harsh learning conditions, the interpretation is that these children have little or no worth in the society, and when there is an effort to improve these conditions, the children are more likely to perceive such as privileges rather than rights. More so, learning is impeded under conditions that are not favourable for learning, as the child is first seeking comfort rather than the new information being transmitted to him or her. About 100% of respondents stated that there is availability of teachers in schools. It was noted in some schools that some teachers teach up to four subjects in secondary schools. There is therefore need for even distribution of efforts to bring to fruition the overall SDG on education.

RQ3: Are there disparities in the boy-girl enrolment in the LG?

Even though the reports from the Local Government office claim that boy-girl enrolment in Ado-Odo Ota Local Government Area is fairly balanced, the questionnaire administered present a conflicting picture. Findings reveal that the number of enrolled male children is 24% higher than the females. This suggests that the girl-child is still significantly denied the right to education, and as such, cannot compete favourably with her male peers; this means that she is still being disempowered, and her future would still be determined by her gender. Though the Local Government reports that the regular sensitisation/ awareness programmes she organises has greatly boosted the enrolment of the girl-child in its schools, with the findings from the questionnaire, it could be hypothesised that if there has been a significant increase in the enrolment of the girl-child, then the enrolment of the girl-child before the introduction of the MDGs was approximately insignificant now that SDGs is in operation. If so, then the efforts of the Local Government with respect to girl-child education are laudable. This output though, is not commensurate with the expectations of the United Nations by 2015, as it is expected that by then, the disparities in boy-girl enrolment in schools would have been eliminated.

RQ4: What are the factors responsible for the non-enrolment of school-aged children in the LG?

The study reveals that not many children are out of school and those who are not in school are likely to have financial constraints. Truly, Universal Basic Education is free, notwithstanding, some parents are faced with economic hardships. As such, they require their children to hawk goods to support their livelihood and sustenance as families, and these children become part of the economic providers in the family.

In an interview with participants, it was pointed out that some parents or guardians probably do not see the need for child education, especially when it is a girl-child. This is buried in the age-long belief of the patriarchal Nigerian society, that the place of the girl-child is the home, and as such any educational investment on her would be a waste. This erroneous belief however, places these children who are not in school at a disadvantaged position.

RQ5: What is the feasibility of achieving child rights, education and gender balance in post- 2015 development agenda?

During an interview session with Mr. Bisi Orebiyi, the MDG Technical Assistant, Ogun State, he had this to say:

In 2005, we got debt relief and following that, the office of the Senior Special Assistant to the President on MDGs was created. It is from that office that the Conditional Grant Scheme was created. The scheme started with the Federal ministries, agencies and departments. We give money to them to execute MDG-related projects like the construction of classroom blocks, roads and streetlights. Later on, the Conditional Grant Scheme moved to the state level. The state was given some money to execute its own MDG-related projects. The budget was divided as: 50% to the Federal Government, 50% to the State Government. In order to fast track the MDG attendance in Nigeria, the Local Government was brought in, learning from the experience of the Millennium Villages. We have two Millennium Villages in Nigeria: the Ikara Millennium Village in Ondo State and the Pampaida Millennium Village in Kaduna State. From these two Local Governments, the MDGs increased and later spread across 113 Local Governments in 2011. Here in this case, one Local Government per senatorial district is selected plus those in the FCT and the two Millennium Villages. After the 2011 set, in 2012, the number of Local Governments increased to 148 out of which we have Abeokuta and this project which Abeokuta is also involved in healthcare, water sanitation and education. In the case of education, we look at access, that is, what we need to do in order to make education accessible to the people; looking at structure improvement, facilities, basic amenities (desks, chairs, water, and toilets). In this case, the Local Government Planning Committee and the Local Government Technical Team were constituted. The Conditional Grant Scheme to the Local Government is a tripartite system supported by the Federal Government, State Government and the Local Government. The Federal Government contributes 50% of the funds, the State Government contributes 30% and the Local Government contributes 20%. We have already selected some schools where renovations would be carried out in the case of building more classroom blocks and constructing toilets, supplying a thousand desks and benches for pupils, that is, double seats benches which would be able to sit 2000 pupils in total. In fact, I was driving some days ago and I passed through Abeokuta where I saw one of these projects being carried out, that is, three classroom blocks were being constructed in one of the selected schools. We have also executed projects like this in other Local Governments. The Local Governments outside this tripartite arrangement are also supporting these school projects on their own. They supply benches, construct classroom blocks, for example, there is a Primary school at Oke Ore which is one of their beneficiaries.

Though from the interview conducted the participants agreed that the government has provided uniforms and also some necessary infrastructure (furniture) and textbooks required for learning (for few schools), these are however insufficient. There are still lots of gaps to be filled as indicated by this study. There is the need for more funds to be directed towards child education in Nigeria. In addition to this, more strident measures in the awarding of contracts alongside monitoring and supervision of allocated projects would go a long way in the realisation of MDGs on child rights and education. Some of the efforts of the government would be siphoned by corrupt individuals if intensive policies are not in place. With the new development agenda (SDGs), it is believed that gender equality and education for all would be a reality.

CONCLUSION AND RECOMMENDATION

The outcomes of the study show that there is limited awareness about children's rights among children in primary schools and secondary schools in Ado-Odo Ota L/G of Ogun State. From further interaction with the students, there is low participation in children's rights activities like seminars and conferences. Children are hardly informed through the mass media on the issue of child rights and SDGs. To remedy these situations, it is recommended that media houses and schools should raise awareness about societal concerns in a balanced way so that some pertinent aspects such as children's rights are not left unaddressed. Most importantly, all partners such as parents, teachers (by giving emphasis to children's rights topics in subjects in the sciences and Mathematics), religious leaders, medical doctors among others need to step up campaigns to supplement existing communication media in promoting awareness about children's rights among children and adults. These campaigns can be in form of designing and using publicity materials such as posters, brochures, branded materials like T-shirts, caps, pens, calendars, umbrellas and billboards among others.

Particularly, attention should be paid to language use in promoting awareness of child rights among the citizenry. The English language which is the language of wider communication, together with Nigerian Pidgin as well as other indigenous languages should be deftly employed to effectively pass across intended messages by the mass media. Careful attention should be paid to the domestication of discursive choices to make mass media messages on child rights and basic mutually intelligible to the masses. This could even include forms of English that reflect the local nuances of the environment where the campaigns occur.

The use of literary materials such as pamphlets, story books, etc. could also effectively serve as tools for publicising the MDGs and SDGs. Language is key to development in any society, therefore, creative writers should weave plots around themes that reflect the unpleasant impacts of gender disparity, lack of education, lack of awareness of child rights, etc. on individuals and the society at large. In doing this though, they should note that simplicity is key, therefore, messages should be comprehensible enough to a large heterogeneous audience, in such a way that can influence change of attitude and perception; this simplicity lies largely in appropriate language choice and its proper use (Onobajo, 2001; Eko, 2002; Owolabi & Nnaji, 2013).

The findings also show that there are still inadequate teachers and teaching materials in many of the schools sampled. Odekunle and Okuwa (2012) are therefore of the opinion that for basic education system to achieve its set goals, the need to put certain educational materials, equipment and facilities in place cannot be over-emphasised. Unfortunately, these facilities are in short supply at all levels of education. In some cases, they are non-existence and where they are available at all, they are in deplorable conditions. To enhance proper teaching-learning processes in basic schools, facilities such as libraries and laboratories must be well-equipped and dilapidated buildings must be reconstructed. Also, teaching aids must be made available to teachers to facilitate learning and the school environment should be made conducive and attractive for proper administration. The findings clearly point out that there are still insufficient teachers in schools and the ones who are available are overburdened. The popular saying that "no educational system can rise above the quality of teachers in the system" is rather apt at this moment in the Nigerian education system. Odekunle and Okuwa (2012) assert that the attainment of the Universal Primary Education component of the MDGs depends largely on the availability of qualified teachers. For UBE to have its intended impact on students and pupils, teachers and instructors as well as school administrators must be capable of imparting permanent literacy and numeracy and some useful communication skills to their pupils and or students. The government is therefore enjoined to provide adequate and qualitative staffing for schools.

FUTURE RESEARCH DIRECTIONS

For future directions, this study recommends that the Local Government should embark on intense rural re-orientation and mobilisation to further sensitise the masses (through campaigns, literature, rallies, television programmes, flyers, among others) on child rights and education, paying particular attention to simplifying the messages. Educational policies that allow for monitoring of roads and streets during school hours would help to ensure that all children are in school when they should, rather than leaving them to disregard their basic rights. The masses too are not left out; it has been observed that many citizens are passive towards the enforcement of child rights. Stakeholders and appropriate authorities likewise should guard against negligence towards quality of education, gender and child rights; there is dire need for government and stakeholders to see basic education as a fundamental tool to attaining national development. Therefore, more funds should be directed towards basic education in order to enhance its quality. If we can achieve quality basic education in Nigeria, then attaining the Sustainable Development Goals will not be a mirage. There is also the need to address the variance between quantity and quality of the education attained by primary school pupils and secondary school students in Nigeria.

At the inception, the focus on education was on increased enrolment. This, however, is at the expense of ensuring the quality of the education attained by the students. Beyond 2015, there is a need to critically reassess this issue with a view to ensuring that the quality of education received by primary school pupils and secondary school students is emphasised beyond the mere numbers.

REFERENCES

Alika, I. H., & Egbochuko, E. O. (2009). Drop Out from School among Girls in Edo State: Implications for Counselling. *Edo Journal of Counselling*, *2*(2), 135–141.

Eko, E. (2002). *Effective Writing*. Heinemann Educational Books.

Kainuwa, A., & Yusuf, N. B. M. (2013). Influence of Socio-Economic and Educational Background of Parents on their Children's Education in Nigeria. *International Journal of Scientific and Research Publications*, *3*(10), 1–7.

Kiprotich, A., & Ong'ondo, C. O. (2013). An Assessment of the Level of Awareness about Children's Rights among Children in Eldoret Municipality, Kenya. *Journal of Emerging Trends in Educational Research and Policy Studies*, *4*(2), 279–287.

Kumar, J. L. (2006). *Fundamentals of Child Rights: Concepts Issues and Challenges*. New Delhi: Mehra Offset Press Delhi.

Nigeria Education Fact Sheet. (2012). The United States Embassy. Retrieved from http://photos.state.gov/libraries/nigeria/487468/pdfs/JanuaryEducationFactSheet.pdf

Nigeria Millennium Development Goals Report 2010. (2010). Government of the Federal Republic of Nigeria.

Nwankwo, E. H., & Okwemba, A. (2002). *How the Kenyan Media Covers the Children's Rights Issues: A Content Analysis*. Paper Presented to International Federation of Journalists (IFJ), Nairobi, Kenya. Retrieved on October 22, 2008 from www.ifj.org/assets/docs/089/021/36f1759- cbfee15.pdf

Odekunle, O., & Okuwa, O. (2012). Enhancing Quality Basic Education for the Attainment of Millennium Development Goals in Nigeria: Suggested Policy Interventions. *Journal of Emerging Trends in Educational Research and Policy Studies*, *3*(1), 92–98.

Okpukpara, B. C., & Chukwuone, N. A. (2001). *Child Schooling in Nigeria: The Role of Gender in Urban, Rural, North and South Nigeria*. Centre for Rural Development and Cooperatives.

Olaniyan, O. (2011). *The determinants of child schooling in Nigeria (No. RP_217)*. African Economic Research Consortium.

Onabajo, O. (2001). Communication Models in Non-Formal Education. *Humanities Review Journal*, *1*(2), 82–88. doi:10.4314/hrj.v1i2.5913

Otive, I. (2006). *The State Of Education in Nigeria*. A Keynote Address Delivered at a Round Table Organized by Civil Society Action Coalition on EFA.

Owolabi, D., & Nnaji, C. I. (2013). The English Language and the Mass Media as Tools for Sustainable Development in Multilingual Nations. *International Journal of Language and Linguistics*, *1*(4), 124–130. doi:10.11648/j.ijll.20130104.16

Oyero, O. (2010). Child Rights and Media Democratization: An Agenda for Realization of the MDGs in Nigeria. *International Journal of Communication*, (11): 202–215.

Oyero, O. S. & Oyesomi, K. (2014). Perceived influence of television cartoons on Nigerian children's social behaviour. *Estudios em Comunicação*, (17), 91-116.

Oyesomi, K. O., Oyero, S., & Okorie, N. (2014). Media Advocacy, Development Journalism and Child Right issues in Nigeria. Scholars Journal of Arts, Humanities and Social Sciences, 2(2).

Republic of Kenya. (2001). *The Children's Act, 2001*. National Legislative Bodies / National Authorities Commonwealth Secretariat.

Republic of Kenya. (2006). *Commonwealth Millennium Development Goals Partnerships*. Commonwealth Secretariat.

Riddell, A. (2003). *The Introduction of Free Primary Education in Sub-Saharan Africa.* Background Paper for EFA Global Monitoring Report, 4.

KEY TERMS AND DEFINITIONS

Child Rights: The rights of children to association with parents, human identity as well as the basic needs for physical protection, food, universal state-paid education, health care, and criminal laws appropriate for the age and development of the child, equal protection of the child's civil rights, and freedom.

Children: Boys and girls between ages 7 and 18.

Education: The process of facilitating learning, or the acquisition of knowledge, skills, values, beliefs, and habits.

Gender: The state of being male or female. It is determined by the conception of tasks, functions and roles attributed to women and men in society and in public and private life.

MDGs: The Millennium Development Goals (MDGs) were the eight international development goals for the year 2015 that had been established following the Millennium Summit of the United Nations in 2000, following the adoption of the United Nations Millennium Declaration.

SDGs: A new, universal set of goals, targets and indicators that UN member states will be expected to use to frame their agendas and political policies over the next 15 years.

Chapter 10
Women Entrepreneurs in Nigeria:
Where Is the Mass Media?

Tolulope Kayode-Adedeji
Covenant University, Nigeria

Oyinkansola Ige
Lagos Business School, Nigeria

Thelma Ekanem
Covenant University, Nigeria

ABSTRACT

For ages, the African culture has limited the activities of women and conditioned, to a large extent, the mentality of most African countries about the place and positioning of women in the society. The mass media have been used as a tool in this. Promoting the abilities and achievements of women in the society is one of the important roles of the media in reducing the rate of gender discrimination. These achievements are becoming noticeable in politics and entrepreneurship; thus, setting a standard for other women in the society to build on. This chapter will explore the role of the media in promoting the woman entrepreneur in Nigeria. The study employed the survey research method for data gathering. Findings showed that women have equal opportunity to grow their businesses. Data shows that this growth is slow as there are no significant differences between respondents who agree or disagree with the availability of enabling environment for women to grow their business as compared to their male counterpart. The study recommends that the mass media need to give more voice to their businesses and activities to pave way and encourage the younger women in the society.

INTRODUCTION

Reeve and Baden (2000) describes gender discrimination as the systematic, unfavourable treatment of individuals on the basis of their gender, which denies them their rights, opportunities or resources." Despite the stand of the Nigerian constitution on equality amongst male and female, some Nigerian cul-

DOI: 10.4018/978-1-5225-1859-4.ch010

tures have constituted a barrier to realising this constitutional provision for gender equality. Explaining further, Anaeme, (2012, p. 1) affirms that "gender discrimination is as old as human history". Socially, "sexual differences have been used to justify cultures in which one sex or the other has been restricted to considerably inferior and secondary roles." Gender equality when discussed by researchers usually draws attention to the African and Asian continent. Gender issues most times have always been about the women: "Female characters devote their primary energy to improving their appearance and taking care of homes and people" (Wood, 1994, p. 32, Sarkar, 2015). Many researchers have supported this argument. (Anaeme, 2012; Kufin & Sinha, 2013). However, in many countries, domestic roles and roles concerning child care remains highly gender specific.

However, the 1999 Constitution of Nigeria (42) stresses that there should be "freedom from discrimination on the grounds of ethnic group, origin, religion, circumstance of birth, disability or political opinion." Despite the constitutional provision, gender discrimination is still identified in Nigeria with both sexes but it is more prominent among women. According to the Nigeria Master Web Citizen News (2012), records show that of the 134 countries in the Equality index, Nigeria ranks 118. While stressing that women need to perform equal roles in the society to aid development, the Secretary General of the United Nations, Ban Ki-Moon, (2014), during the International women's day titled "equality for women is progress for all" argued that "countries with more gender equality have better economic growth. Companies with more women leaders perform better. Peace agreements that include women are more durable. Parliaments with more women enact more legislation on key social issues such as health, education, anti-discrimination and child support." Analysing the situation further, the World Bank (2005) believes that "Nigeria is a highly patriarchal society, where men dominate all spheres of women's lives.

Furthermore, Moemeka (2000) describes the media as the "reality media of mass information", because, according to him, "they are second to none in the ability to disseminate information far and fast- to reach millions of people in different parts of the world at virtually "no time". This is how effective media messages can be, in Nigeria and across the globe. In recent times, the mass media have been used to promote women achievements in Nigeria especially in the area of politics.

As the participation of women continue to increase in Nigeria politics, most studies (Adenle & Oso, 2014; Ajaji, 2014; Mojaye, 2011) on gender discrimination have focused on women in politics and the role of different forms of mass media in promoting the activities of these women. To determine and compare media coverage of both genders in Nigeria by the various media, a research was carried out by the Institute for Media and Society, titled 'Media Coverage of Women Politicians in the 2011 Elections in Nigeria'. The research compared the level of coverage of both genders in Nigerian television, radio and newspapers during election campaign. Result shows that male participants in the elections were more visible in the media than the female participants. The visibility of male aspirants in the media during the campaigns was not just higher than that of female aspirants; it was extremely higher (Institute for Media and Society, 2011). This study through survey, takes a step further to know the audience's view on the promotion of women entrepreneurs in the Nigerian mass media.

THEORITICAL FRAMEWORK

Framing theory was developed by a sociologist Evrving Goffman in 1974. According to Baran and Davis (2003), "Goffman argues that we constantly and often radically change the way we define or typify situations, actions and other people as we move through time and space. In other words, our experience of the world is constantly shifting. We can step from one world to another without noticing it." Goffman explains this theory in relation to the role of the media in shaping our idea and thinking about ideology. Goffman's theory provides an intriguing way of assessing how media can elaborate and reinforce a dominant public culture". In this research, the theory is used to critically explain how the media can reinforce the culture of gender inequality in Nigeria. Different programmes in the media these days portray women as taking up some culture-defined roles. By so doing, the media reemphasizes those cultural beliefs in the mind the audience. Young women begin to take to this mentality and thoughts that the media feed them with. In most cases, people hold on to information derived from the mass media.

Social construction of reality is another theory that can be used to explain media role in gender discrimination. Peter Berger and Thomos Luckman wrote a book in 1966 titled "The Social Construction of Reality". According to Baran (2008 p.427*)*, "the two authors in this book offered a compelling theory to explain how cultures use signs and symbols to construct and maintain a uniform reality". Social construction then "is the idea that all presumptions of the real and the good are constructed within relations among people" (Baran, 2008 p.427). Social construction of reality theory explains that the way we present ourselves to other people is shaped partly by our interactions with others, as well as by our life experiences. How we were raised and what we were raised to believe affect how we present ourselves, how we perceive others, and how others perceive us. In short, our perceptions of reality are coloured by our beliefs and background. Our reality is also a complicated negotiation. What is real depends on what is socially acceptable. Social construction of reality theory argues that the building blocks for the construction of these realities come primarily from the mass media. The mass media here is seen as an instrument used through its information process and messages in forming the ideology that today exists about the male and female gender. These realities have therefore become the norm and belief of the society when it comes to gender. This general culture of the female gender being inferior to the male gender begins to transmit in all aspects of life in the African society because of the societal construct that has been built overtime.

LITERATURE REVIEW

Media scholars amongst other scholars from diverse disciplines have come to accept the power of the mass media in the society (Van Dijk 1991; Shoemaker, & Reese 1991). The role of the media is not limited to providing timely information to the public. The media also possess the ability to structure a society, contribute, to a large extent, in enhancing activities, influencing people and enhancing changes in any society. However, in resent time it has been argued through research conducted "during the late 1940s, 1950s and 1960s that the mass media have only a very limited influence" (Curran, Gurevitch & Woollacott, 2005 p.7). According to McQuail, (1998), the power of the mass media is growing continually. The arena of media has seen a significant shift within the last fifteen years. It wouldn't be wrong then to say that we have shifted from media 1.0 to media 2.0. Although, some scholars tend to point out the adverse effects of the new media, there is no denying their impact. Kaul, (2013, p.127) notes that "In the age of the Internet and other new media technologies, change has occurred not only in the technology itself, but also in how individuals interact, define, and organize their world." This makes journalists moulder of the society through the information they provide. This is why their professional activities are guided by ethical codes of conduct.

Discrimination occurs when a person or persons is deprived certain rights based on colour, age, sex or anything that deals with culture, attributes and values. It manifests when a member of a group is treated differently or deprived of the benefits other members of the group enjoy. Gender discrimination in Africa is culture-based, having been passed down from one generation to another. Arowolo and Aluko (2010) note that the major role of a woman in Africa is in the kitchen, where she is expected to provide daily meals for her family. However, Esidence, and Abdul (2013) believe that "gender bias especially in the Nigeria's political system is often traced to the onset of colonialism in Nigeria. Thus, the western cultural notion of colonialism woven around male superiority reflected in their relations with Nigerians." The discrimination that exists in homes against women manifests in all areas of the society. For example, young African girl is expected to help her mother in the kitchen in preparing food for the family. Contrarily, the boy child may be allowed to play football at that time. This mentality builds up in the mind of young girls and as they grow older, they assimilate such beliefs and passing it on. Agbalajobi (2010 p.78) likewise argues that the "nature of our society celebrates men as being unique, stronger, and fit for the public space while women are seen as feeble and weak and meant to stay within the confluence of private space". All these happen despite that The 1999 Nigerian Constitution of Nigeria (42, 1a & b, 2 & 3) states rightly that there should be no discrimination on the basis of gender, religion, ethnicity, age or circumstances of birth against any citizen.

Gender Discrimination in Africa and Asia

According to the Fact Sheet from the International Labour Office (2007), discrimination in Africa is characterized by sharp contrasts. The facts state that "while there have been major advancements against racial segregations in some countries, gender inequalities remain widespread". The fact sheet also states that gender discrimination in Africa is also expressed through "social isolation, violence, verbal harassment and verbal abuse". Obasola (2013, p. 1027) believes that "women bear the brunt of hardship which manifests in various forms". He believes also that "women are key agents for effective grassroots implementation of poverty reduction programmes and economic regeneration". In other words, with the presence of gender discrimination in Africa, it will be almost impossible for there to be a feasible reduction in the level of poverty in Africa. Obasola says that the efforts of developing countries to "modernize discriminatory laws and galvanize women's participation have been frustrated by deep-rooted cultural barriers that so often run parallel with poverty" (Obasola, 2013, p.1027). In relation to education, Tuwor and Sossoh (2008, p. 367) believe that girls do not enrol in schools due to "gender inequality and discrimination, which takes its root in patriarchy". For them, " one of the causes of the high illiteracy rate in 42% of West African countries is lack of access to quality formal education and gender inequality in the educational systems within the sub-regions" (Tuwor & Sossoh, 2008, p. 364). Ojalammi (2011, p.1) supports this position that "many studies have shown that women in Sub-Saharan Africa are more disadvantaged than [women] in any other region of the world".

Gender discrimination still exists in Asia also and manifests especially in politics, labour and education. Labani, Zabaleta, and Ruiz (2010, p.6) believe that "tackling governance, leadership and political participation" from a "gender transforming perspective" entails considering women as responsible individuals, capable of taking decisions and indispensable for their society's democratization processes".

According to the International Labour Office (2013), the Asian and Pacific region "continues to experience traditional forms of discrimination such as those based on gender and ethnic origin and is increasingly confronted with new forms of discrimination brought about by structural economic reforms, economic openness and greater movement of people". Smith and Byron (2005, p.1) believe that gender discrimination has affected many areas of the life of Asian women and that it has led to "inadequate medical care and nutritionally inferior diets for many women", resulting in "poor health and babies with low birth weight which affects future generations". They also believe that there is a preference for sons over daughters in Asia and that this has led to "infanticide in some cases and in recent years by abortion of female children, leading to shortage of women in population statistics". They add that at the national level, "efforts to reduce poverty and increase the pace of economic development are hindered by gender discrimination" in Asia.

Labani, Zabaleta, and Ruiz (2010, p.7) believe that Asia "presents a high level of poverty, illiteracy, social and economic discrimination primarily affecting women". In areas like Pakistan, gender discrimination appears paradoxical. In their words: "on (sic) one hand, women have prominent political leadership and on the other hand, there is excess female mortality with an estimated 13% of women missing…" In other words, in some areas of Asia, there is pronounced women participation in politics while in some other parts of Asia, women are still discriminated against.

In another sense, Sekhar and Hatti (2005, p.3) observe that in India, males "decisively outnumber the females and women (sic) which constitute less than half of the total population". This most likely is another reason for gender discrimination in Asia. They also observe that "fertility decline in East and South Asia has been relatively rapid at both higher and lower levels of economic development".

Balatchandirane (2007, p.3) is of the opinion that "gender discrimination and disparities", especially in the areas of education in South Asia, are still the largest in the world. She says that "more that 100 million children of primary school age are not in school in Asia with South Asia accounting for one of the worst shortfalls with girls being disproportionately affected" (Balatchandirane, 2007, p.35).

Esteve-Volart (2004, p.8) notes that female labour participation figures in India are still low. For her, "despite the government's attempts at improving women's working conditions, inequalities remain". She argues further that "men are more likely to get promotions than women" in the Asian labour force.

In a document focused on *Gender and Rural Development*, it was stated that Asia still faces key gender gaps in the areas of economic empowerment, voice and unequal workloads. Women's voices and influence remain low, as women are expected to take full responsibility for domestic and caregiving activities, which limit their economic opportunities. In another document titled *Gender Equality in East Asia: Progress and the challenges of Economic Growth and Political Change*, it is also stated that "girls educational attainment has seen significant achievements in East Asia, be it in increased literacy rates, enrolment rates or completion rates".

Smith and Byron (2005, p.1) also are of the opinion that there are "circumstances under which increasing women's power may reduce discrimination against girl-children" in Asia. They argue that such powers may be used "altruistically" to "improve the well-being of all their children regardless of their gender". They propose that "women who are better educated and spend more time outside the home may become aware of gender inequalities and subsequently have a growing sense of responsibility for remedying them in their households". They add that "if women have same sex preferences and thus favour their own gender, they may tend to allocate increased resources under the control of girls".

Media and Gender Discrimination

As pointed out earlier, the mass media plays a significant role in promoting gender discrimination and the society in one way or the other is exposed to the mass media contents. The mass media sets the agenda for the audience and has the power to draw attention to the different genders through story slant.

Convention on the Elimination of All Forms of Discriminations against Women (CEDAW) has defined "gender discrimination" as any distinction, exclusion or restriction made on the basis of sex which has the effect or purpose of impairing or nullifying the recognition, enjoyment or exercise by women, irrespective of their marital status, on a basis of equality of man and woman, of human rights and fundamental freedoms in the political economic, social, cultural, civil or any other field.

The Convention on the Elimination of All forms of discrimination against women (CEDAW) was created in 1979 by the UN General Assembly. It was set up to ensure equality between men and women. This they believe will bring about respect for human rights. However, Adedeji and Ajibogun (2013) state that "it has been described as a bill of right for women; as it spells out the area in which women experience discrimination and commits countries to amend their laws, construct national gender policies and create institutions to deliver them". Member nations are expected to follow the following objectives: (United Nation Women, 2009).

- To incorporate the principle of equality of men and women in their legal system, abolish all discriminatory laws and adopt appropriate ones prohibiting discrimination against women;
- To establish tribunals and other public institutions to ensure the effective protection of women against discrimination; and
- To ensure elimination of all acts of discrimination against women by persons, organizations or enterprises.

Gender discrimination has no limit in areas of life. Sylvia and Ogugua (2012) stress that women "are discriminated against in virtually all spheres of life in the home front, in educational institutions, in working place, and in political appointments". "The worrisome aspect of this discrimination is that it has persisted for so long that it is deeply rooted and institutionalized in Nigerian system that uprooting same would definitely pose an uphill task" (Sylvia and Ogugua, 2012). Statistics presented by Gender statistics Newsletter (2011) reveal that women are still lagging behind in occupying equal positions with men in areas such as education. Although there have been recorded growth of women participation in politics, only a few have been able to occupy political seats. Black and Juhn (2000) also attest to the increase of female professionalism. According to them, this increase is as a result of "a demand shift favouring women over men in these highly skilled occupations. They stress further that the "notion of a "gender-specific" demand shift is compelling in the case of high-school graduate men and women who work in very different industries and occupations" (Black & Juhn, 2000 p.450). The rights of women in Nigeria have been trampled on in the society even though according to Nnabuenyi and Otinche (2015 p.33), "Section 42 (1) (2) of the Constitution of Nigeria makes provision for gender equality among male and female children. But as a result of culture and religious beliefs in the country, female children are often deprived of their fundamental rights" and women do not have the same right as their male counterparts.

Consequently, the sustainable development goals on women and equality set by the United nations notes that, "though the world has achieved progress towards gender equality and women's empowerment progress, providing women and girls with equal access to education, health care, decent work, and representation in political and economic decision-making processes will fuel sustainable economies and benefit societies and humanity at large" (United Nations 2014 p.7).

Women in Entrepreneurship

The growth of women entrepreneurship in the recent history of Nigeria has proved that women who are said to be the home keepers naturally possess the ability to manage money-making ventures and multi-task effectively. Adewunmi (n.d) affirms that "women in general are endowed with some exceptional abilities, if which properly harnessed, for entrepreneurial purpose could result in positive and enviable results". This is a quality identified with all women in the world. However, Chinedum, Dephne, Paul, and Sylvia (2009) have identified Nigeria as one of the African countries where women play an increasing important role in entrepreneurial activities. According to Ukonu, and Tafamel (2011, p. 226), "Many women are becoming more educated and the idea that a woman should stay at home, baby sit, cook, go to market, take care of the children and the home is no longer current." Through empirical study Aterido and Hallward-Driemeier (2011) notice that even with the increase in the population of academically qualified women, fewer women are taking leadership positions in firms for which they work when compared to men.

Women entrepreneurship has attracted attention for decades because it creates job for the owners and others, thereby contributing to economic growth. Despite that, Organisation for Economic Co-operation and Development, (2004, p.5) notes that "women still represent the minority in entrepreneurs." Ukonu and Tafamel (2011, p. 230-231) affirm this that, although women are well recognised in informal business sector, they face a lot of restrains when it comes to the formal sector. Food and Agricultural Organisation of the UN (2014, p.1) stresses that "women increasingly run their own enterprises, yet their socio-economic contributions and entrepreneurial potential remain largely unrecognized and untapped." According to the International Labour Organisation (2014), the societal attitude and norms stand as a barrier to most women entrepreneurs especially in moving from a small scale business which is informal to the formal stage. This, the Organisation believes affect the "contribution of women to socio-economic development, job creation and environmental stewardship" (International Labour Organisation, 2014 p.1). This has limited entrepreneurial development in businesses run by Nigerian women. To curb this problem, Aderemi, Ilori, Siyabola, Adegbite and Abereijo (2008, p.174) note that there is need for "special effort and initiatives by the government to remove the socio-cultural bias about the type of business that women or men should engage through the promotion of self-employment and community-based enterprises of women in technological ventures, Government agencies supporting MSSEs (medium and small scale entrepreneur) should also undergo guidance and capacity building on how to create a level playing field for women engaged in small enterprises and government and NGOs should provide and facilitate access to credit for women entrepreneurs, as lack of credit is a major barrier to their innovative capability, performance and growth." With these solutions given by various authors on the issue of women entrepreneurship in Nigeria, there is need for the various mass media to promote women entrepreneurship in Nigeria.

The Role of the Mass Media

The mass media have continued to grow with the emergence of new media. According to a survey carried out in 2010 by Harvard Business Review Analysis Services, "More than three-quarters (79%) of the

2,100 organizations surveyed were either currently using social media channels (58%) or preparing to launch social media initiatives (21%)." New media are a mix of mass media communication. However, the mass media all over the world have the social responsibility of disseminating information and also for the purpose of interaction, amongst other functions. The role of the mass media, according to Moemeka (2000), and Wright (1979), involves surveillance, correlation, socialization and entertainment. This is why they are referred to as the "forth estate of the realm". In the same vein, the mass media has been used in so many areas. With this overwhelming role the media plays in the society, it is inevitable to acknowledge its power and prominent position in the society: the media influences change, shapes the society and enhances situations in any society.

In most African countries, discrimination against women is identified as part of the culture and values of the African people. The information dissemination activities of the media therefore matters in addressing gender inequality as a societal issue. The contribution of women activities and education of the girl child are viewed as helping to aid and enhance development in Nigeria. This purpose is far from being achieved as women still strive and struggle to prove their capability amongst men. Media of mass communication are designed to reach large audience. It is "an array of communication which reaches large numbers of people through radio, television, newspapers and magazine. Ukoha (2013) affirms that "radio, television, movies, videos, videos games, computer network and the Internet have assumed central roles in our daily lives." From 1983 till date research carried out by many scholars has consistently shown that women are either portrayed in advertisements as either sex symbols or housewives. Media of communication are expected to package their messages to suite the values of their audience. This is the reason specific areas of the mass media such as advertisements have suffered criticism from their target audience. Most of these adverts targeted at Nigerian audience portray the female gender in a discriminative light. Children, young people and adults begin to see themselves in such a light. This usually results in lack of confidence and lack of self-drive to move the women forward. According to Deputy Secretary General of the United Nations Asha-Rose Migiro, "mass media sways how people view what it means to be masculine and feminine, sadly, often resorting to destructive gender stereotypes in the process," such stereotyping, in turn, feeds into the gender discrimination which is a root cause of violence against women and girls, mass media should put an end to its one-dimensional gender portrayal and its negative depictions of women's role in society."

Gender Discrimination in Present Time

The issue of gender discrimination is still stands as a concern to the society. This is with regards to the argument of many scholars specialized in the area of study. According to Ifemeje and Ikepezer (2012, p.51), Gender discrimination "existed from the dawn of civilization and has continued over centuries." Also "previous studies in Nigeria show that gender discrimination against men and women based on their ethnic, religious and geographical identity is common" (Gberevbie, Osibanjo, Adeniji, & Oludayo 2014, p.98). Consequently, scholars have looked into different forms of gender discrimination in different aspects of life. Owoyemi and Olusanya (2014, p.10) looked into gender discrimination in the workplace and discovered that "despite all the initiatives aimed at eradicating the discrimination of women in paid employment, evidence is still showing that these constitutions theoretical commitments and acknowledgment of women's crucial roles remained violated and perniciously undermined, therefore more awareness will help address the problem." Further empirical study such as Gender bias in Legal profession by Levinson and Young (2010, p.1) shows first "that the power of implicit gender biases persists, even in the next generation of lawyers; and second, that the emergence of a new generation of egalitarian law students may offer some hope for the future."

However, a further study on "gender discrimination and employee performance among staff" by Gberevbie, Osibanjo, Adeniji, and Oludayo (2014, p.102) finds that "discrimination does exist though in mild and subtle manner. Put differently, gender stereotyping in the society has relegated female children into second fiddle in the society, and also the behaviour of male counterparts does not help in any way." Having dwelled on different aspects of gender discrimination, scholars have proffered various solutions to the problems gender discrimination is likely to cause the society. While commending international organisations for laying emphasis on the subject of gender discrimination, Brummet (2008, p.20) discovers from his study that "underinvestment in women's education has a negative effect on growth" and there is need to carry out more research in the area as it will be of great benefit to the society in terms of societal development. To achieve this development through gender equality, Njogu and Orchardson-Mazrui (n.d p.18) believe that "it is worth reconfiguring the negative aspects of culture, deconstructing and reconstructing them so that they can be more enhancing of humanity." In simpler terms, the unfavourable aspects of the African culture for economic development have to be reviewed or modified to sustain economic development.

According to Ifemeje and Ikpezer (2015, p.51), "gender based discrimination is an all pervasive while identifying that although in the present age the law has given women a platform to participate in all aspects of life where they were not previously permitted to thrive, "women are hardly heard in decision making process but are always used as instrument of voting in political palace and child bearing and rearing of socio-cultural milieu" (Ayeni & Ajibogun, 2013 p.117). However, Abara (2012, p.4) stresses that solution to gender discrimination is still out of reach because "even though the 1999 constitution of Nigeria prohibits discrimination on the ground of gender, the customary and religious laws continue to restrict women's right". Explaining the barriers to achieving the solutions proffered by most scholars who have addressed the issue of gender discrimination, Abara (2012, p.4), notes further that "the combination of federation and a tripartite system of civil, customary and religious law makes it very difficult to harmonize legislation and remove discrimination measures".

METHODOLOGY

The study used primary and secondary data as required to determine the promotion of women entrepreneur in the mass media. The survey technique was used. According to Sobowale (2008, p.27), the survey technique "involves drawing up a set of questions on various aspects of a subject to which selected members of a population are requested to react." A total of 200 respondents were selected through the random sampling method from 3 Local Government Areas in Lagos, namely Abuleade, Surulere, and Amuwo-Odofin. These Local Government Areas were randomly selected. Questionnaire was administered as a survey tool in 2015 to ascertain gender issues in entrepreneurship.

Findings

Table 1 shows 46% of the respondents agree to the fact that women entrepreneurship has grown in Nigeria. This growth draws attention to the enabling environment that is now beginning to identify with the existence or presence of women in the society, no matter their areas of interest. Consequently, the

Table 1. Women entrepreneurial growth in Nigeria

Distribution	Percent (%)
To a very large extent	20%
To a large extent	34%
To a moderate extent	30%
To a small extent	11%
To an extremely small extent	5%
Total	100% n=200

Table 2. Provision of enabling environment for women in entrepreneurship

Distribution	Percent (%)
Yes	46%
No	44%
I don't know	10%
Total	100% n=200

female gender were said to have shown more interest in growth, self-development and change as they have started getting more involved in entrepreneurship in recent times to support the family.

Almost half of the respondents agreed that Nigerian women are provided with the enabling environment to excel just as their male counterparts. However, almost the same percentage of the respondents also disagreed that the women were still far behind in terms of opportunities to compete. Although it is just a 2% difference, it could be deduced that the opportunity for women to grow their business is increasing although slowly.

Table 3 shows that the television and the Internet are the most frequently used media when it comes to promoting women entrepreneurship in Nigeria. The Internet topped as the most used medium which could be of its high rate of penetration. It is seconded by the television whose audio-visual power has the ability to attract and retain the attention of its audience. However, it must be noted that in as much as the Internet platform is a mass medium, its contents are not provided only by media personnel but also by business persons who are seeking recognition for their brands in other to make sales or provide services and increase customer base.

Table 4 shows that the media promotes mostly entertainment programmes and advertisements. The table further indicates that the media concentrates a lot on entertainment as compared to other media categories. The Nigerian society is becoming interested in entertainment contents. As a result of this,

Table 3. Medium of communication mostly used in promoting women entrepreneurship

Distribution	Percent (%)
Television	34%
Radio	6%
Internet	38%
Newspapers	4%
None	3%
Other	15%
Total	100 n=200

Table 4. Categories of content most promoted by the mass media

Distribution	Percent (%)
Education	8%
Entertainment	33%
Current Affairs	17%
Advertisements	21%
Religion	4%
Sports	6%
Entrepreneur	11%
Total	100 n=200

organisations and product manufacturers are ready to sponsor entertainment programmes to create awareness for their products and services. This form of sponsorship and advert placements was said to have in turn helped to sustain media organisations.

DISCUSSION

Gender discrimination originates from societal cultures. These cultural heritages have been known to affect various aspects of the life of the woman in particular, especially, in African societies. As women in Nigeria continue to engage themselves more in entrepreneurial activities, which could be a competitive environment, there is need for the society to look at women beyond what culture prescribes in order to drive economic growth for development. To achieve this, the society needs to assess the effectiveness of both male and female gender equally in business and provide equal enabling environment for their businesses to grow.

The result of the research shows that there is a little level of growth in providing enabling environment for women as compared with men in growing their businesses. However, this growth is very small, 2%, as the survey revealed. These are concern indicators that probably in some parts of the society or businesses, women are not trusted enough to handle businesses at a higher level. It could also be deduced that changes in this area of discrimination is not far-fetched. That change could create a privileged platform for the women folk to be encouraged to take up high entrepreneurship roles if given the opportunity and the support irrespective of their gender.

Furthermore, the Internet has proved to be an effective medium in promoting entrepreneurial activities because of its wide spread and opportunity to sell and promote their brands conveniently. Consequently, it can be argued that the promotion of women entrepreneurship that exists on this platform is driven not by the media itself but by women entrepreneurs. This is because the Internet is not solely controlled by media personnel as the medium provides opportunity for personal and organisational use.

It has become common place through diverse research in present times that the mass media generally have concentrated on featuring entertainment content. Brown (1992 p.253) confirms that "audience exposure to entertainment television programs worldwide is rapidly increasing, and several nations are infusing pro-social messages within the content of entertainment television programmes to promote development." This dominant media function is changing media operations mostly in a negative light. Consequently, other media operators especially those in the management level believe that it is only through the platform of entertainment and advertisements that the media organisations can survive.

However, with this dominance, the television medium as found in this research ranks high in promoting women entrepreneurs in Nigeria. It is effective in promoting women entrepreneur because of its visual power. The finding that the rate at which women entrepreneurs in Nigeria are being promoted is slow and could still be challenging obviously means that the reality and acceptance of women in the business world is still slow. This is likely to affect the economy of the country, especially considering the down turn of the Nigerian economy at the moment. This could further lead to backwardness and limit the nation in competing with other industrialized nations. Also, this form of slow acceptance of women in the business world could affect foreign investment negatively. There is need to keep educating the public especially the media to give opportunities to both genders in the world of business for economic involvement and politics leading to development and nation building. As opposed to Curran, Gurevitch and Woollacott (2005) who believe that the media has limited influence on the culture and socio-economic dynamics of a country, this study shows that there is so much the media can do in promoting a way of life, and in changing the orientations of members of a particular culture especially in relation to women entrepreneurship.

FUTURE RESEARCH DIRECTION

A further research on gender discrimination can focus on comparing gender discrimination as portrayed by the media in different continents as it affects development. This will establish if any relationship exists between gender discrimination and underdevelopment. The study can do a comparison of gender discrimination or gender representation in diverse mass media. This will help know how both genders are represented by the mass media and the societal implications of such media presentation. Also, a discourse analysis can be done in determining how live TV show presenters portray genders through their presentations. Another research can analyse the present effects of societal gender discrimination on the mentality of adolescents.

CONCLUSION

The mass media has as part of its roles educating and teaching people. This is the reason the society looks up to the media for information. Media programmes are very influential as they have high effect on the people. Media contents however do reflect the culture of the environments. These are cultures whose view of women sometimes makes it difficult for women to aspire to a high level or for the world to embrace their promotion to a high level.

Although in the recent times, there is a notable increase in the promotion of businesses run by women especially with the use of the television medium and the Internet. However, most of these businesses promoted through these media listed above are paid for either through the method of trade by barter adopted by a few of these media establishments or monetarily.

RECOMMENDATIONS

There is the need for media and other non-governmental organisations to seek ways of promoting the activities of businessmen and women as the mass media still has a strong agenda setting ability. Gupta, and Sinha (2010) confirm that there is a growing reliance on the mass media as main source of information. Media contents, especially that concern women, do reinforce cultural bias against women. This act has further depicted and emphasised how the society sees women. This could enhance gender inequality leading to discrimination. Care should therefore be taken especially by media practitioners and presenters. The danger in this is that young girls begin to grow with the idea that their fellow male counterparts are not equal to them. There is need for the mass media to begin to present educational messages in a neutral form that will not depict women as weak and incapable of achieving what we refer to as "big goals". In Nigerian, it is clear that women are great achievers and have the ability to affect the society in a positive way if given the opportunity. A good example is in the Nigerian political system. Though it is believed that the male candidates are given more opportunity than the female counterpart during election campaign, researchers do acknowledge the fact that more women are now participating in politics.

Women's activities in business should be promoted by the media and more opportunities be given for them to grow to the top. Rather than media organisations seeking financial gain, stories that concern women's business growth should be promoted on their medium without charges as their activities also result in absolute development as well as their male counterparts'.

REFERENCES

Abara, C. J. (2012). *Inequality and discrimination in Nigeria tradition and religion as negative factors affecting gender.* Paper presented at the Federation of International Human Rights Museums.

Adedeji, A., & Ajibogun, G. (2012). Gender issues and democracy: The Nigeria Experience. *International Journal of Academic Research in Progressive Education and Development., 2*(1), 117–127.

Adenle, G., & Oso, L. (2014). *Muted gender: How Nigeria press portrays female politicians. In Women's political visibility and media access: The case of Nigeria.* Cambridge Scholar.

Aderemi, H. O., Ilori, M. O., Siyanbola, W. O., Adegbite, S. A., & Aberiojo, I. O. (2008). A an assessment of the choice and performance of women entrepreneurs in technological and non-technological enterprises in Southwest Nigeria. *African Journal of Business Management., 2*(10), 165–176.

Adewunmi, A. (n.d.). *Nigerian women Agro-entrepreneurship development: Issues and challenges.* Retrieved on 2 August, 2015 from http://peda.cc/nigerian-women-agro-entrepreneurship-development-issues-and-challenges/

Agbalajobi, D., (2010) Women's participation and the political process in Nigeria: Problems and prospects. *African Journal of Political Science and International Relations, 4*(2), 75-82.

Ajala, V. (2014). *Exploring the indifference of women and minorities in Nigeria's politics. In Women's political visibility and media access: The case of Nigeria.* Cambridge Scholar.

Anaeme, F. (2012). Reducing gender discrimination and violence against women through library and information service. *Library philosophy and practice.* Retrieved on 12 January, 2015 from https://www.researchgate.net/profile/Francis_Anaeme/publication/267264193

Aririguzoh, S. (2005). The function and structure of communication in entrepreneurship. *Nssuka Journal of the Humanities*, (15), 160-178.

Aterido, R., & Hallward-Driemeier, M. (2011). Whose business is it anyway? *Small Business Economics, Springer, 37*(4), 443–464. doi:10.1007/s11187-011-9375-y

Ayeni, M. D., & Ajibogun, A. G. (2013). Gender issues and democracy: The Nigeria Experience. *International Journal of Academic Research in Progressive Education and Development., 2*(1).

Baden and Reeves. (2000). *Gender and development: concepts and definitions.* Bridge (the development-gender) Report No 55. Retrieved on 27 May, 2015 from www.bridge.ids.ac.uk/report

Balatchandirane, G. (2007). Gender discrimination in education and economic Development: A study of Asia. *International Studies*, *40*(4), 349–378. doi:10.1177/002088170304000403

Baran, S. (2008). *Introduction to Mass communication. In Media literacy and culture* (5th ed.). London: McGrill Hill.

Baren, S. J., & Davis, D. K. (2003). *Communication theory. In Foundation, ferment, and future* (3rd ed.). Wadsworth.

Black, S., & Juhn, C. (2000). The rise of female professionals: Are women responding to skill demand? American Economic Association.

Brummet, Q. (2008). The effect of Gender Inequality on growth: A cross-country Empirical Study. *The Park Place Economist, 16*(1), 13-23.

Ceulemans, M., & Fauconnier, G. (1979). Mass media: The Image, role and Social conditions of women. A collection and analysis of research materials. *Global Media Journal.*

Chinedum, N., Dephne, H., Paul, T., & Sylvia, C. (2009). African Journal of Business and Economic Research. *Business and Social Profiles of Female Immigrant Entrepreneurs in Nigeria*, *4*(2-3), 85–92.

Das, J. K. (2003). *Education broadcast through radio*. Retrieved on 21 July, 2015 from http://www.aiaer.net/ejournal/vol19107/8.htm

Drakapoulos, S. A., Theodossiou, I., & Dodd, S. (2008). Gender Discrimination and Institutional Frameworks: Evidence from Four European Union Countries. *International Journal of Economic Sciences and Applied Research*, *1*(1), 6–36.

Du toit, L. (2005). A Phenomenology of Rape: Forging a new vocabulary for action. In *(Un) thinking Citizenship*. Burlington, VT: Ashgate Publishing.

Ekundayo, A., & Ama, B. (2014). Nigerian women and political participation: The way forward. *International Journal of Educational Foundations and Management*, *2*(1), 71–81.

Esidene, E. C., & Abdul, S. U. (2013). The role of women politics: Interrogating the gender question for an enhanced political representation in the fourth republic. *Afro Asian Journal of Social Sciences.*, *4*(4), 1–17.

Esteve-Volart, B. (2004). *Gender discrimination and growth: Theory and evidence from India.* Retrieved on 10th March, 2015 from http://eprints.lse.ac.uk/6641/1/Gender_Discrimination_and_Growth_Theory_and_Evidence_from_India.pdf

Ezeocha, J., & Mojaye, E. (2014). Mass Media influence on discrimination and violence against women in Delta State, Nigeria. *Journal of Communication and Media Research*, *3*(1), 75–92.

Food and Agricultural Organisation of UN. (2014). The State of food and agriculture. *Innovation in Family Farming,* 95-139.

Gberevbie, D., Osibanjo, A., Adeniji, A., & Oludayo, O. (2014). An Empirical performance among Academic staff of government University in Lagos State, Nigeria. *International Journal of Social, Education, Economics and Management Engineering, 8*(1). Retrieved on 10th March, 2015 from http://www.google.com.ng/url?sa=

Gupta, A., & Sinha, K. A. (2010). Health coverage in mass media: A content analysis. *Journal of Communication, 1*(1), 9–25. Retrieved from www. krepublishers.com/02-Journals/JC/JC-

International Labor Office. (n.d.). *Discrimination at work in Africa*. Retrieved on 10th March, 2014 from http://www.ilo.org/wcmsp5/groups/public/@ed_norm/@declaration/documents/publication/wcms_decl_fs_89_en.pdf

Kaul, V. (2013). Journalism in the Age of Digital Technology. *Journal of Communication and Media Technologies., 3*(1), 125–143.

Kufin, P., & Sinha, K. A. (2013). Gender discrimination in job Ads: Evidence from China. *The Quarterly Journal of Economics, 128*(1), 287–336. doi:10.1093/qje/qjs046

Labani, S., Zabaleta, C., & Ruiz, L. (2010). *Gender analysis of women's political participation in seven south-east Asian countries: Bangladesh, Cambodia, the Philippines, Indonesia, Sri-Lanka, East Timor and Vietnam (2008-2009)*. Retrieved on 10th March, 2015 from https://en.wikipedia.org/wiki/Gender_inequality_in_Bangladesh

Levinson, J., & Young, D. (2010). Implicit Gender Bais in the Legal Profession: An Empirical study. *Duke Journal of Gender Law & Policy, 18*(1), 1–9.

McQuail, D. (1998). *Mass Communication Theory. An Introduction* (3rd ed.). London: Sage.

Moemeka, A. A. (2000). Reporter's handbook: An Introduction to effective journalism. Morris Publication.

Motsaathebe, G. (2009). Gendered roles, images and behavioural patterns in the soap opera generations. *Journal of Africa Media Studies, 1*(3), 429-435.

National Coalition of Affirmative Action. (2009). *National gender policy*. Retrieved on 7, March, 2015 from www.aacoalition.org/national-policy-women

Nigeria Masterweb Citizen News. (2012). *Gender inequality: The Nigerian case*. Retrieved on 2 March, 2015 from www.nigeriamasterweb.com

Nnabuenyi, U. M., & Otinche, S. I. (2015). Culture and the right of the rural women in Nigeria: an overview. *International Journal of Basic, Applied and Innovative Research, 4*(2), 30-43.

Obasola, K. E. (2013). Gender discrimination and cultural Anachronisms in African Societies: Implication for Girl Child Education. *International Journal of Social Science and Education, 3*(4), 1027–1034.

Ojalammi, S. (2011). *Gender and development in Sub-Saharan Africa*. Retrieved on 10th March, 2015 from http://www.bing.com/search?FORM=SK216DF&PC=SK216&q=Ojalammi%2C+S.+(2011).Gender+and+development+in+Sub-Saharan+Africa

Okaka, W. (2014) *Effectiveness of ICT for Gender Equality and Women empowerment in small Island Developing States (SIDS).* Third International Conference on small Island Developing States Stakeholders Forum. Retrieved from www.researchgate.net

Okpe, O. (2005). *Mainstreaming Gender in the African Development Process: a Critic of NEPAD and the Women Question*. Nigeria: BSU Press.

Organisation for Economic Co-Operation and Development. (n.d.). *Women's entrepreneurship: issues and policies.* Retrieved from http://www.oecd.org/cfe/smes/31919215.pdf

Owoyemi, O., & Olusanya, O. (2014). Gender: A precursor for discriminating against women in paid employment in Nigeria. *American Journal of Business and Management, 3*(1).

Price, R. (2011). *What is the history of entrepreneurship?* Retrieved from www.blog gcase.org/2011/02/04/what-is-the-history-of-entrepreneurship

Reeves, H., & Baden, S. (2000). *Gender and development concepts and definitions.* Bridge Institution of Development Studies. Retrieved from http://www.bridge.ids.ac.uk/reports/re55.pdf

Sarkar, S. (2015). Gender Stereotyping in Product Promotion: A Dominant Tool of Advertising in Indian Media. *International Journal of Media Journalism and Mass Communications, 1*(2), 1–12.

Sekhar, T., & Hatti, N. (2005). *Discrimination of Female Children in Modern India: From conception through childhood.* Retrieved on 10th March, 2015 from /www.researchgate.net/publication/215875262_Disappearing_Daughters_and_Intensification_of_Gender_Bias_Evidences_from_Two_Village_Studies_in_South_India

Smith, L., & Byron, E. (2005). *Is greater decision making power of women associated with reduced gender discrimination in South Asia?* Retrieved on 10th March, 2014 from http://link.springer.com/article/10.1007/s10680-013-9288-3

Sobowale, I. (2008). *Scientific Journalism* (2nd ed.). Lagos: Idosan Consult.

Tuwor, T., & Sossoh, M. A. (2008). Gender discrimination and education in West Africa: Strategies for maintaining girls in school. *International Journal of Inclusive Education, 12*(4), 363–379. doi:10.1080/13603110601183115

Ugwulebo, B. (2011). Female gender in professional education. *Marketing and Management of Professional Survey Education.* Retrieved from www.fig.net/pub/fig2011

Ukoha, K. E. (2013). *Media violence and violent behaviour of Nigeria youth: Intervention strategies.* Nigeria Educational Research and Development Council. Retrieved from www.web.a.ebscohost.com

Ukonu, O. I., & Tafamel, A. E. (2011). Problems, challenges and prospects of female Entrepreneurs in Gwagwalada, Abuja. *Africa Research Review, 5*(3), 226–246. doi:10.4314/afrrev.v5i3.67354

United Nations. (2014). *Equality for women in progress for all women watch.* Retrieved on 5th March, 2015 from http://www.un.org/womenwatch/future/iwd

Wood, J. (1994) Gender Media: The influence of media on views of gender. In *Communication, gender and gender.* Thomson Learning.

World Bank. (2005). *Gender –responsive social analysis: A guidance note. Incorporating social dimension into bank-supported projects.* Retrieved form http://web.worldbank.org/archive/website01028/WEB/IMAGES/GENDERGU.PD

Worthington, N. (2011). Gender discourse and Ubuntu Media Philosophy. *Journalism Studies*, *12*(5), 608–623. doi:10.1080/1461670X.2010.545205

Wright, C. R. (1979). Sociology of mass communication. *Annual Review of Sociology*, *5*(1), 193–247. doi:10.1146/annurev.so.05.080179.001205

Zoonen, L. (1991). Feminist perspectives on the media. In J. Curran & M. Gurevitch (Eds.), *Mass Media and society*. London: Arnold Ltd.

KEY TERMS AND DEFINITIONS

Discrimination: A situation whereby, a segment of people in a society is treated unjustly compared to others in the same society because of their colour, age, class, sex, ethnicity, etc.

Entrepreneur: A person or group of persons who engages in a business for the purpose of serving customers and making profit.

Framing: The act of drawing specific meaning or interpretation through written or spoken contents.

Gender Inequality: A situation where a male or female is treaded unjustly because of his or her gender.

Gender: A biological identification of the state of being a male or a female.

Mass Media: Technological device or printed documents used in the delivering of information about societal issues and happenings to a large heterogeneous audience at the same time in different places.

Women Empowerment: A form of freedom given to women in a society, to freely contribute to the development of themselves and the society in which they belong.

Section 3

Issues in African Mass Media

Discourse on the prevalent issues affecting the African mass media.

Chapter 11
Nigerian Government's Attempts at Removing Fuel Subsidy and the Attendant Mass Protests:
A Content Analysis of National Newspapers' Coverage

Adeyemi Aderogba
Osun State University, Nigeria

Olanipekun Gbenga
Straightgate Publishers, Nigeria

ABSTRACT

The Federal Government of Nigeria made several unsuccessful attempts at stopping petrol subsidy by arguing that the policy is unsustainable. Seven unsuccessful attempts were made between 1999, when the country returned to democratic rule, and 2012. Though the government might have good reasons for removing petrol subsidy, it was clear that there was an obvious disconnection between the government and the people as the people were not in support of the policy. This chapter examined coverage of petrol subsidy removal by three national dailies. Content analysis was adopted as the research method. It was discovered that 49.4% of the stories published by the selected newspapers in the study period were not in favour of petrol subsidy removal. It was therefore recommended that government should endeavour to carry Nigerian citizens along ever before introducing such kind of policy and that media practitioners should set an agenda for fighting corruption in the oil industry.

DOI: 10.4018/978-1-5225-1859-4.ch011

INTRODUCTION

Nigeria is reputed as the world's eighth largest producer of crude oil and sixth largest exporter of same. Petroleum remains the backbone of the Nigerian economy. The country derives 80% of her revenue as well as 90% of her foreign earnings from petroleum with obvious consequences. The country is also the largest nation on the continent of Africa with about 170 million people. According to Falola and Heaton (2008:224), "the success of the country has been undermined in recent decades by ethnic and religious conflict, political instability, rampant official corruption and an ailing economy." With the drastic fall in oil prices, subsidy removal have resurfaced on pages of the national dallies.

Oil subsidy or petrol subsidy is perhaps one of the most contentious issues in Nigerian history. It is an issue that creates misunderstanding between successive leaders and the led, as the leaders made series of efforts to deregulate the downstream sector of the petroleum industry during the last three decades. Each time such an effort was made, people vehemently resisted it to the point of mass protests.

Nigerians have witnessed changes in pump price of petrol 26 times since 1973. Nineteen of those changes were hike in price, while the country experienced price reductions seven times within the same period. The latest price adjustment was announced by the Federal Government on May 11, 2016. The price was moved from ₦86.50 to ₦145, that is 67.6 percent increase in price.

Nigerian masses are very suspicious of any hike in the price of petrol as increase in the price of the product would automatically translate to increase in prices of other goods and services. Also, a huge chunk of the population lives in abject poverty. Odusola (1997:121) asserts that:

Sub-Saharan African countries are noted for prevalence of absolute poverty in all its characteristic features. Poverty in these countries is massive, pervasive and chronic, engulfing a large proportion of the society.

In Nigeria, seventy percent of the population lives in abject poverty. Onyeizugbe and Onwuka (2012) believe that petrol subsidy removal would seriously worsen the economic situation of the country's poor masses. They affirm that:Many Nigerians see the subsidy, which gives them the cheapest petrol price in the region, as the only benefit of being an oil producing country. Most live in grinding poverty and on less than $2 a day. There is little infrastructure, high unemployment and only intermittent electric power…

In all of these increments, the media, notably the newspapers, played a significant and salient role in portraying or reporting the struggle between the ruled and the rulers.

What is Subsidy?

There are various descriptions of subsidy by experts. Momoh, in Nwosu (1996:81), enunciates that:

A subsidy is any public money or other resources out of the wealth of the entire nation which government makes available to any enterprise it owns for the purpose of enabling it to sell any of its goods or services or both at any price below what it would have sold it if its aim was to make any profit at all.

Organization for Economic Co-operation and Development (OECD, 2006:7) describes subsidies as: "*a* measure that keeps prices for consumers below market levels, or keeps prices for producers above market levels or that reduces costs for both producers and consumers by giving direct or indirect support".

To Rolph Earl, in Robinson (1967:86), subsidy is described "as transfer payments because by definition they refer to payments by government for which the government receives no product, service or asset…"

Berge (1951) is of the opinion that: "subsidy is a gift; a one-way transactions, without provision for payment." Subsidies could come in the form of grants and other direct payments. Based on this fact, tax concessions, in-kind subsidies, cross subsidies, credit subsidies and government guarantees, hybrid subsidies, subsidies through government procurement, derivates- subsidies and market price support are all identified forms of subsidies (Steenblik nd: 18-26).

The summary of these descriptions and definitions is that oil subsidy comes under government grants or direct payments. It is a situation whereby government decides to shoulder the responsibility of paying certain percentage of the cost of the pump-price of petrol for every buyer, be it rich or poor, male or female, young or old.

How Sustainable is Petrol-Subsidy?

Supporters of petrol-subsidy removal argue that oil subsidy is unsustainable because subsidy bills are ever-growing and could have adverse effects on other sectors of the economy. Ascher (1999:406-407) draws attention to "…the absurd mismanagement of Nigerian oil…, the under-pricing of petroleum in Nigeria which does not always benefit consumers." He notes:

Petroleum under-pricing, coupled with deliberate neglect of refineries, created shortages and smuggling opportunities that brought scarcity rents to those in the military that controlled oil flows.

Nigeria is a poor nation though with great potentials. Nwachukwu (2012) affirms that global finance ranking of the richest or poorest countries of the world confirms that Nigeria is the 43rd poorest nation in the world with a GDP purchasing-power parity per head of \$2,358. The public affairs analyst believes that Nigeria requires all the money it could get to provide needed infrastructural facilities that are grossly inadequate in the country.

Corruption is one of the major vices attributed to petrol subsidy and total collapse of local refineries. Nwachukwu (2012) observes that petrol subsidy makes local refineries unprofitable for investors and that subsidy is the reason for incessant fuel shortages at fuel stations across the country as corrupt and unscrupulous petrol-marketers usually divert subsidized fuel to neighbouring countries for higher profit. Okogu (1995:385) supports this claim. The expert believes that:

The neighbouring countries of Chad and Cameroon have much higher fuel prices than Nigeria, and the brisk smuggling activities across the borders with these countries testify to this. In fact, the Cameroonian refinery at Cap Limboh is closed down from time to time because of cheap, surplus supplies of smuggled Nigerian products.

Ben Nanaghan supported these arguments in a newspaper article published in the *Punch* of January (2012, 1, p. 6). He pleaded with Nigerians to ignore the hardship petrol removal had caused them in the interim; he believed that the problem was not peculiar to Nigeria. According to him, Singapore went through this same stage in the 1960's when that country's foremost leader, Late Lee Khuan Yew, who ruled Singapore from 1959 to 1990 as Prime Minister, took over the leadership mantle of Singapore and through his transformation agenda transformed the country to what it is today - a pride of the whole world.

Why Should Petrol-Subsidy Policy be Retained?

Antagonists of petrol-subsidy removal are of the opinion that removal of subsidy would rather kill Nigerians than build them. They strongly believe that what the government needs to do is to kill corruption and not the people. They are of the opinion that corruption is a major issue in Nigeria and it must be tackled headlong (James, Lawrence and Weber, 1999); Ekpo, Chime and Enor, 2016). "Kill corruption and not Nigeria", this was a popular inscription on some of the protesters' placards during the 2012 nation-wide strike. Opposition politicians, human right activists, the academia, students, artistes and other groups gathered to protest during the strike.

In an article titled "The President breaks social contract with the people", one of the national leaders of an opposition political party at the time, Senator Bola Tinubu, accused President Jonathan of betraying people's hope. According to the politician, "the very first thing this government needs to do is to clean up and throw away the salad of corruption in the Nigerian National Petroleum Corporation (NNPC)" *(The Punch*, 12th of January, 2012 pg 18).

So many people believe that Nigerian National Petroleum Corporation (NNPC) is heavily corrupt. The organization has been accused of corrupt practices such as non-remittance of huge revenue to the federation's account. The figure was initially put at ₦450billion but the Nigerian Extractive Industries Transparency Initiative (NEITI) suggested that the figure could be as high as ₦842 billion, an allegation which the NNPC debunked (see the *Guardian* Editorial of 5th January, 2012 pg 16).

However, findings of the Federal Legislators' Ad Hoc Committee on petrol subsidy during President Goodluck Jonathan's regime revealed a rot in the NNPC as quite a number of oil marketing firms and some individuals got paid for petrol they never imported to the country in the first instance. A number of companies and people were indicted. But Nigerians waited in vain to see justice done against corrupt people that diverted proceeds from the petrol subsidy remittal into their private bank accounts. President Muhammadu Buhari's administration has made a number of implicating discoveries against the NNPC. In an article titled "What next for NNPC after board dissolution?", Emeka Ugwuanyi listed some of the corruption allegations against the corporation as "sharp practices such as oil swap, offshore processing agreement, violation of regulations, illegal deductions and non-remittance of funds…by withholding ₦3.8trillion from the ₦8.1trillion it generated in the past three years" (see *The Nation*, 1st July, 2015 pg 2-3).

Late Chinua Achebe, a world-renowned novelist, advised government to seriously consider what he called "scale of corruption in Nigeria". According to the writer, the country looses at least $10 billion to corruption annually. He believed tackling corruption should be the focus of the present President Jonathan's administration. *(The Punch*, 12th January, 2012 pg 3).

Apart from corruption, the manufacturers have concerns for huge cost of generating electricity at their respective factories for production processes. They lamented that the sector spends over N2bn a week on alternative electricity source before the removal of fuel subsidy on January 1, 2012. According to Roseline Okere – a public affairs analyst, Small and Medium Enterprises (SMEs) were the worse hit. (See *The Guardian*, 4th January, 2012, pg. 44).

Is Petrol Really Being Subsidized by the Nigerian Government?

Another school of thought is of the opinion that Nigerian government is not in anyway subsidizing petrol. In an article titled "The Real Cost of Nigeria Petrol", Agbon Izielan, a former head of Department of Petroleum Engineering, University of Ibadan, argued that there was no petrol subsidy in Nigeria. He claimed that the reverse was the case. According to him, there was high sales tax of 91.2 percent when petrol sold for N65 per litre before the January 2012 hike in price. The university don further explained:

We have about 170,000 barrels of refined products for about 13.26 million litres of petrol, 6.8 million litres of diesel and 2.72 million litres of kerosene/ jet fuel: This is not enough to meet internal national demand. So we send the remaining of our non-export crude oil volume (270,000 barrels per day) to be refined abroad and import the petroleum product back into the country. We will just pay for shipping and refining.

Agbon compared the price of petrol in the USA with that of Nigeria against the background that one barrel of crude oil is 42 gallons or 168 litres of crude oil, which when it is refined will give 45 gallons of petroleum products consisting of 4 gallons of LPG, 19.5 gallons of gasoline, 10 gallons of diesel, 4 gallons of jet fuel/Kerosene, 2.5 gallons of fuel oil and 5 gallons of bottoms. The researcher put the cost of 1 litre anywhere in Nigeria at N33.36 per litre as against N65 in 2011 or N87 in 2015. This calculation is based on the fact that the government earmarked 445,000 barrels per day for meeting domestic refinery needs. But Agbon believed that government ought to have based its calculation on the price of swapped oil and not on the price of petrol in the international market. He concluded, just like the former Petroleum Minister in General Ibrahim Babangida's administration, Prof. Tam David West, that there was no petrol subsidy in Nigeria because the actual price of a litre of swapped petrol at the time was N34.45. The price of a litre of petrol at N65 before the 1st of January, 2012 price hike was already too high.

History of Petrol Subsidy Removal in Nigeria

Subsidy on petrol became government policy in Nigeria in 1973, when government made the first attempt at regulating fuel price. General Yakubu Gowon became the first Nigerian leader to increase the pump price of petrol in Nigeria by jacking it up from 6K to 8.45K. In 1976, General Murtala Mohammed adjusted the price to 9K. In 1978, General Olusegun Obasanjo adjusted the price from 9K to 15.3K.

General Obasanjo's successor, President Shehu Shagari, moved the price to 20K (25 cents), while General Ibrahim Babangida raised it to 39.5kobo on 31 March, 1986 (the currency exchange market was also liberalized during the same period and the Naira was allowed to float freely against major world currencies). On 10 April 1988, the Babangida administration increased the price again to 42K per litre, the price was to be further increased to 60K for private cars on 1 January, 1989. This latest increase was the first in the nation's history to instigate nation-wide protest. General Babangida jacked up the price again on 6 March, 1991 to 70K. He had the intention of further price hike but he exercised restraint because of the political situation in the country. Okogu (nd: 388) asserts that:

In 1993, many attempts were made to raise the price, but the government's timing was clumsy at best and mischievous at worst. The price was supposed to be raised in May 1993 — just weeks before the Presidential elections — when it was clear that such action at that time would almost certainly lead to social unrest. In August, just days before leaving office, the former military government announced new prices for petroleum products.

The then newly appointed Ernest Shonekan's Interim National Government raised the price to ₦5 on 8th November, 1993, a move that caused mass outrage against the illegitimate government and eventually led to its fall. General Abacha that took over from Mr. Shonekan calmed people's nerves by reducing the price to ₦3.25 kobo on 22 November, 1993, only to increase it to ₦15.00 on 2 October, 1994 and reduce it to ₦11.00 two days later. After the demise of General Abacha, it was the turn of General Abdusalami Abubakar to rule. He increased the price on 2 December, 1998 to ₦25.00 before reducing it to ₦20.00 on 6 January, 1999 (The *Punch* of 2nd of January 2012 page 9).

Right from 1973, petrol prices have either been increased or decreased 26 times by successive government. The prices were increased 19 times and reduced five times. Right from 1989 when General Babangida increased the price and the very first nation-wide protest was recorded, Nigerian people have responded violently to upward petrol price adjustment. Former President Jonathan announced a change in pump price of petrol on 1st January, 2012 to ₦141.00 per litre, but the nation-wide strike that ensued and the fear that it could lead to total breakdown of law and order in the country forced the President to revert to ₦97.00 per litre. Fall in the price of crude oil in the international market coupled with 2015 general elections that were around the corner then forced former President Jonathan to reduce the pump price to ₦87.

However, subsidy on energy is not peculiar to Nigeria because so many nations around the globe are having difficulties in tackling the problem as a result of political considerations (Widodo et al, 2012; A Guidebook to Fossil-fuel Subsidy Reform, 2013).

The objectives of the study were as follows:

1. To examine the prominence of reportage of petrol price hike in Nigerian newspapers during the period under study.
2. To determine the shade of opinion of reportage of petrol price hike by Nigerian newspapers.
3. To identify the predominant sources of news reportage on petrol price hike in Nigeria during the study period.
4. To examine major themes of news reports on petrol price hike.

Research Questions

A number of research questions were formulated in examining major issues highlighted in the research objectives above. The following research questions were therefore treated in the study:

1. How prominent was reportage of petrol price hike in Nigerian newspapers during the period under study?
2. What were the shades of opinion in the reportage of petrol price hike by Nigerian newspapers?
3. Which were the predominant sources of news reportage of petrol price hike in Nigeria during the study period?
4. What were the major themes of news reports on petrol price hike?

Table 1. Showing the number of time petrol prices were adjusted in Nigeria since 1973

S/N	Date	Administration	Price			Percentage Change	Remark
			Old Price	New Price	Amount Settled for		
1	1973	Gowon	6K	8.5K	8.45K	40.8	
2	1976	Murtala	8.45K	9K	9K	6.5	
3	1978	Obasanjo	9K	15.3K	15.3K	15.9	
4	1982	Shagari	15.3K	20K	20K	30.7	
5	1986	Babangida	20K	39.5K	39.5K	97.5	
6	1988	Babangida	39.5K	42K	42K	6.3	
7	1989	Babangida	42K	60k	60K	42.9	First nation-wide protest
8	1991	Babangida	60k	70K	70k	16.7	
9	1993	Ernest Shonekan	70K	₦5.00K	₦5.00K	614	
10	1993	Abacha	₦5.00K	₦3.25K	₦3.25K	-35	Reduction in price
11	1994	Abacha	₦3.25K	₦15.00 K	₦15.00K	361	
12	1994	Abacha	₦15.00 K	₦15.00 K	₦11.00K	-26.6	**Reduction in price**
13	1998	Abubakar	₦11.00 K	₦25.00 K	₦25.00K	127.3	
14	1999	Abubakar	₦25.00 K	₦25.00 K	₦20.00K	-20	Reduction in price
15	2000	Obasanjo	₦20.00 K	₦30.00 K	₦30.00K	50	
16	2000	Obasanjo	₦30.00k	₦22.00 K	₦22.00K	-26.7	Reduction in price
17	2002	Obasanjo	₦22.00 K	₦26.00 K	₦26.00K	15.3	
18	2003	Obasanjo	₦26.00 K	₦42.00 K	₦42.00K	61.5	
19	2004	Obasanjo	₦42.00 K	₦50.00 K	₦50.00K	19	
20	2004	Obasanjo	₦50.00 K	₦65.00 K	₦65.00K	30	
21	2007	Obasanjo	₦65.00 K	₦75.00 K	₦75.00K	15.4	
22	2007	Yar' Adua	₦75.00 K	₦65.00 K	₦65.00K	-13.3	Reduction in price
23	2012	Jonathan	₦65.00 K	₦141.00 K	₦97.00K	49.2	
24	2015	Jonathan	₦97.00 K	₦86.50 K	₦86.50K	-10.8	Reduction in price
25	2016	Buhari	₦86.50 K	₦85.50 K	₦85.50K	-1.2	Reduction in price
26	2016	Buhari	₦86.50 K	₦145.00 K	₦145.00K	67.6	

THEORETICAL FRAMEWORK

The theory selected for this study is Agenda Setting Theory.

Some communication scholars are of the opinion that Agenda-Setting Theory is somehow controversial (McQuail, 1993:276). But be that as it may, nobody can out rightly dispute the fact that treatment of issues by the media determines the level of importance society will ascribe to such issues. Even though mass media may not be able to exercise the power to determine what people actually think, they have the power to determine what we think about (Folarin 1998:68). Folarin also identifies the following elements of the agenda-setting theory: the quantity or frequency of reporting; prominence given to reports; the degree of conflict generated in the reports; and cumulative media - specific effects over time.

Method

Content analysis method was employed for this study for a number of reasons.

Content analysis is one of today's most extensively employed analytical tools. The research method has been applied fruitfully in many empirical studies (Allen & Reser, 1990). According to Berelson (1952:18), content analysis is a research technique for the systematic, objective, and quantitative description of the manifest content of communication. Three national newspapers (the Punch, the Guardian and This Day) were selected for this study using purposive sampling technique. It was expedient that newspapers with national outlook and wide circulation were chosen in line with the objectives of the study. Various fuel price hike incidents were randomly selected.

The following content categories were derived for the study: prominence, story slant, story type, sources of story and themes of story.

Study Period

There were 7 occurrences of fuel-price increments in Nigeria from 1999 to 2012, out of which four occurrences were randomly selected for the study. The following fuel-hike incidents were selected as samples: June 1, 2000, May 29, 2004, May 27, 2007 and January 1, 2012. The three selected newspapers- the *Punch*, the *Guardian* and *This Day*- were analysed for one week each for the four study periods. The study periods were from June 1 to 7, 2000 (study period 1), May 29 – June 4, 2004 (study period 2), May 27 – June 2 2007 (study period 3) and January 7 to January 12 (study period 4).

Study Universe

As earlier mentioned, three national dailies (the Punch, The Guardian and This Day) were content analyzed. A total of 883 articles relevant to the topic were sampled across four periods of increment in petrol prices in Nigeria. Of this number, *Punch* newspaper published 338 articles of different genres on fuel increment. This constituted 38.3% of the total number of articles considered relevant to the subject matter. The *Guardian* and *This Day* shares of the total publications were 251 and 258 or 28.4% and 29.2% respectively.

Unit of Analysis

This includes story headlines, illustrations and entire story.

Operational Definitions

- **Prominence:** This refers to placement of stories/illustrations on different pages of the newspapers which tells much about the prominence of such items. Front-page as well as back-page materials of the selected newspapers were considered to be more prominent than those that were buried in the belly of the newspapers.
- **Story Type:** This is a content category that captures all forms of newspaper genres such as news story, interview, pictures/illustrations, etc.
- **Sources of Story:** Content category that identifies sources where stories emanated from.
- **Pro-Subsidy Removal:** Newspaper contents (stories) in support of subsidy removal.
- **Anti-Subsidy Removal:** Newspaper contents not in favour of subsidy removal.
- **Corruption in the Oil Industry:** Newspaper publications on corruption in the oil industry.
- **Industrial Action/Strike:** Newspaper contents on industrial/strike action.
- **Slant:** This refers to tone of stories towards the government. The slant could either be positive, negative and neutral.
- **World Oil Market Reports:** Newspaper publications on situation in the world oil market.

Data Presentation

Data gathered through content analysis of the three selected newspapers are presented in Tables 2a to 6b.

Tables 2a and 2b indicates that 483(54.7%) of all the published contents by the three newspapers selected for the study were news stories, feature stories took the second position with just 171(19.4%), followed by illustrations with 149(16.9%).

Petrol subsidy removal dominated front pages of all the newspapers selected for this study. Four periods were examined as stated earlier, 38 (4.3%) of all the stories were front-page stories. All the newspapers that were studied almost paid equal attention to the matter. Petrol subsidy also dominated the inside pages of the newspapers.

Tables 4a and 4b show that huge chunk (67.5%) of contents published on petrol subsidy were negative towards the government. This clearly showed that overwhelming majority of the citizens of Nigeria were not in favour of government's policy of subsidy removal. Only 13.14% of the materials were in favour of the government policy.

Tables 5a and 5b show that 32.8% of all the relevant publications in the three newspapers selected for this study were from government sources, this is followed by non-governmental organization (NGO) with 27.6%, while journalists contributed the third largest share of all the relevant materials, that is, 16.9%.

Tables 6a and 6b show themes derived for the study;49.4 percent were against the removal of petrol subsidy. Only 16 percent of the total publications were in support of petrol subsidy removal, while 7 percent of the materials published in the three newspapers were on corruption, and 7.2 percent were on the refineries.

Table 2a. Story type

		News Story		FEATURE		EDITORIAL		INTERVIEW		ILLUSTRATION		TOTAL	
		Freq	%	Freq	%	Freq	%	Freq	%	Freq	%	Freq	%
Punch	Study Period 1.	29	3.3	10	1.1	1	0.1	6	0.7	11	1.3	57	6.5
	Study Period 2.	25	2.8	8	0.9	2	0.2	6	0.7	4	0.5	45	5.1
	Study Period 3.	22	2.5	9	1.0	-	-	4	0.5	2	0.2	35	4.0
	Study Period 4.	119	13.5	26	2.9	2	0.2	-	-	54	6.1	201	22.8
Guardian	Study Period 1.	30	3.4	11	1.3	2	0.2	7	0.8	8	0.9	58	6.6
	Study Period 2.	29	3.3	9	10.2	3	0.3	5	0.6	6	0.7	51	5.8
	Study Period 3.	24	2.7	6	0.7	2	0.2	5	0.6	2	0.2	39	4.4
	Study Period 4.	38	4.3	34	3.9	2	0.2	4	0.5	25	2.8	103	11.7
This Day	Study Period 1.	27	3.1	13	1.5	3	0.3	6	0.7	10	1.1	60	6.8
	Study Period 2.	32	3.6	10	1.1	1	0.1	7	0.8	8	1.0	58	6.7
	Study Period 3.	21	2.4	8	0.9	-	-	6	0.7	2	0.2	36	4.1
	Study Period 4.	87	9.9	27	3.1	2	0.2	4	0.5	17	2.0	140	15.9
	Total	483	54.7	171	19.4	20	2.3	60	6.8	149	16.9	883/100	

Table 2b. Story type: Summary of Table 2a

	NEWS STORY		FEATURE		EDITORIAL		INTERVIEW		ILLUSTRATION		TOTAL	
	Freq.	%	Freq.	%	Freq.	%	Freq.	%	Freq.	%	Freq.	%
THE PUNCH	195	22.1	53	6.0	5	0.6	16	1.8	71	8.0	340	38.5
THE GUARDIAN	121	13.7	60	6.8	9	1.0	21	2.4	41	4.6	252	28.5
THIS DAY	167	18.9	58	6.7	6	0.7	23	2.6	37	4.2	291	33.0
TOTAL	483	54.7	171	19.4	20	2.3	60	6.8	149	16.9	883/100	

Table 3a. Prominence

	The Punch						The Guardian						This Day					
	Front Page		**Inside Page**		**Back Page**		**Front Page**		**Inside Page**		**Back Page**		**Front Page**		**Inside Page**		**Back Page**	
	FREQ.	%	FREQ.	%	FREQ.	%	FREQ.	%	FREQ.	%	FREQ.	%	FREQ.	%	FREQ.	%	FREQ.	%
Study period 1	10	1.13	45	5.1	6	0.68	10	1.13	39	4.42	3	0.34	11	1.26	25	2.83	2	0.23
Study period 2	8	0.91	30	3.4	7	0.79	9	1.02	18	2.04	4	0.45	8	0.91	35	3.96	3	0.34
Study period 3	6	0.68	19	2.2	3	0.34	5	0.57	21	2.38	2	0.23	6	0.68	24	2.72	4	0.45
Study period 4	14	1.59	201	22.8	2	0.23	13	1..47	143	16.19	3	0.34	13	1.47	129	14.61	2	0.23
TOTAL	38	4.3	295	33.4	18	2.01	37	4.19	221	25.03	12	1.36	38	4.30	213	24.12	11	1.25

Table 3b. Prominence: Summary of Table 3a

NEWSPAPER	FRONT PAGE		BACK page		INSIDE PAGE		TOTAL	
	Freq.	**%**	**Freq.**	**%**	**Freq.**	**%**	**Freq.**	**%**
THE PUNCH	38	4.3	18	2.01	295	33.4	351	39.75
THE GUARDIAN	37	4.19	12	1.36	221	25.03	270	30.58
THIS DAY	38	4.3	11	1.25	213	24.12	262	29.67
TOTAL	113	12.79	41	4.6	729	82.56	883	100

Table 4a. Slant

Period	Punch Slant						Guardian Slant						This Day Slant					
	FA		NU		NF		FA		NU		NF		FA		NU		NF	
	Freq.	%	Freq.	%	Freq.	%	Freq.	%	Freq.	%	Freq.	%	Freq.	%	Freq.	%	Freq.	%
1	13	1.47	9	1.02	39	4.45	12	1.36	8	0.91	32	3.62	10	1.13	7	0.79	21	2.38
2	15	1.7	9	1.02	21	2.38	9	1.02	3	0.94	19	2.15	12	1.36	10	1.13	24	2.72
3	7	0.79	4	0.45	17	1.93	9	1.02	4	0.45	15	1.7	7	0.79	5	0.57	22	2.49
4	7	0.79	65	7.36	145	16.42	7	0.79	30	3.4	122	13.82	8	0.91	17	1.93	119	13.48
Total	42	4.75	87	9.85	222	25.14	37	4.19	45	5.1	188	21.29	37	4.19	39	4.45	186	21.01

Table 4b. Slant: Summary of Table 4a

Newspaper	Positive		Negative		Neutral		Total	
	Freq.	%	Freq.	%	Freq.	%	Freq.	%
The Punch	42	4.75	222	25.14	87	9.85	351	39.75
The Guardian	37	4.19	188	21.29	45	5.1	270	30.58
This Day	37	4.19	186	21.01	39	4.45	262	29.67
Total	116	13.14	596	67.5	171	19.37	883/100	

Table 5a. Sources of story

Newspaper	Study Period	Govt		Foreign		Individual		NGO		Journalist		Others		Total	
		Freq.	%	Freq.	%	Freq.	%	Freq.	%	Freq.	%	Freq.	%	Freq.	%
Punch	Study Period 1	15	1.7	7	0.8	12	1.36	15	1.7	16	1.81	1	0.11	66	7.5
	Study Period 2	16	1.8	3	0.3	7	0.8	9	1.0	13	1.5	1	0.1	49	5.5
	Study Period 3	12	1.4	5	0.6	8	0.9	10	1.1	9	1.0	-	-	44	5.0
	Study Period 4	47	5.3	4	0.5	14	1.6	56	6.3	21	2.4	3	0.3	145	16.4
Guardian	Study Period 1	19	2.2	3	0.3	13	1.5	12	1.4	15	1.7	2	0.2	64	7.2
	Study Period 2	15	1.7	4	0.5	11	1.3	11	1.3	13	1.5	1	0.1	55	6.2
	Study Period 3	16	1.8	1	0.1	7	0.8	9	1.0	11	1.3	2	0.2	46	5.2
	Study Period 4	56	6.3	4	0.5	13	1.5	44	5.0	21	2.4	2	0.2	140	15.9
This Day	Study Period 1	16	1.8	7	0.8	14	1.6	13	1.5	11	1.3	-	-	61	7.0
	Study Period 2	13	1.5	3	0.3	6	0.7	14	1.6	13	1.5	1	0.1	50	5.7
	Study Period 3	15	1.7	-	-	2	0.2	9	1.0	10	1.13	3	0.3	39	4.4
	Study Period 4	50	5.7	3	0.3	13	1.5	42	4.8	14	1.6	2	0.2	124	14.0
	Total	290	32.8	44	5.0	120	13.6	244	27.6	167	18.9	18	2.0	883/ 100	

Table 5b. Source of story: Summary of Table 5a

Newspaper	Government		Foreign		Individual		NGO		Journalist		Others		Total	
	Freq.	%	Freq.	%	Freq.	%	Freq.	%	Freq.	%	Freq.	%	Freq.	%
Punch	90	10.2	19	2.2	41	4.6	90	10.2	59	6.7	5	0.6	304	34.4
Guardian	106	12.0	12	1.4	44	5.0	76	8.7	60	6.8	7	0.8	305	34.5
This Day	94	10.6	13	1.5	35	4.0	78	8.8	48	5.4	6	0.7	274	31
Total	290	32.8	44	5.0	120	14.0	244	27.6	167	18.9	18	2.0	883/100	

Table 6a. Story themes

Newspaper	Study Period	Corruption Related		Refinery-Related Reports		Industrial Action		World Oil Market Report		Anti-Subsidy Removal		Pro-Subsidy Removal		Others		Total	
		Freq.	%	Freq.	%	Freq.	%	Freq.	%	Freq.	%	Freq.	%	Freq.	%	Freq.	%
Punch	Study Period 1	10	1.1	10	1.1	12	1.6	2	0.2	18	2.0	8	0.91	5	0.6	65	7.4
	Study Period 2	3	0.3	5	0.6	12	1.6	3	0.3	20	2.3	7	0.8	3	0.3	53	6
	Study Period 3	4	0.5	3	0.3	5	0.6	1	0.1	17	1.9	9	1.0	-	-	39	4.4
	Study Period 4	4	0.5	4	0.5	12	1.6	4	0.5	115	13	21	2.4	3	0.3	163	18.5
Guardian	Study Period 1	11	1.3	4	0.5	12	1.6	5	0.6	11	1.3	4	0.5	2	0.2	49	5.5
	Study Period 2	4	0.5	10	1.1	11	1.3	3	0.3	9	1.0	5	0.6	3	0.3	45	5.1
	Study Period 3	3	0.3	4	0.5	5	0.6	2	0.2	13	1.5	6	0.7	2	0.2	35	4
	Study Period 4	3	0.3	2	0.2	6	0.7	3	0.3	107	12.1	26	2.9	1	0.1	148	16.8
This Day	Study Period 1	11	1.3	6	0.7	14	1.6	9	1.0	15	1.7	7	0.8	6	0.7	68	7.7
	Study Period 2	5	0.6	9	1.0	12	1.6	1	0.1	16	1.8	6	0.7	3	0.3	52	5.9
	Study Period 3	3	0.3	4	0.5	5	0.6	2	0.2	18	2.0	8	0.9	-	-	40	4.5
	Study Period 4	1	0.1	3	0.3	6	0.7	3	0.3	77	8.7	34	3.9	2	0.2	126	14.3
	Total	62	7.0	64	7.3	112	12.7	38	4.3	436	49.4	141	16	30	3.4	883/100	

Table 6b. Story themes: Summary of Table 6a

Newspaper	Corruption Related		Refinery Related Reports		Industrial Action/ Strike		World Oil Market Report		Anti-Subsidy Removal		Pro-Subsidy Removal		Others		Total	
	Freq.	%	Freq.	%	Freq.	%	Freq.	%	Freq.	%	Freq.	%	Freq.	%	Freq.	%
Punch	21	2.4	22	2.5		4.6	10	1.1	170	19.3	45	5.1	11	1.3	320	36.2
Guardian	21	2.4	20	2.3	34	3.9	13	1.5	140	15.9	41	4.6	8	0.9	277	31.4
This Day	20	2.3	22	2.5	37	4.2	15	1.7	126	14.3	55	6.2	11	1.3	286	32.4
Total	62	7.0	64	7.2	112	12.7	38	4.3	436	49.4	141	16	30	3.4	883/100	

DISCUSSION

Based on the data presented on Tables 1 to 6, the research questions developed for the study will be answered. The research questions are:

1. How prominent was the reportage of petrol price hike in Nigerian newspapers during the period under study?
2. What were the shades of opinion in the reportage of petrol price hike by Nigerian newspapers?
3. Which were the predominant sources of news reportage of petrol price hike in Nigeria during the study period?
4. What were the major themes of news reports on petrol price hike?

Research Question 1: How prominent was the reportage of petrol price hike in Nigerian newspapers during the period under study?

Petrol price hike was an important issue in Nigeria during the study period. The issue still remains so important to date. Petrol price hike adorned the pages of the selected national dallies all through the study period. Table 3 provides an answer to the question on prominence. A total of 113 stories made it to the front pages of the three newspapers selected for the study within the four periods (28 days) this study covered. This figure constitutes 12.79% of all the reports that were considered relevant to this study. This implies that an average of four story-titles were published on the front pages of the newspapers each day the issue lasted. So also, as much as 41 stories on the issue, that is 4.6% of all the stories, made the back pages of the newspapers, while 729 (82.56%) adorned the inside pages of the newspapers.

Research Question 2: What were the shades of opinion in the reportage of petrol price hike by Nigerian newspapers?

Information provided in Table 4 shows that government was scored low all through the period under study. Huge number of published contents of the newspapers (596= 67.5%) was not in favour of the government. Only 116 (13.14%) contents favoured the government, while 171 (19.37%) were considered neutral.

Research Question 3: Which were the predominant sources of news reportage of petrol price hike in Nigeria during the study period?

Government sources were the most active during the periods under study. A total of 290 news stories published in the three newspapers came from government sources. This figure constituted 32.85% of all the relevant publications. This is followed by non-governmental organizations with 244 (27.6) of all the relevant news stories. One hundred and twenty (14%) news stories came from individuals, while journalists' share was 167 (18.9%).

Research Question 4: What were the major themes of the news reports on petrol price hike?

The most important theme identified in the study, according to Table 6, was anti-fuel subsidy removal. Four hundred and thirty-six (49.4%) of all the relevant contents were on this theme. This is followed by pro-fuel subsidy removal with 141(16%); industrial action/strike's share of all the relevant materials was 112 (12.7%), and refinery-related issues were 64 (7.2%), while corruption's share was only 62 (7%).

SUGGESTION FOR FURTHER STUDY

Petrol subsidy removal is a developing issue in Nigeria, and it will remain so until that sector of the economy is fully liberalized. Muhammadu Buhari-led administration recently moved the price to ₦145 per litre though petrol importers are still agitating for another price increase. As the event is unfolding, communication and media researchers may need to focus attention on the media coverage of the oil industry with a view to examining problems and prospects of the industry, as well as audience assessment of the effects of deregulation policy on ordinary Nigerians and on the nation's economy.

CONCLUSION

Petrol price hike is a sensitive issue that touches virtually all the citizens of Nigeria. Findings in this study show that the issue was widely reported. The topic dominated the front pages as well as back pages of the three newspapers selected for the study with an average of four news stories per day for each newspaper. The front pages of the newspapers were devoted to 12.79% of all the stories. Nigerian press set the agenda for discussion throughout the four periods.

Stories got to the press through various sources such as government which had the biggest chunk of the news with 32.85%, NGO with 27.6%, 18.9% of the news was traced to journalists and individual accounted for 14% of all the stories considered relevant to the study.

The themes derived for the study were also diverse. News items that were against the removal of petrol subsidy termed (anti-fuel subsidy removal) were huge. 49.4% of all the stories fell under this category, this was followed by stories regarded as pro-fuel subsidy removal (16%). The other theme was "industrial action/strike" with 12.7%. Surprisingly, corruption was not considered a serious issue as just 7% of all the contents were on corruption in oil industry. This may be the consequence of majority of the editorial content flowing into the media through government sources, which is 32.85% of all the published contents. Government basically focused on how to convince the masses to support the price increase and not how to open up the skeleton in its cupboard by exposing the rot in the oil industry.

The media on the other hand failed to pursue an agenda different from that of the government, even though all the selected newspapers are privately owned. Going by the data, NGO and journalists constituted 46.5% of the sources of information; these two sources would have made a significant difference had journalists set an anti-corruption agenda.

The questions to be answered are the following:

1. Is the government really subsidizing petrol?
2. What is the level of subsidy?

Answers to these two questions are not that straight forward. Even though content analysis as a research method might not have provided direct and generalizable answers to these questions, literature review for this study gave some indications. We can authoritatively say that Nigerian oil industry is shrouded in mysteries. There is dire need for transparency and accountability. Lack of transparency and accountability has made the industry a nest for corruption. As shown through the literature review, corruption strives well in NNPC.

Until corruption in the industry is dealt with, one cannot ascertain the level of subsidy in the industry.

RECOMMENDATIONS

Government actively dialogued with the citizenry as 32.85% of all the published contents during the crises came from government sources. This kind of engagement should continue at all times. Government should explain its programmes to the people so that they (the people) can support good programmes. As laudable as petrol subsidy removal might seem, statistics revealed that people were not in support of it. Nigerian press should set the agenda to fight corruption. It is ridiculous that a meagre 7% of all the contents were on corruption in the oil industry. The press needs to place fighting against corruption on the front burner.

REFERENCES

Adeyemi, A. (2013). Nigerian media and corrupt practices: The need for paradigm shift. *European Scientific Journal*, *9*(1), 125–126.

Agbon, I. (2012, January 8). The Real Cost of Nigeria Petrol. *The Guardian*, p. 7.

Allen, B., & Reser, D. (1990). Content Analysis in Library and Information Science Research. *Library & Information Science Research*, *12*(3), 251–260.

Ascher, W. (1999). *Why governments waste natural resources: Policy failures in developing countries*. Baltimore, MD: Johns Hopkins University Press.

Balogun, M. (2012, January 1). Bridging the Wealth gap. *The Punch*, p. 16.

Berelson, B. (1952). *Content Analysis in communication Research*. Glencoe, IL: The Free Press.

Berge, S. (1951). Subsidies and Competition as Factors in Air Transport Policy. *The American Economic Review*, *41*(2), 519–529. Retrieved from http://www.jstor.org/stable/1910826

Clements, M. B. J., Coady, D., Fabrizio, M. S., Gupta, M. S., Alleyne, M. T. S. C., & Sdralevich, M. C. A. (2013). *Energy subsidy reform: lessons and implications*. International Monetary Fund.

Ekpo, C. E., Chime, J., & Enor, F. N. (2016). The Irony of Nigeria's Fight Against Corruption: An Appraisal of President Muhammadu Buhari's First Eight Months in Office. *International Journal of History and Philosophical Research, 4*(1). Retrieved from http://www.eajournals.org/journals/international-journal-of-history-and-philosophical-research-ijhphr/vol-4-issue-1-march-2016/the-irony-of-nigerias-fight-against-corruption-an-appraisal-of-president-muhammadu-buharis-first-eight-months-in-office/

European Union Trade Policy. (n.d.) Retrieved September 25, 2015 from http://ec.europa.eu/trade/policy/accessing-markets/trade-defence/actions-against-imports-into-the-eu/anti-subsidy/

Falola, T., & Heaton, M. M. (2008). *A History of Nigeria*. Cambridge University Press. doi:10.1017/CBO9780511819711

International Institute for Sustainable Development. (2013). *A Guidebook to Fossil-fuel Subsidy Reform for Policy-Makers in South East Asia*. Manitoba, Canada: Author.

Lasisi, A., & Augoye, J. (2012, January 1). A history of protests against subsidy removal. *The Punch*. p. 16.

McQuail, D. (1993). Mass Communication Theory: An Introduction (2nd ed.). New Delhi: Sage Publication.

Nanaghan, B. (2012, January 1). Still on Fuel Subsidy Removal. *The Punch*, p. 16.

Nwachukwu, E. (2012, January 13). Fuel Subsidy: Examining the facts. *The Punch*, p. 18.

Nwosu, N. I. (1996). The Politics of Oil Subsidy in Nigeria. *Africa: RivistaTrimestrale Di Studi E Documentazione Dell'istituto Italiano Per L'africa E L'oriente, 51*(1), 80–94. Retrieved from http://www.jstor.org/stable/40761059

Odusola, A. F. (1997). Poverty in Nigeria: An Eclectic Appraisal. Poverty Alleviation in Nigeria. *Nigerian Economic Society,1997 Annual Conference*. Ibadan: Pat Mag Press.

Okogu, B. E. (1995, January01). Issues in Petroleum Product Pricing in Nigeria. *Journal of African Economies, 4*(3), 378–405. Retrieved from http://jae.oxfordjournals.org/

Onyeizugbe, C. U., & Onwuka, E. M. (2012). Fuel Subsidy Removal as an Imperative for Enhancing Business Development in Nigeria. *VSRD International Journal of Business and Management Research.* Retrieved on May 18, 2013 www.vsrdjournals.com

Post, J. E., Lawrence, A. T., & Weber, J. (1999). *Business and Society: Corporate Strategy, Public Policy, Ethics* (9th ed.). Boston: McGraw Hill.

Post, J. E., Lawrence, A. T., & Weber, J. (1999). Business and Society: Corporate Strategy, Public Policy, Ethics (9th ed.). Boston: McGraw Hill.

Robinson, W. C. (1967). What is a government subsidy? *National Tax Journal, 20*(1), 86–92. Retrieved from http://www.jstor.org/stable/41792130

Steenblik, R. (n.d.). *A Subsidy Primer. Global Subsidies Initiative of the IISD*. Retrieved September 25, 2015 from https://www.iisd.org/gsi/subsidy-primer/

Ugwuanyi, E. (2015, July1). What next for NNPC after board dissolution? *Nation (New York, N.Y.)*, 2–3.

Widodo, T. G. A., Sahadewo, S. U. S., & Chaerriyah, M. (2012). Cambodia's Electricity Sector in the Context of Regional Electricity Market Integration. In Y. Wu, X. Shi, & F. Kimura (Eds.), *Energy Market Integration in East Asia: Theories, Electricity Sector and Subsidies* (pp. 173–206). Jakarta: ERIA.

Zahariadis, N. (2008). *State subsidies in the global economy*. New York: Palgrave Macmillan. doi:10.1057/9780230610514

KEY TERMS AND DEFINITIONS

Corruption: An act that exhibits lack of integrity, virtue, transparency and accountability, aimed at unjustly enriching oneself or a group of people through un- ethical means. It is the misuse of power/and office for private aggrandizement.

Deregulation: An action that allows businesses to run without government's interference by allowing the industry to open up to competition and normal market forces of demand and supply.

Government Policy: The direction the government decides to move on certain issues of national importance.

Nigerian National Petroleum Corporation (NNPC): An oil company owned by the Federal government of Nigeria. The company is involved in all levels of oil-business activities in Nigeria.

Petrol: Premium motor spirit. It is one of the major by-products of petroleum. It is used especially as fuel for vehicles and to power generating sets.

Petroleum Industry: The industry that has to do with up-stream and down-stream sector of oil and gas business.

Protest: Open objection of the people to an increase in the price of petrol.

Subsidy Removal: Government anti-subsidy policy on the price of petrol. It is the withdrawal of the money paid by the Federal Government of Nigeria to importers of petrol to reduce the cost price of the product.

Chapter 12
Press Freedom and Socio-Economic Issues in the Nigerian and Ugandan Democratic Landscape

Okorie Nelson
Covenant University, Nigeria

ABSTRACT

Press freedom is a global concept that encourages the freedom of the press from external influences or interferences. In Africa, press freedom is enshrined in the constitutions of most countries. However, there are several factors and issues that limit press freedom in Africa. Using Nigeria and Uganda as examples, this chapter explores press freedom and socio-economic issues in a democratic media landscape. Furthermore, the chapter was organized in a number of rubrics; it x-rayed the concept of press freedom. It also examined the constitutional basis of press freedom in Uganda and Nigeria. The chapter concludes that Press freedom promotes transparency and flourishes more in countries where there are constitutional and legal provisions are greatly promoted because of the democratic principles and ideals which are conducted with openness.

INTRODUCTION

The concept of press freedom has become widely accepted and encouraged among policy makers, activists, politicians, media scholars and professional journalists across the globe. Press freedom has been globally accepted because it is either implied or explicitly enshrined in national constitutions all over the world (Alabi, 2003; Okunna, 2003, Okoro, 2004; Ufuophu-Biri, 2006; Okorie, 2010). In Africa, for instance, press freedom as implied in the constitution guarantee that "every person shall be entitled to freedom of expression, including the freedom to hold opinions and to receive and impart ideas and information without interference." Okorie (2010, p.159) reasoned that:

DOI: 10.4018/978-1-5225-1859-4.ch012

In the international community, press freedom has been a core issue discussed among media professionals and government officials in the 21st century. The issue and debate of press freedom has been prevalent due to its importance in the transparency and governance of nations across the globe. Also, the concept of press freedom is at the heart of human capital development as well as sustainable development of a nation. Basically, press freedom denotes the right of every person to own a printing press, to publish whatever information or ideas that pleases him, to decide the editorial policy of a publication and to enforce it upon his staff and to distribute freely without having to obtain license from the authorities or face suppression or proscription.

However, studies have shown that African journalists experience socio-economic issues and factors which have crippling effects on their professional practice as the 'Fourth Estate of the Realm'(Wilcox, 1975; Okunna, 2003, Ufuophu-Biri, 2006; Okorie, 2010). Countries like Nigeria and Uganda are typical examples of African countries in term of socio-economic controls or interferences of the press activities and operation in African societies. Nevertheless, the press struggles with these socio-economic issues and factors to operate as the watchdog of the populace in making the government accountable to the citizens of the nation. This chapter seeks to examine the issues and challenges confronting the concept of free press among media professionals in the African society. Furthermore, the chapter will examine how free the press is in a liberalized media landscape, placing emphasis on Nigeria, Uganda and Namibia.

CONCEPT OF PRESS FREEDOM

Freedom of the press has been said to be a cornerstone to judge maturity of democracy in many societies. Freedom of the press is a concept that has often been explained as the freedom of the press to publish whatever information they feel is necessary for the public to know. According to Oster (2013), 'the media is to be accepted as a legal institution with specific rights and responsibilities'. He also added that 'the media is to be defined as 'a natural or legal person gathering and disseminating to a mass audience information and ideas pertaining to matters of public interest on a periodical basis and according to certain standards of conduct governing the newsgathering and editorial process'. In the same vein, Sani (2014) believes that the mass media as a source of information is expected to be reliable, but in some countries, dictatorship adulterated the information emanated from media to control public opinion and ensure conformity.

Press freedom is a topical issue of great relevance not only to private individuals and the media industry, but also to governments in their efforts to achieve a democratic society. A free press plays a critical role in informing electorates and serving as a check on government's activity, as well as giving meaning to public opinion. It provides a framework for public authorities to be held accountable and as a means for citizens to freely express and exchange their views on public issues. Conceivably, freedom of the press and of expression constitute the foundation on which a democratic system is built (Hardy, 2008, p.80).

There are some misconceptions about what press freedom really is, for example, according to Weaver (1977), press freedom is the relative absence of governmental and nongovernmental restrictions on the media, and that of Piccard (1985), who subscribes to lack of legal controls in his definition of press freedom. Pearson (2013) explains that press freedom should be seen as a media system that is free, pluralistic and independent. This dispensation is essential for democracy and development.

Pearson (2014) used the term 'press freedom warrior' in his article *Press freedom, social media and the citizen* to describe a person who has made substantial sacrifice in the name of free expression and free media. Pearson also went ahead to include in his definition of a 'press freedom warrior' as those who have suffered in other ways because of their commitment to truth-seeking and truth-telling. Some have been the victims of lawsuits and have had to pay damages to those who have set out to gag them. Others have forsaken lucrative positions in government or public relations so they can continue as Fourth Estate watchdogs in preference to becoming political or corporate lapdogs. He also added the term 'press freedom worrier' and he explained that a press freedom worrier was a person who worry about a continuing array of regulatory, technological, economic, corporate and ethical threats to free expression and a free media.

Theorizing a Free Press

There are different theories that have tried to explain media freedom. These theories portray media freedom in different ways. Some of these theories are:

1. The Free press theory (most fully developed in the United States of America, but applying elsewhere) proclaims complete freedom of public expression and of economic operation of the media and rejects any interference by government in any aspect of the press. A well- functioning market should resolve all issues of media obligation and social need.
2. The Social responsibility theory (found more in Europe and countries under European influence) is a modified version of free press theory placing greater emphasis upon the accountability of the media (especially broadcasting) to society.

Media are free but they should accept obligations to serve the public good. The means of ensuring compliance with these obligations can either be through professional self-regulation or public intervention (or both). Social responsibility theory allows free press without any censorship but at the same time the content of the press should be discussed in public panel and media should accept any obligation from public interference or professional self -regulations or both. The theory helped in creating professionalism in media by setting up a high level of accuracy, truth, and information. This theory however can only be applied in certain areas as it is not applicable to all forms of government practiced across the globe.

Constitutional Bases of Press Freedom in Nigeria and Uganda

In Nigeria, according to Okorie (2010, p.160), the 1999 Constitution is more elaborate in its provision, not only on the general freedom of expression or of the press but also on the obligations of the media to uphold the fundamental objectives and directive principles of the State policy. For example, S. 39 provides as follows:

Sub section 1: every person is entitled to freedom of expression, including the freedom to hold opinions and receive and impart ideas and information without interference.

Sub section 2: without prejudice to the generality of subsection (1) of this section, every person shall be entitled to own, establish and operate a medium for the dissemination of information, ideas and opinions.

Provided that no person other than the Government of the federation or of a state or any other person authorized by the president on the fulfillment of conditions laid down by an Act of the National Assembly, shall own, establish or operate a television or wireless broadcasting station of any purpose whatsoever.

While Section 39 deals with general provisions on freedom of expression, conditions for owning and operating the media and restrictions on such freedom, the Constitution also makes similar broad provisions on the obligations of the press (media) to the government and people of Nigeria. The functions and obligations of the press are well stated in the section dealing with the fundamental objectives and directive principles of state policy which is found in chapter II of the 1999 Constitution. For example, S. 22 provides that:

The press, radio and television shall at all time be free to uphold the fundamental objectives contained in this chapter (chapter II) and uphold the responsibility and accountability of the Government to the people.

Also, the press is required among other things, to maintain the national ethics and articulate the duties of the citizen as contained in section 23 and 24 of the Constitution (Ewelukwa, 2004; Okoro, 2004;Okorie, 2010). The above cited provisions especially the entire S. 39 are in line with international law, co

In Uganda, Article 29 (1) (a) of the Ugandan Constitution 1995 states that: "Every person shall have the right to freedom of speech and expression, which shall include freedom of the Press and the Media".

Article 41: *Every citizen has a right of access to information in the possession of the state or any other organ or agency of the State except where the release of the information is likely to prejudice the security or sovereignty of the state or interfere with the right to the privacy of any other person.*

Article 43: *In the enforcement of the rights and freedom prescribed in this chapter, no person shall prejudice the fundamental and other human rights and freedom of othersor the public interest.*

In consonance with Article 29 of the Constitution – the Press and Journalist Act CAP 105 was enacted mainly to fulfill this Constitutional obligation, one of its main objective is stated: "an Act to ensure the freedom of the Press". (a) The Electronic Media Act CAP 104. This Act was set up by the Broadcasting Council to license and regulate Radio and Television Stations and to provide for the licensing of Television sets. (b) The Access to Information Act 2005. In furtherance to Article 41, this Law was enacted in 2005.

Socio-Economic Factors Limiting Press Freedom

There are several socio-economic issues and factors facing the press in a democratic media landscape. These issues and factors are in different ramifications such as controls, interferences, pressures and regulations. Some of these challenges are:

Government Control and Measures

The government uses certain preferential treatments to 'buy' the most influential journalists in the country... appointing these influential critics in the media to top posts within the government. 'Coopting' of journalists ensures that they are reduced to being mere stooges of government officials (Uche, 1989). For example, Reuben Abati, who is an influential media expert was appointed as Special Adviser on Media and Publicity to President Goodluck Jonathan of Nigeria. Prior to his appointment, he was previously a newspaper columnist and the chairman of the editorial board of the Nigerian newspaper 'The Guardian' from 2001 to 2011. Many Nigerian reacted negatively about the appointment because they feared that Reuben may be influenced negatively and would discontinue his anti-corruption crusade

These pernicious laws and decrees against the media gave government officials legal backing to persecute, fine, detain and imprison journalists, and to proscribe media houses. For instance, the Offensive Publications (Proscription) Decree 35 1993, made it possible for the government to clamp down on six media houses across the nation. Even government owned media were not spared. In Uganda, The 2000 Press and Journalist Act requires journalists to register with the government-affiliated National Institute of Journalists of Uganda (NIJU) and obtain a license from the Media Council, which has been criticized for lacking independence. Journalists must also meet certain standards, including the possession of a university degree, to be full members of NIJU. This situation also exist in Nigeria, scholars such as Okwuchukwu (2014,p.39) observed that:

... regulatory bodies set up by the government can be a source of media control. Where there are defects or loopholes in the decree that set up such regulatory bodies, these can be used for repressing freedom of expression. It is thought that government may intentionally leave loopholes in laws and decrees so as to exploit such in silencing any opposition. Media scholar believe that one pitfall in the decree that set up the National Broadcasting Commission (NBC) is the power given to the commission to revoke the licenses of stations which do not operate in accordance with the code and in the public interest. The decree did not specify either how to seek redress or what the public interest is, as in the American Federal Communication Commission. Thus the decree allows the NBC to provide licenses in perpetuity only to withdraw them at whim.

Other measures of government control include denying journalists access to places and persons for information, refusing to give government advertisements and dubious labelling of documents containing valuable information. Okwuchukwu (2014, p.38) observed that "In Nigeria three types of government (colonial, civilian and military) had been in power at one time or the other in the country and all three types of government have implemented policies that restrained freedom of the press in Nigeria. To achieve control, media outfits have been deliberately denied advertising revenue, while journalists have been killed, maimed, harassed, detained, jailed, and repressive laws and decrees enacted to cow the media." All these measures have been used to harass journalists in Nigeria and Uganda.

1. **Proprietor Influence:** The ownership of a media outlet has significant control on the freedom of the press. Some politicians establish newspaper, radio and television houses to serve as a vehicle for political advertising as well as a megaphone for their political party. For example, Chief Gbenga Daniels established the *National Mirror*, which is a national tabloid used to propagate his interest and ambition. Chief UzorKalu established the *Sun* newspaper which he uses to expose the evils, drawbacks and weakness of his political enemies. Similarly, Chief Ahmed Bola Tinubu established the Television Continental (TVC), Nation Newspaper and Radio Continental, which serve as a communication platform to propagate the ideology of All-Progressive Congress (APC). Thus, proprietor influence is a limiting factor for the socio-economic development of press freedom in Nigeria.

2. **Advertisers Influence:** Most media outlets survive on advertising as a major source of revenue. Some scholars argue that advertising is the financial nexus that enhance the operations of any media organization. Many advertisers know this and take undue advantage of media organization. For example i*n Uganda,* media owners are somewhat complicit in the erosion of press freedom in Uganda. To safeguard their investments, they reportedly comply with government requests, including onerous instructions as to which journalists they may employ. The sustainability of newspapers is compromised by declining circulation rates, which leave outlets more dependent on advertising. Threatened or actual advertising boycotts by corporations and the government, which are especially problematic for smaller media outlets, limit media diversity and pluralism.
3. **Editorial Policy and Orientation:** An editorial policy can be referred to as the beliefs of a media organization which it would like to manifest in its news coverage, Ate (2006). A policy is a definite course of action selected from among alternatives to guide and shape present and future actions. Orientation concerns itself with the aims, objectives or interests of a particular activity or organization or establishment. Policies and orientations in the views, of Ate (2006) can be influenced by the socio-political and economic environment in which the media establishment is operating. A lot of newspapers in Nigeria and Uganda are bifurcated along political lines. For example, *Tribune* and *Nation* newspapers in Nigeria are polarized towards the South-Western part of Nigeria, propagating the ideology of Awolowo; *Champion* and *Sun* Newspapers are polarized towards the South-Eastern part of Nigeria, propagating the interest of the Ibo people; *Guardian* and *Thisday* newspapers are elitist in nature. The editorial policy and orientation of a newspaper is a crucial factor that affects the socio-economic development in Africa.

SUGGESTIONS FOR RESEARCH

For better representation and comprehensive study on press freedom in Africa, an empirical review of press freedom violation in North, East and West Africa should be conducted. Also, studies on journalism ethics and press freedom need to be extensively investigated in line with international best practices and inquiries.

CONCLUSION

Press freedom is essential for the socio-economic development of the African continent. Press freedom promotes transparency and flourishes more in countries where there are constitutional and legal provisions are greatly promoted because of the democratic principles and ideals which are conducted with openness.

Countries like Nigeria and Uganda face several challenges that negatively affect press freedom, which have socio-economic and political undertones. It is suggested that journalists in Uganda should spearhear a media campaign to expunge draconian press laws from the statute books. Also, Ugandan journalists should seek help from human rights organizations to support their quest to expunge draconian press laws from the statute books. In the same vein, The Nigerian press shouldn't relent in its quest in the passage of the Freedom of Information bill into law. The media should capitalize on its agenda-setting function in providing adequate coverage and information about the Freedom of Information bill. The passage of this bill into law would ensure that the media have legal access to public records as well as discard the conduct of government affairs in secret.

REFERENCES

Ake, C. (1985). *Political economy of Nigeria*. Lagos: Longman.

Alabi, S. (2003). Press freedom: Background, legal bases and constraints. *UNILAG Communication Review*, *4*(1), 52–65.

Barton, F. (1979). *The press in Africa: Persecution and perseverance*. London: Macmillan. doi:10.1007/978-1-349-04075-9

Constitution of the Federal Republic of Nigeria. (1999). Lagos: Federal Government Press.

Constitution of the Federal Republic of Uganda 1995

Croteau, D., & Hornes, W. (2000). *Media society. Thousand Daks*. Pine Forge Press.

Ewelukwa, N. (2001). *Introduction to press law*. Maranatha Press ltd.

Folarin, B. (1998). *Theories of mass communication*. Ibadan: Stirling-Horden Publisher.

Gambo, D. (2006). Legal and ethical challenges for Nigerian journalists in a Democracy. *An Interdisciplinary Journal of Communication Studies*, *20*(4), 86–95.

Ojo, E. (2003). The mass media and the challenges of sustainable democratic values in Nigeria: Possibilities and limitations. *Media Culture & Society*, *25*(1), 62–76.

Okorie, N. (2010). Press freedom in Nigeria: Legal constraints and possible solutions. *International Journal of Communication*, *11*(1), 159–173.

Okoro, N. (2004). Constitutional limitations of press freedom in Nigeria: Cases, indices and implication. *An Interdisciplinary Journal of Communication Studies*, *18*(1), 36–70.

Okunna, C. (2003). The press, the law and the society: A practical and ethical approach'. *UNILAG Communication Review*, *4*(1), 65–79.

Oster, J. (2013). Theory and doctrine of 'media freedom' as a legal concept. *Journal of Media Law, 5*, 57-78.

Perrottet, A. &Robie, D. (2011). Pacific media freedom 2011: A status report. *Pacific Journalism Review, 17*, 148-186.

Sani, A. A. (2014). Press freedom: Analyzing the portrayal of Nigerian state on National dailiescover page. *Journal of Yasar University*, *35*(35), 6255–6260. doi:10.19168/jyu.71007

Ufuophu-Biri, E. (2006). Post-Mortem of the Nigerian press freedom. *An Interdisciplinary Journal of Communication Studies*, *20*(4), 143–157.

Ukonu, M. (2004). Issues in the Nigerian press laws versus press freedom. *An Interdisciplinary Journal of Communication Studies*, *18*(1), 14–28.

Wilcox, D. (1975). *Mass media in black Africa: Philosophy and Control*. New York: Praeger.

KEY TERMS AND DEFINITIONS

Democracy: Government by the people; a form of government in which the supreme power is vested in the people and exercised directly by them or by their elected agents under a free electoral system.

Government: The political direction and control exercised over the actions of the members, citizens or inhabitants of communities, societies, and states. It is the form or system of rule by which a state, community, etc. is governed.

Issues: Issues are things people are thinking about and talking about that affect their life in the society. Issues can also be problems or situations that affect the public interest of a nation.

Mass Media: Channels used for delivering of information about societal issues and happenings to a large heterogeneous audience.

Nigeria: A country in West Africa, it has about 160 million people; it is regarded as the largest Black Country in the world.

Press Freedom: The right to public newspapers, magazines and other printed matter without governmental restriction and subject only to the laws of libel, obscenity, sedition, etc.

Uganda: A country in East Africa.

Chapter 13
A Discourse on Mass Media and Society

Nneoma A. Anozie
Chukwuemeka Odimegwu Ojukwu University, Nigeria

ABSTRACT

Mass media and society, a popular concept in media studies, has constituted much discourse due to roles media play in society and perceived effects that can result thereof. This chapter is inspired by the term 'medicalization of the society' whereby ailments are regarded as medical issues and subjected to medical diagnosis and treatment, regardless of their true causes. Similarly, the violence, moral decadence and ethno-religious crises witnessed in the society are largely ascribed to the media. This chapter examined the said effects of mass media with society's social systems, cultures and values, with a view to finding a relationship. It argues that these societal makeups especially ones as formidable as Africa's also affect largely members' conducts and reactions to media contents. However, it advocates children's news segment, adherence to media ethics, and use of media programmes to enhance learning, proper socialization, abolishment of negative cultures, media literacy among others.

INTRODUCTION

'Medicalization of the society' is a term that describes a process by which non-medical problems become defined and treated as medical problems, usually in terms of illnesses and disorders (Cornard, 2007). According to Cornard, illnesses or "syndromes" that relate to behaviour, a psychic state, or a bodily condition that now has a medical diagnosis and medical treatment, are numerous. Obesity, reproduction, sleep problems, countless addictions, depression or post-traumatic stress disorder, just to name a few — are now subjected to medical diagnosis and treatment. Cornard (2007:3) stated thus;

Clearly, the number of life problems that are defined as medical has increased enormously. Does this mean that there is a new epidemic of medical problems or that medicine is better able to identify and treat already existing problems? Or does it mean that a whole range of life's problems have now received medical diagnoses and are subject to medical treatment, despite dubious evidence of their medical nature?

DOI: 10.4018/978-1-5225-1859-4.ch013

Karp (1996), in Ehernberg (2010), believes that depression in the United States stems from a cultural context that results in collective vulnerability to emotional distress. His thesis, as explained by Ehernberg (2010), connotes that there has been a personal dislocation due to the triple movement of medicalization, disconnection and post-modernization. These increasingly loose human connections according to him contribute to an increase in depression; the loss of social bond of the community gives depression its deep meaning. Horwitz & Wakefield (1980) in Ehrenberg (2010), discuss main issues and debates regarding medicalization of depression and blurring of borders between what is normal and what is pathological. They (Horwitz & Wakefield) show these three processes as factors that drive individuals to distress:

1. Low social status,
2. Loss of valued attachments,
3. Inability to reach important goals.

The implication is that certain illnesses and mental disorders have causative agents that spring from abnormalities in the social system – our society, and the medicals take the fall for them.

A similar issue occurs when society ascribes the violence, ills and problems it witnesses to the mass media as the main causative agent. The media as the fourth estate of the realm in the society are saddled with numerous if not heavy responsibilities, ranging from surveillance to educating, informing and entertaining members of the society. They are expected to function effectively and live up to expectations. Yet, when they discharge these duties expected of them, they are accused of inciting violence among children and youths (Gerbner, 1980; Huesmann, Moise-Titus, Podolski, & Eron, 2003), raising fear of insecurity among adults through news contents, and reporting incidents of insecurity (such as terrorist attacks) especially in Nigeria. Ifeduba (2013) contends that there have been complaints of newly licensed media using their media platforms to incite ethnic animosity (Tettey, 2006), bias (Olorunnisola, 2009), irresponsibility and unethical practices (Skjerdal and Lule, 2009; Ifeduba, 2011). Media imperialism (Schiller, 1973, in Toholde, Olaniran and Aliede, 2015) is blamed for cultural deterioration in the society. Ifeduba (2013) asserts that there is a new genre of sport newspapers that devote their pages to European football leagues, a situation that seems to promote cultural imperialism since the publishers have access to local leagues and other sporting events but choose to publicise European leagues, clubs and stars.

However, the mass media are part of an already existing society with established value systems, cultures, religious ideologies, ethnic groups and political structure. There has to be a society before any mass medium is established. The media cannot operate in a vacuum or in isolation; and actually, those who are termed media practitioners emanate from the society. The main aim of any medium is to report occurrences in the society; therefore, messages contained in the media are reflections of events occurring in the society. Societies which enjoy positive value systems and cultures, stabilised political and economic structure, steady growth and development and good standard of living, will ultimately enjoy positive image from the media situated in them. When the reverse is the case however, there are bound to be negative media contents. It is pertinent to understand what and who constitute the society and the media especially as it pertains to Africa (with particular reference to Nigeria). Understanding the societal makeup will reveal the value systems, and cultures, which the members live by, and how these affect their conducts and their response to media messages. This will help establish a relationship between the said effects of mass media and the effects of negative cultures, value systems, etc. on society.

BACKGROUND

Klapper (1960) posits that concern about the possible influence of mass communication has been prevalent for several decades and has continued in a less intense way as long as there is quasi- mass communication. Social observers, according to Klapper, have modernized the popular saying, "the pen is mightier than the sword", claiming that the mass media are more powerful than the atomic bomb. Parents, teachers, religious leaders and lawmakers all point accusing fingers at the media as the force driving the malpractices witnessed in the society that affect children and youths the most. Similar concerns have been raised in Nigeria too. Mercy (2009) in Udende & Ugande (2015) posits that in Nigeria, since independence, there have been series of reported cases of violent crimes all over many university campuses and various communities in many states. They include incidences of rape, clashes among cult members, political clashes, land disputes, religious violence, and tribal problems among others. Even media scholars themselves wonder if they are really at fault. This has prompted a lot of scholars to conduct research on media effects on society members, and propound theories and models to this respect.

Despite the volume of research, theories and models on exposure to media contents and its subsequent effects on audience, the issue still remains contentious. Some scholars argue that the media alone are not capable of asserting such influence on audience as proposed (Klapper, 1960; Burton, 2010; Ekeanyanwu & Peter, 2015). Such magnitude of influence requires other factors inherent in the social system and the societal makeup to take effect. It should be noted that before this glamour about mass form of communication emerged, the society, the human race, has survived and has been able to extend its control over the environment so successfully due to the major fact that the human brain is distinctively capable of communicating (Cronkhite, 1976). If communication is distinctively human, that means it is societal since the society in question is the human society. This means that before communication in mass forms began, man has always communicated personally, interpersonally and in social groups. In the process of communication, not only goods or services were exchanged; both positive and negative ideas, beliefs, values, norms were equally exchanged. Agbanu (2014) posits that beliefs are the information an individual has about objects and actions, the foundation of attitudes and opinions. He noted that in different parts of the world, individual and societal beliefs vary on many issues, and they determine how people perceive things around them.

As noted earlier, society begets the media and there can be no media without the society. The media do not manufacture news, nor did the practitioners themselves fall from the sky. They only reflect happenings in the society. This implies that the negative media reports or conducts are products of events occurring in or value systems inherent in the society. Hence, the focus of this chapter is to bring to light the corresponding traits that induce or magnify the said effects of mass media in the society. Subsequent subtopics are dedicated to buttressing this point. Equally, suggestions that are deemed tangible to ameliorate the situation are outlined thereafter.

THE MASS MEDIA

Mass media are generally regarded as those media that make use of technology in order to reach their audiences who are scattered, far, unknown, of different age groups, classes and cut across gender and cultures (Anyanwu, in Aliede, 2015: 30). Nwabueze (2014:23) describes mass media as "channels of communication through which information are disseminated to a large, heterogeneous, anonymous and scattered audience". Mass media saturate our society: the television in the living room, the newspaper stands on the streets, the radio in the car and sometimes in the bedroom, the computer at work and at home, the fliers and most predominantly, the smart phones on our hands are some of the media channels daily delivering advertisements, news, opinion, music, and other forms of mass communication. According to Akin (2005), there is no standard for how large the audience needs to be before communication becomes "mass" communication. There are also no constraints on the type of information being presented as a car advertisement and a report on Boko Haram terrorists' attacks are both examples of mass media messages.

African media practitioners are men and women who are trained or who have the skills required to manage news in the communication-driven world. They are the 'gatekeepers', a term that was first employed by the Austrian psychologist Kurt Lewin to refer to individuals who govern the travel of news items in the communication channel (Bittner, 1989). Bittner further defined gatekeepers as any person or formally organised group directly involved in relaying or transferring information from one individual to another through a mass medium. These gatekeepers determine what kind of news gets to the audience. Cross (1983), as paraphrased in Agbanu (2014:2), posits that the responsibility which the society places on the media gatekeepers as they determine how people think and live their lives, not only made them the managers of the various media organisations, but also managers of human minds as they "tell the rest of us how to live, what to be afraid of, what to be proud of, how to be successful and how to be loved."

As noted earlier, the media have a heavy responsibility of effectively serving the society in which they exist. When such society is as heterogeneous as Nigeria, their responsibilities get more tasking as each individual or group tends to interpret media messages based on the value system they are familiar with or their need satisfaction. Benson-Eluwa (2013) in her work "mass media and society" detailed roles media play for the society. They include;

1. **Mobilization:** Promoting national interests especially in the areas of economic and political development.
2. Providing support for established authority and norms;
3. Serving as a market place where society can market goods or services through advertisements;
4. **Cultural Transmission:** Communicating society's norms and values, and transferring them from one generation to another or to society's visitors.
5. **Socialisation:** Advancing societal values and promoting patriotic attitude towards the state, and a host of others.

In a developing nation such as Nigeria, media roles are oftentimes highly demanding. They are expected to contribute immensely to the economic, political, individual and national development. Through educational programmes, awareness campaigns and so on, they conduct mass education on issues regarding health, personal hygiene, environmental sanitation, skill acquisition, proper patriotic behaviours and so on. Most importantly, they serve as intermediaries between the government and the citizens. Through them, the citizens can communicate opinions and concerns to the government. They also serve to keep the government in check by effectively playing their watchdog role. It is evident then that media's responsibilities are numerous, crucial, as well as tasking. It will only take an impeccable and extraordinary set of media practitioners to perform such huge tasks perfectly and not err at one point or the other, or run the risk of encountering criticisms however harsh from observers - the society.

The Society

Ordinarily, a society can be said to be a group of people inhabiting a particular area who share common customs, laws, norms, and are bound by common interests. In a society like Nigeria however, the meaning and characteristics of a society may vary. This is due to the convergence of different cultures, traditions, languages, values, religions and even political structures. The UNESCO's (1981) definition of community in Anyanwu & Aliede (2015) relates closely to what is obtainable in Nigeria. Community was defined as an aggregation of groups which vary in social class and economic status, often in political or religious affiliation and also in outlook and opinion. Each culture or tradition differs from the other hence what is acceptable in a culture can be prohibited in another. The interests and value systems of each culture may differ. The mode of dressing and interaction also differ. These differences can be integrated to form a common interest or can result in clashes of interests which can translate into full blown crises when not managed properly.

The Nigerian society comprises different social groups. The home, school, different religious, ethnic, political and otherwise groups, government and media all make up the society. Socialization and education begin from the home and extend to the school. The different religious groups also partake in inculcating good morals and patriotic behaviors in members. The ethnic groups serve as means of identity; each ethnic group has characteristics unique to them and for which they are known. These characteristics whether inherent or cultivated could be found in the culture, traditions, beliefs, and value system that members identify with. The government, political, economic and other social institutions are supposed to be avenues for members of the society to contribute to the growth and development of their society by utilising the good morals, good upbringing, sound value system and educational prowess inculcated in them since childhood. The government especially is charged with the duty of piloting the affairs of the society. It is set up to keep the society organised, use the resources of the state to better the lives of everyone, ensure the security of lives and properties of members, ensure that peace and unity thrive, provide basic amenities such as good motorable roads, drinking water, constant power supply, good health care services. When these various social groups function properly, there is bound to be harmony, tangible development and subsequent positive media reportage. However, when the reverse is the case, there are bound to be negative outcomes.

Unfortunately, the Nigerian society is characterised by bribery and corruption, low level of education and poor educational policies, sometimes negative or adverse cultures, traditions and value systems, lack of basic social amenities, avarice, chronic bad governance, government ownership of and otherwise imposition on mass media outlets, to mention a few. The different religions and ethnic or political groups in the country exhibit 'mutually exclusive' features, and have a high tendency to discriminate against or take less cognizance of others. The "us and them" ideology is very popular among them. Some of the violence witnessed in the country have been religiously justified by their perpetrators. A vivid instance is the Danish Cartoon Crisis of 2006 where about 139 people were killed in the attacks and demonstrations which lasted weeks mainly in Nigeria, Libya, Pakistan and Afghanistan (Jorgensen, 2006 in Akdulut, 2014). Each of these characteristics has its own effects and determines how the society is portrayed in the media. Just as depression, which has largely been medicalised, derives its deep meaning from loss of social bond (loneliness), low social status (which means poverty in the Nigerian sense) and inability to accomplish important goals, so also can youth violence, ethno-religious crises, and deterioration of cultures which the media take blame for be associated with troubled homes or neighbourhoods, inability to attain basic education and enlightenment, unemployment, hatred, conflict of interest, bad governance, and Nigerians' inordinate preference for foreign products. Hence, when the society malfunctions, degrades in value, gives in to violence, condones or practises adverse cultures and traditions, the media reflect these vices in their news contents.

Media Effect/Influence Studies

Scholarly research on media effects abound. Livingstone (1996), in Curran & Gurevitch (1996), posits that since the 1920s, thousands of studies on mass media effects have been conducted. Notable is the George Gerbner's (1980) research that kept track of how violent television contents affected children and the generality of the society. The research gave birth to cultivation theory that suggests that heavy viewing of violent contents on television literally cultivates a view of the world, thus making children grow up to cultivate and accept violence as a natural part of the society. There is also Leonard Eron's longitudinal research on the effect of television on children (Ekeanyanwu & Peter, 2015). Eron observed children that watched more violent contents on the television periodically at ages 8-9, 19 and 30, and found them to be more aggressive and violent with higher tendency to be convicted of a crime than those who watched less violent contents on the television.

Nigerian media scholars have equally conducted series of research on influence of the media on the audience. Asemah and Edegoh's (2014) "Viewers' perception of ritualism in Nigerian home videos" shows that portrayal of ritualism in Nigeria home videos affects viewers who now begin to see ritualism as means of solving life's problems. Also, their findings show that such movies affect Nigeria as a nation negatively in the international community. Gever's (2014) work, "Assessing the psychological effects of horrible pictures on newspaper readers" establishes significant relationships between exposure to horrible pictures and newspaper readers' emotional trauma and aggressive behaviour. However, his findings noted family relations, friendship, political association and others as extraneous factors that contribute to the said effects. The social media is not left out. Studies have shown a correlation between frequent exposure to social media platforms and poor academic performance, indecent dressing, addiction and other negative influences (Ashiekpe, 2015; Danladi, 2015; Ijeh, Umukoro & Amune, 2015) among social media users. Udeze, Nwodu and Nwodu (2015) posit that the quantity of information coming from the less developed nations has increase considerably because of access to social media. Unfortunately, such information contains more negative contents than good thereby increasing the negative portrayal of the developing nations. However, some studies on social media effects have shown positive influences on users who use them (social media) for academic purposes and social interaction. Among them are Danladi (2015), Dike and Anikwudike (2015), Tsegyu and Danjuma (2014), and Ochonogor, Ohiagu and Nwammuo (2015).

Some Relevant Media Theories and Models

Media theories and models are propounded to help understand how the media function and to answer questions posed by the society on the functions of the media. Media/Cultural Imperialism by Herbert Schiller in 1973 suggests that the Western media, which are more sophisticated, dominate other media around the world especially the third world. They (western media) diffuse the western values into the third world cultures through their media contents. Since the media are said to have powerful effects on their audience and that they react to media messages, audience in the third world are said to imbibe these foreign media contents. The proposition of this theory coincides with that of Attitude Change theory that suggests that the media change, shape, form and generally influence people's behaviour (Nwabueze, 2015:43). The dependency model propounded by Sandra Ball-Rocheach and Melvin DeFleur in 1976 focuses on the relationship between the mass media entity (information system) and society itself (social system). The model proposes that individuals in modern society have become increasingly dependent on the media for information (Bryant and Thompson, 2002). The Comstock psychological model developed by G. Comstock and his associates in 1978 shows that the behaviour of an individual viewer may be influenced by televised actions; a person learns some behaviour by watching it presented on television, and may adopt the learned behaviour depending upon the salience and personal excitement or motivation gained by engaging in such behaviour (Bryant & Thompson, 2002). The agenda setting theory by McCombs and Donald Shaw in 1992 posits that the media determine what the public discuss and think about through frequent reporting of certain issues, thereby giving such issues prominence and setting them as public agenda of discuss (Owuamalam, 2014).

Reviewing Media Effects and Correlating Them with Society Equivalent

According to Bryant and Thompson (2002), the level of dependency relationship and the strength of the media effects hinge on the stability or instability of the society and the degree of societal importance placed upon mass media as an information source. Hence, the notion of *cultural imperialism* is not easily attainable as has been advertised especially among Africans (who can be said to be strong-willed and unwavering in their belief system). This is because of their strong belief in their customs and traditions handed over to them by their ancestors. Ogan (1988), in White (2001), posits that while the economic components of media imperialism may be measured statistically, the cultural components can be difficult to measure. Despite the concern about western media or culture influx especially in Nigeria media, it has not extinct the people's native cultures. In Igbo land for instance, the *Iri Ji Ohuru* – new yam festival, *Igba Mkpe* or *Iru Uju* – the rigorous traditional burial rites and (sometimes severe) widowhood practices, and the breaking of Kola nut (*Iwa Oji*) reserved only for men have not been replaced by any alternative culture from the west. Such imperialism is mostly witnessed among youths who have access to formal education. Regardless, they still go back to their roots to perform some academic activities and school functions such as Igbo day and even prohibited practices, chiefly the secret cult activities (which tallies with the African traditional cults and rituals). Also, cultural imperialism that has thrived especially in Nigeria can be linked to the society's preference and demand for foreign products owing perhaps to their lack of faith in indigenous ones.

Concerning *mediated violence*, it is commonly held that television viewing does more harm than good, especially to young audiences. Particular attention has been focused on the negative effects watching violent media programmes have on social behaviour. Paik and Comstock (1994 in Mares, 1996) reviewed 217 studies that linked television viewing to aggression. However, Ekeanyanwu and Peter (2015:68-69) believe that no single factor no matter how potent, can cause humans to behave violently. Hence, a child who grew up in a violent home may likely exhibit violent behaviour with minimal exposure to mediated violence. Mares (1996) found positive results in studies she reviewed to include self-control, positive interaction with peers; and she suggested that future studies should consider the positive angle of television viewing for young audiences. Livingstone (1996), in Curran & Gurevitch (1996), opined that researchers must ask whether more aggressive people choose to watch violent television programmes (i.e. selective exposure), whether violent television programmes make viewers aggressive (i.e. media effects), or whether certain social circumstances (societal make-up) both make people more aggressive and lead them to watch more violent television contents (i.e. a common third cause).

There is also the issue of *agenda setting* as theory and function of the mass media. The media set agenda for their public. This agenda is set precisely by media practitioners because they have been made the gatekeepers. It is difficult if not impossible for each medium to exhaustively report every occurrence in the society hence, the need to select the perceived important or most pressing issues. In doing this, other events which the medium deems irrelevant are neglected. This can be detrimental since there is huge likelihood of media gatekeepers giving more publicity to news reports that are more profitable to their media organisations. Equally, journalists tend to cover stories of sources that pay them. Therefore, *news commercialisation* and *brown envelope syndrome* can affect media gate-keeping, which in turn affects agenda setting. However, this practice is vividly witnessed in public offices whereby officials meant to provide services to the public are tipped in order to do their job and prioritize the needs of those who tipped them at the detriment of those who do not. Besides, Ngoa (2012) noted that the media do not set agenda; therefore, opinions, attitudes and behaviours of people may change or be altered not necessarily because the media have caused an issue to be elevated in importance to the public; but rather, because people manipulate the media (media content) selectively to meet their individual needs and dispositions. This coincides with Klapper's (1960) selective theory whereby audience of mass media selectively expose themselves to, perceive and retain media messages. Furthermore, media, according to Burton (2010:71), are part of capitalism and its interests, and they inevitably promote views of the social elites – the ruling class that is, the rich members of the society. They do this through news commercialisation and brown envelope syndrome, thereby making the media accessible to those who can afford to pay. Olorunnisola (2009:90) in Ifeduba(2013), stated that rather than critique Nigerian Government's propagandistic offer of neo-liberal economic policy for possible inconsistencies with Nigeria's socio-political and economic antecedents, media managers focused on ways in which the businesses of their respective media houses will thrive in the evolving economic environment, leaving Nigerian citizens unprotected by their calibre of surveillance. Ifeduba (2013) agrees that it is now a common practice to charge money for news coverage. Except the bill is paid up-front, public interest notwithstanding, reporters will not be sent to an event. However, the same practice is witnessed in the society. The rich in the society are more respected and honoured than the poor. No one wants to associate with poverty thus, the poor are often disregarded. Even the poor themselves can afford to grovel at the feet of the rich just to receive favour from them. In a society where inordinate adoration is ascribed to money, it does not matter how genuine their riches were attained, or how honourable, truthful and just they are, as long as money lies in their pockets, they call the shots in the society. They are the ones who buy airtime and advert placements in the media. Therefore, it is not out of place for the media to promote the ideology since they are obviously business enterprises whose interest just like any other business enterprise is to maximize profit, generate revenues, gain popularity and so on in order to stay in business. And so just as the poor worship the rich for personal gain at the detriment of their own self, so do the media run the interest of the rich for their personal gain (Burton, 2010:72).

There are concerns about *media influence on political violence and voting pattern of audience* (society members). Some studies correlate a relationship between partisan media contents or political campaign advertisements and political strife cum audience voting pattern (Kenya election of 2007 in Semetko, 2010; Udeze, & Uwem, 2013). Some are controversial (M'Bayo, Amobi & Danjuma, 2014) while some others found no or minimal relationship (Owuamalam, 2014; Ojekwe, 2016). The mass media are the most common source of information about election campaigns in democracies and societies in transition around the world hence concerns about political bias in the mass media dominate debates about the roles and responsibilities of the media at election time (Semetko, 2010). Semetko noted also that behind these concerns is the assumption that there may be effects, intended or unintended, on public opinion and political behaviour and, ultimately, electoral outcomes. However, real life experience suggests that other societal factors such as religious and ethnic affiliations, bribery, corruption and countless electoral malpractices affect political behaviour, voters' decision, and electoral outcomes intensively and extensively especially in developing nations. The last presidential election conducted in Nigeria attests to this fact. The election results showed that Goodluck Ebele Jonathan (GEJ) recorded massive votes in the eastern region as opposed to his major opponent General Muhammadu Buhari (GMB) who recorded massive votes in the northern region. The negative campaign advert (defined by Goldstein and Freeman, 2002 in Alawode & Adesanya, 2016) publicized by AIT (African Independent Television) against GMB did not affect the election outcome. Nor did the praise campaign in favour of Goodluck Jonathan by NTA (Nigerian Television Authority) win him the election. Though it could be argued that Nigerians voted for GMB because he promised them change, the fact still remains that religion and ethnicity contributed heavily to voters' choice of candidate.

In addition, media influence on behavioural patterns of audience or otherwise mass communications intended for attitude change, are most times successful or unsuccessful depending on the favourability or not of the existing culture of the beneficiaries (that is, the target audience) or the disposition of the beneficiaries themselves (Nwodu, personal communication May 7, 2015). Klapper (1960) noted too that mass communication do not directly influence people, but just reinforce people's predisposition hence, mass communications play a role as a mediator in persuasive communication. The exclusive breastfeeding campaign succeeded because breastfeeding was already practiced among targeted audiences. The killing of twins was successfully aborted, not only because the agents of change canvassed laudably, but also because the victims (especially the mothers), their relatives and friends (which when calculated, amounted to the entire community) grieved each loss greatly. Over the years, there have been vigorous feminist activism and laudable outcry for empowerment, emancipation and equality for women. The media takes active part in this. Yet, there are still cultures in Nigeria where women face unwholesome practices such as girl-child marriage, adverse widowhood practices, and female circumcision, and have a number of restrictions to their human rights. The law and government have not made it any easier. Recently, the Senate passed a bill into law allowing under-age marriage. Education and media exposure may have somewhat liberated those who have access to them. However, a gap begging to be filled is always revealed when the number of educated females is crosschecked with the number of the uneducated ones. It should be noted too that given the economic stance of Africa, a lot of people are too busy trying to secure their daily meals to adopt media messages. Some cannot even afford media devices such as radio and television sets, smart phones to mention a few. Baran and Davies (2003:9), cited in Ekeanyanwu and Peter (2015, pp.60-61), explains thus;

If we have learned anything about the media over the past century, it is that they are not demonic forces that inevitably precipitate societal or personal disasters. Media alone do not create couch potatoes and cyber-addicts, or foster massive political demonstrations.... People using media have the power to create division or community. Media technology alone is powerless to initiate useful change. But technology can augment and amplify the actions of individuals and groups and in so doing, facilitate rapid and widespread social change on an important scale.

There are theories too that makes reference to other factors responsible for media effects on the society. The two step flow and the multi-step flow theories suggest that media messages pass through one or multiple stages before getting to the audience (Nwabueze, 2015). These media messages may have been coloured or adjusted in content by the carriers especially the opinion leaders or social groups which the audience belong to and source information from. These theories and some other studies recognise the roles society's social groups play in inducing or amplifying the effects of media messages on the audience.

The ideas related here are not meant to disprove years of meticulous efforts made by media scholars to find a lasting solution to questions they posed. That will be sheer heartlessness. The aim is also not to excuse the outright bad conducts of the mass media and their practitioners. The aim therefore, is to establish the fact that social groups, societal values, beliefs and cultures especially ones as strong and formidable as the Nigeria's also largely affect audience's reaction to media contents and the conduct of media practitioners. Every member of the society including media operators belongs to one or more of these social systems. Whether at the management, editorial, or reportorial level, they all come from different homes, have different norms and values inculcated in them, and are affiliated to different religious or ethnic groups, not to mention the different cultures and traditions. This variety of people fuse together to form media gate-keepers who determine what kind of news gets to the society. They too live in the society that is constantly plagued by chronic bad governance, insecurity, economic and political instability. It is conceivable how these factors can affect media messages and conduct of practitioners as each person/medium will tend to act according to the value system they are familiar with or look out for their best interest or the interest of whichever group they fall. Mihailidis (2009:8) have the following to say of media messages:

All media messages, beyond context, are bound to a specific perspective. They are created by a specific organization or group, within a platform and with a certain audience in mind. The details in creating messages often define the scope, angle, and theme of the information presented. Whether on war, climate change, or entertainment, media messages in specific countries will adhere to an agenda, often dictated by the political, social, and ideological systems in place.

SUGGESTIONS FOR THE MEDIA AND THE SOCIETY

Having considered several facets to 'media effects and society effects', it is pointless apportioning blames especially without a corresponding suggestion to make the society a beautiful place in which everyone can live happily ever after. Because concern about the possible influence of mass communication has prevailed and will continue as long as there is quasi mass communication (Klapper, 1960), the following suggestions are deemed tangible to ameliorate the situation.

For the Mass Media

- **Using Mass Media Programmes as Instruments of Learning:** Abimbade (2006), in Adegbija, Fakomogbon & Adebayo (2013), confirmed that in 1957, the Nigeria Broadcasting Service was established which eventually took over the transmission of educational programmes. The import is that the media since their inception in Nigeria have been used for educational purposes. Since the media play major role at a child's formation stage, the school board can introduce a learning period that shows movies and dubbed television programmes with sound educational qualities and moral values for students' benefit under the directives of an instructor. Students are expected to discuss and practice lessons learnt from such programmes. Also, students can organize dramas, talk shows and record them to be stored for learning purposes or publish newspapers and magazines of educational significance with the aid of the school management board. Where such is not attainable, students should be encouraged to contribute to existing newspapers and other media outlets through letters to the editor, opinion articles, contribution to phone-in programmes etc. This way, the mass media become instrument of learning instead of evil forces that instill fear and violent behaviour in society members.

- **Advocating "Children's News Segment":** This implies dedicating some minutes of the day's schedule to reporting happenings in the society specifically for children, especially ones that contain violence. This will be achieved by considering their psychological makeup and level of understanding while packaging and reporting news with violent contents to enable such news to have no or minimal negative effects on them.
- **Diversifying Avenues of Revenue Generation:** Despite the fact that the media operate to inform, educate and entertain the audience, they still have to consider revenue generation for the remuneration of staff, maintenance and even profit making. Since the media have to consider the business part of their establishment, they should engage in other means of revenue generation such as event planning (for broadcast), book editing and publishing (for newspaper), web designing and management (for social media).
- **A Redress of the Editorial Policies and Mission Statements of Media Outlets:** As Adesoji & Hahn (2011) rightly said, most newspapers (and indeed all media outlets) in Nigeria at their inception profess laudable editorial policies and ideologies which often give the impression that the collective good of the country and the well-beings of its citizens are the primary purposes for their establishment. Sticking to such policies and ideologies will make for more balanced and objective news stories in the media. It will also create a sense of neutrality and restore or at least increase considerably audience's confidence in the media.
- **Adherence to Ethical Standards and Rules of the Profession:** This issue is difficult to discuss and can be tongue-tying when considering a lot of factors (chiefly economic and political factors in Africa). It is difficult to ask media owners and practitioners to adhere to the code of ethics, set goals that seek to advance development, peace and unity in the society and implement them, be fair and objective and not report only terrorism, violence, conflicts; advance our culture, norms, traditions and summarily be positive, when the society itself is hardly positive. Day by day we witness incidences of terrorism, violence and conflicts. We experience hunger, unemployment, bad governance and other social vices. If a country like Nigeria can pass girl-child marriage into law, what culture then should the media advance? How can you ask a reporter who has not received his poor salary for three months to be fair and objective and shun brown envelope? How do you tell media owners to adhere to the code of ethics when they have to pay their staff and run other expenses? It is really tongue-tying but equally a truth that should be told because once we decide to neglect the ethics of media profession due to economic uncertainties, we might as well shut down the African media and everyone go home to continue the crude ways of communication.

For the Society

- **Proper Socialization:** Charity begins at home. As mother Theresa of Calcutta (1910-1997) rightly said; "peace and war begin at home. If we truly want peace in the world, let us begin by loving one another in our families. If we want to spread joy, we need for every family to have joy." Parents should raise their children properly, inculcate good morals, and monitor their activities, their association with school mates, friends and neighbours, and their exposure to television and other media of mass communication so that when they detect any change in their behaviour, they can decipher its true source. In cases where children do not have homes, foster homes can be set up for this purpose.
- **Quality Education:** Teachers should take their duties seriously so that children entrusted in their care do not go astray. The government and other institutions can assist to enhance the standard of education by increasing teacher's salary, recruiting more teachers to ease the workload of the existing ones and updating the educational system. Education should be made more lucrative through establishment of workshops in schools for carpentry, cloth designing etc., standard scientific laboratories for food processing in order to promote agriculture, provision of adequate sports equipments, and award of scholarships to brilliant students etc.
- **A Responsible Government:** Udeze, Nwodu and Nwodu (2015) posit that the vices witnessed in underdeveloped countries such as Nigeria are fueled by bad governance – bad governance increases criminal activities and social strife, economic and political instability which are reported in the mass media. The government should shun corruption and perform the duties expected of them. They can assist in reducing vices in the society by creating jobs or equipping youth with loans or machineries to enable them establish their own businesses. They can also go into partnership with private firms and small businesses to enable them grow extensively. All it takes is to become humane, understand people's suffering and identify with them. Nigeria is blessed with enough resources that can go around if the leaders will allow it. The moment African political leaders shun corruption and seek the interest of their nation first before theirs, hunger, unemployment, economic and political unrest will become history.
- **Media Literacy:** This has been suggested by other scholars in the field of communication to be introduced by parents to their children and be included in school curriculum beginning from secondary schools (Ende, 2011; Jusoff & Sahimi, 2009). The reason is to enable the public to analyse media messages critically so as to control how the media affect their lives especially negatively. Sparrow (2007), in Jusoff & Sahimi (2009), has suggested several ways parents could teach their young children to be media literate to include setting limits, reducing and carefully choosing channels for young children and school age children's viewing. It is suggested that parents discuss with their young children about what is right and what is wrong, especially on aggressive behaviours (Jusoff & Sahimi, 2009). Adults should express disapproval of on-screen behaviour, raise questions about the realism of televised information and encourage children to discuss them and engage children in moral reasoning, teach children to evaluate television content rather than accept it uncritically and by extension, engage children in social problem-solving and pro social acts, such as helping, sharing and defending victims of injustice (Berks, 1994 in Jusoff & Sahimi, 2009).
- **The Mass Media Should Be Viewed from a Positive Perspective:** The media equally have positive effects on the society and that aspect should not be neglected. Through their educational and awareness programmes, they afford their audience the opportunity to acquire informal education.

They alert the society of outbreak of disease if any, and inform them of proper preventive measures (the outbreak of Ebola virus disease is a good example). Without the media, it will be difficult for society to be abreast of happenings in their own environment and their government, not to talk of the world at large. The media afford businesses avenues to market their goods and services to a larger audience. There are many other positive and interesting uses and effects of the media. The society should harness them instead for their own benefit and avoid pointing accusing fingers.

- **Abolish Negative Cultures, Traditions, Beliefs, and Value System:** The African nation is notable for her rich cultures and traditions, rich both in quality and quantity. A good number of them serve to inform, entertain and educate. Akuko Ifo – tales by moonlight are used to inculcate good morals in children. The different dance groups – atiliogwu, egedege, entertain at traditional functions. The new yam festival and indeed many others are rich and commendable. Such cultures should be promoted and modernised to satisfy the yearnings of this generation positively. Unfortunately, there are some cultures and traditions that are better heard-of than experienced. Traditions that terminate human life through ritual killings or that bury their dead along with living humans; cultures that condone girl-child marriage, female circumcision and unwholesome widowhood practices; value systems that sacrifice human dignity at the altar of money and riches, beliefs that are inconsequential and that mar both individual and national development, and summarily negative practices that dehumanize and retard growth and development at all levels should be abolished in the society.
- **Individual Decision:** We are all masters of our minds. The human mind is capable of making and executing decisions. Every normal individual is equipped with common sense and conscience to decipher what is right and wrong. It is the refusal to choose the right, the embrace of the wrong that destroys the society. Though it is difficult to choose right in the face of hunger, some of the societal vices are done to advance the selfish interests of the perpetrators or their sponsors. The society should learn to say NO to vices that encourage strife in the society. All should endeavour to shun laziness, idleness, corruption, hatred, greed, jealousy so as to make the society a better place to live in.

CONCLUSION

The black nation is notable for her richness in cultural heritage, strong religious observance, assorted ethnicities and languages. The members (of the black nation) consist of men and women linked by social groups. Each of these groups professes different values and norms. The mass media are part of this society. An indisputable, indispensable union of mass media and society has continued to exist; complement each other through the various roles each play in ensuring an organized, progressive environment. The social system, cultures, values, norms inherent in the African society have to be positive and members must exhibit positive actions for the media to report positively. If there is good governance for instance, it will translate to better living conditions, reduction in crime rate, and lead to economic and political stability. If the society has positive cultures, beliefs, values and norms, then female children will not be subjected to early marriage and female circumcision, nor widows to severe (sometimes inhuman) widowhood practices. The society will not trade integrity for money; ritual killings, and other such vices will be eradicated. If there is peaceful co-existence, understanding and harmonization of interests among the different ethnic, religious and political groups, conflicts will be reduced to the barest minimum.

Thus, there will be no need to discuss media effects since there are no negative societal issues to report. Subsequently, activities and conducts of the mass media and their practitioners will be enhanced for the better. In all, the media and the society both make up the society and should work hand in hand to achieve peace and unity. After all, when the society becomes inhabitable due to their actions or inactions, both parties will be affected greatly and equally.

REFERENCES

Adegbija, M. V., Fakomogbon, M. A., & Adebayo, M. S. (2013). Roles of Broadcast Media for Instructional Delivery in Open and Distance Learning: Nigeria as a Case Study. *European Scientific Journal, 9*(23), 279–290.

Adesoji, A. O., & Hahn, H. P. (2011). When (not) to be a proprietor: Nigerian newspaper ownership in a changing polity. *African Study Monographs, 32*(4), 177–203.

Agbanu, V. N. (2014). *Propaganda and public opinion: A discourse on political communication and mind management*. Enugu: Rhyce Kerex.

Akbulut, A. (2014). Conflict in the global age: The case of danish cartoon crisis. *European Scientific Journal, 10*(14), 13–23.

Akin, J. (2005). *Mass media*. Retrieved from http://www.beyondintractability.org/essay/mass-communication

Alawode, S. O., & Adesanya, O. O. (2016). Content Analysis of 2015 election political advertisements in selected national dailies of Nigeria. *European Scientific Journal, 12*(5), 234–259.

Anyanwu, C. (2015). Communication and society an introductory overview. In J. E. Aliede (Ed.), *Studies in communication, mass media and society: Discourses and debates* (pp. 28–37). Lagos: PR-MIX.

Asemah, E. S. & Edegoh, L. O. N. (2014). Viewers' perception of ritualism in Nigerian home videos. *ANSU Journal of Arts and Social Sciences, 3*(1), 18-30.

Ashikpe, J. A. (2015). Perceived influence of social media use patterns on the social development of Nigerian adolescents and youths. In J. E. Aliede (Ed.), *Studies in communication, mass media and society: Discourses and debates* (pp. 204–222). Lagos: PR-MIX.

Benson-Eluwa, V. (2013). *Media and society*. Enugu: Virgin Creation.

Bittner, J. R. (1989). *Mass communication an introduction* (5th ed.). Prentice-Hall.

Bryant, J., & Thompson, S. (2002). *Fundamentals of media effects*. New York: McGraw Hill.

Burton, G. (2010). *Media and society: Critical perspectives* (2nd ed.). McGraw-Hill.

Conard, P. (2007). *The medicalization of society: On the transformation of human conditions into treatable disorders*. Maryland University Press.

Cronkhite, G. (1976). *Communication and Awareness*. Cummings.

Danladi, K. (2015). Assessment of the use of social media by secondary school students in Zaria. In D. Wilson (Ed.), *Communication, society and the Nigerian child: Issues and emerging trends in the 21st century* (pp. 227–236). Nigeria: ACCE.

Dike, K., & Anikwudike, C. N. (2015). Evaluation of the influence of social media use on the academic performances of federal polytechnic, Bida students. In D. Wilson (Ed.), *Communication, society and the Nigerian child: Issues and emerging trends in the 21st century* (pp. 91–102). ACCE.

Ehrenberg, A. (2010). *Weariness of the self: Diagnosing the history of depression in the contemporary age.* McGill Queen's University Press.

Ekeanyanwu, N. T., & Peter, A. S. (2015). Children, youths and mediated violence: A reflective evaluation of some selected theoretical models. *Journal of the African Council for Communication Education, 12*(1), 50–71.

Ende, T. S. (2011). Investing in Young People through Media Literacy. *The Journal of the African Council for Communication Education, 9*(1) 54-81.

Gever, C. V. (2014, November 25-27). *Assessing the psychological effects of horrible pictures on newspaper readers.* Paper presented at the Nigerian Psychological Association Conference, Nsukka, Nigeria. Retrieved from http://www.academia.edu/.../assessing_the_psychological_effect

Huesmann, R., Moise-Titus, J., Podolski, C., & Eron, L. (2003). Longitudinal relations between childrens exposure to TV violence and their aggressive and violent behavior in young adulthood: 1977–1992. *Developmental Psychology: American Psychological Association, 39*(2), 201–221. doi:10.1037/0012-1649.39.2.201 PMID:12661882

Ifeduba, E. (2013). Media entrepreneurship in Nigeria: A study of new patterns, orientations and coping strategies. *Covenant Journal of Communication, 1*(2), 145–162.

Ijeh, N. P., Umukoro, E., & Amune, P. (2015). Use of social media by youths in selected cities of south-south, Nigeria. In D. Wilson (Ed.), *Communication, society and the Nigerian child: Issues and emerging trends in the 21st century* (pp. 61–74). ACCE.

Jusoff, K. & Sahimi, N. N. (2009). Television and Media Literacy in Young Children: Issues and Effects in Early Childhood. *International Education Studies, 3*(2), 151-157.

Klapper, J. T. (1960). *The effects of mass communication.* New York: The Free Press.

Livingstone, S. (1996). On the continuing problems of media effects research. In J. Curran & M. Gurevitch (Eds.), *Mass Media and Society* (2nd ed.). London: Edward Arnold.

M'Bayo, R., Amobi, I., & Danjuma, Y. M. (2014). Influence of political campaign advertising on selected voters in Lagos, Nigeria. *KSU Journal of Communication*, (3), 59-76.

Mares, M. (1996). Positive effects of television on social behavior: A meta-analysis. *The Annenberg Public Policy Center, 3*, 1–26.

Mihailidis, P. (2009). *The 5 – A's of media literacy.* Retrieved from https://hofstramass112.files.wordpress.com/2009/09/mihailidis-the-5as-of-media-literacy.pdf

Ngoa, S. N. (2012). Agenda-setting: A theory in process. In Critique and application of communication theories (pp. 1-25). Ota: Covenant University Press.

Nwabueze, C. D. (2014). *Introduction to mass communication: Media ecology in the global village*. Imo: Top Shelve.

Ochonogor, C. I., Ohiagu, O. P., & Nwammuo, A. (2015). Social media: Connecting or disconnecting Nigeria youths in the 21st century? In D. Wilson (Ed.), *Communication, society and the Nigerian child: Issues and emerging trends in the 21st century* (pp. 1–10). ACCE.

Ojekwe, G. I. (2015). *Political advert campaigns and voting behaviour: A study of Akinwunmi Ambode's election ad campaigns in Lagos state*. Retrieved from http://www.inecnigeria.org/wp-content/uploads/2015/07/Conference-Paper-by-Grace-Ojekwe.pdf

Owuamalam, E. (2014). Voter perception of unique selling proposition in Nigeria's presidential political advertisements. *International Journal of International Relation, Media and Mass Communication Studies*, *1*(1), 1–16.

Semetko, H. A. (2010). Election campaigns, partisan balance, and the news media. In P. Norris (Ed.), *Public sentinel: News media & governance reform*. Washington, DC: World Bank.

Toholde, L. Z., & Olaniran, S. (2015). Cultural imperialism and national image: The role of the media. In J. E. Aliede (Ed.), *Studies in communication, mass media and society: Discourses and debates* (pp. 38–52). Lagos: PR-MIX.

Tsegyu, S., & Danjuma, Y. M. (2014). Impact of social media networks on the academic performance of mass communication students of Ibrahim Badamasi Babangida University, Lapai Nigeria. *KSU Journal of Mass Communication*, *3*, 47–58.

Udende, P., & Ugande, G. B. (2015). The puzzle of media violence and actual exhibition of violence by Nigerian youths. In D. Wilson (Ed.), *Communication, society and the Nigerian child: Issues and emerging trends in the 21st century* (pp. 355–370). Nigeria: ACCE.

Udeze, S., & Uwem, A. (2013). The influence of political advertising on Nigerian electorate. *Journal of Communication*, *4*(1), 49–55.

Udeze, S. E., Nwodu, L. C., & Nwodu, E. G. (2015). Social media and global news flow: What has changed? In J. E. Aliede (Ed.), *Studies in communication, mass media and society: Discourses and debates* (pp. 247–254). Lagos: PR-MIX.

White, L. (2001). *Reconsidering cultural imperialism theory*. Retrieved from http://www.tbsjournal.com

KEY TERMS AND DEFINITIONS

Mass Media: Establishments operating in a society (in this case the African society) who are registered and charged with the duty of reporting events and contributing to the development of the society.

Media Effect: Perceived effects (often negative) of media on the audience through their news contents, conducts or activities.

The Society: This can be referred to as the African nation. It comprises a group of people linked by various social groups. Their beliefs, value system and norms determine the behavior of their members which can have negative or positive effects on the nation. The social groups include the home, schools, peer groups, religious communities, ethnic groups, the government and political and economic institutions.

Society Effect: Societal vices such as, negative values systems and cultures, bad governance and clash of interests which affect the actions and conducts of members of the society negatively.

Chapter 14
Broadcasting Policy in Botswana:
The Case of Botswana Television

Seamogano Mosanako
University of Botswana, Botswana

ABSTRACT

The media has been considered an essential tool in propelling social change. Nonetheless, in Botswana, while the media is expected to play a role in the country's development, there exists a broadcasting policy vacuum in that there is no formal media policy to guide the operations of the media. A national television service, Botswana Television (Btv), is used as a case study in this chapter to demonstrate that the lack of media policy in Botswana and possibly other developing countries is generally responsible for poor performance of the broadcast media regarding its role in national development. Such poor performance of the media is reflected by domination of government originated content, domination of foreign content on local television, state control of the media as well as increasing pressure on the media to be more accessible to opposition political parties.

INTRODUCTION

The notion of media policy in this chapter largely refers to principles and decisions of both the government and media professionals that direct the actions of the media. Media policy sets parameters and guidelines through which the media operates (Freedman, 2008). It must be noted that whereas no formal document exists on broadcasting or media policy in Botswana, statutory instruments such as acts and regulations are the main guiding frameworks. Such include the repealed Broadcasting Act (1998), Botswana Communications Regulatory Act (BOCRA, 2012) and Broadcasting Regulations (2004). Additionally, there are actions and decisions of the Botswana Government that guide the operations of the media in Botswana which include directives on coverage of government activities, and national elections to mention but a few. Despite these regulatory frameworks, it can be argued that the media in Botswana operates in a policy vacuum, as there is no national media or broadcasting policy. In view of this, the aim of this chapter is

DOI: 10.4018/978-1-5225-1859-4.ch014

to evaluate the extent to which a lack of formal broadcasting policy in Botswana has impacted the successful operations of the media. Drawing on the development communication framework, this chapter focuses specifically on the national television service, Botswana Television (Btv). Within development communication, it is envisaged that the media plays an informational, educational and a unifying role in the process of national development (Amienyi, 2004; Lerner, 1958; Vilanilam, 2009). Subsequently, it is expected that media policies in developing countries' contexts should be framed with the intention to aid social change. Given the lack of a formal media policy in Botswana, this chapter explores possible policy considerations that could guide broadcast media in Botswana, particularly the national television service, Botswana Television (Btv), with the view to improving the role of the broadcast media in national development.

Television broadcasting in Botswana provides an interesting area of study in that the country is considered a democratic exemplar in Africa, as well as an economic success story. Moreover, a national television service was introduced comparatively late in Botswana in 2000. In light of this context, it is intriguing to explore the impact of lack of a broadcasting policy reflective of the democratic and developmental guiding principles of national development in Botswana. Thus the overarching aim of this chapter is to explore the impact of the media policy vacuum in Botswana on the successful operations of the media's role in national development.

This chapter will provide a background on the media sector in Botswana, with the view to articulating a context within which the national television operates. The chapter also explores policy initiatives guiding the operations of the media in Botswana as outlined in the national development plans and other official documentation. The purpose is to explore the policy issues framing "silent" media policy in Botswana, and the continuities and changes in those issues. The evidence, based on a review of various documents, suggests that national planning mostly focused on provision of infrastructure and technology, while content issues and the structure of the media were largely ignored. The chapter also provides an overview of discernable policy initiatives in the communications sector in Botswana. Such relate to media liberalization in the 1990s, to radical changes in the regulatory system that involve rejection of a national broadcasting policy, and to exemption of state media from oversight of an independent regulatory body.

MEDIA SECTOR IN BOTSWANA: AN OVERVIEW

The media is ascribed the role of information dissemination and in this section, the author considers the anatomy of the media in Botswana- broadcasting in general, and Btv in particular, to explain how the media are connected to national development in Botswana. Botswana has a traditional system of communication, *bogosi* or chieftainship, which has had a central role in the country's communication space, by facilitating national development communication. Through *bogosi*, it has been a practice that government officials, incumbent members of parliament, council representatives and bureaucrats mostly from state owned enterprises, address *kgotla* meetings to consult with the public and disseminate information on a non-political basis. The *kgotla* meetings are community- level gatherings presided over by the *Kgosi,* who ensures the issues discussed are non-partisan. This is a challenge in that by their nature, national development discourses are political issues. However, the *bogosi* system faces limitations of audience reach compared to the electronic media in that communication through the kgotla is inaccessible to those attending the gatherings. Nonetheless, the system remains a strong consultative forum and an information dissemination channel. In recent years (2010-2016), the current government has used the *kgotla* to make major public pronouncements, which traditionally would be made through the mass media. For example, in 2012, President Khama announced a public servants' salary increase during a *kgotla* meeting instead of the Budget Speech or a directive through the Directorate of Public Service Management as is the tradition. This is because the kgotla gatherings also get media coverage.

Regarding press freedom in Botswana, there have been significant changes relating to perceptions of the constraints within which the media in Botswana operate. In the 1990s, the Botswana press was considered to be relatively free in the context of Africa (Kijeski, 1995; Zaffiro, 2000). The press was capable of writing critical stories of the government. However, in recent years, scholars are noting the increasing efforts by the government to highly limit media freedom. The Media Practitioners Act of 2008, and defamation cases and sedition charges against a Sunday Standard editor are among the commonly highlighted examples of government efforts to muzzle the press. Recent reports by Freedom House that categorise the media in Botswana as being partially free also suggest the diminishing positive perceptions of press freedom in Botswana.

The print media is very vibrant in Botswana. There are both public and private publications. The Government owns one national paper, the *Botswana Daily News*, a semi-commercialized daily that circulates free throughout the country, and is by far the most widely circulated paper (Mochaba, et al., 2004). The Department of Information Services manages The Daily News. It carries mainly government and development-oriented stories. These are stories about government projects and policy initiatives. The Department of Information Services also publishes *Kutlwano*, a semi-commercial magazine, which has been published prior to independence in 1962. The magazine is distributed nationally, published in both Setswana and English, and covers cultural news, entertainment and feature articles. The department has a news agency, Botswana Press Agency (BOPA), which supplies news mainly to *Daily News* and Radio Botswana. The Department also has a publicity unit, which provides publicity services like public address systems. The private print media has grown significantly since independence. There are major publications such as *Mmegi*, *Monitor* (Monday paper, owned by the publisher who owns *Mmegi*), *The Gazette, Guardian, The Midweek Sun* (same publisher as *Guardian*), *The Voice, The Echo, Weekend Post* and *Sunday Standard to mention but a few prominent ones*. There are also a number of privately owned monthly magazines. The commercial print media, compared to the one owned by government, is more critical of government developmental efforts, and carries critical stories about government projects and policies. Thus, the public and commercial media provide a somewhat balanced view of developmental stories by respectively providing positive and critical elements in their coverage. The greatest challenge facing the private print media and to a lesser extent commercial radio is that the government, as one of the main advertisers in Botswana, has recently cut advertising in a large number of private newspapers. While the government cites financial constraints for this decision, commentators are arguing that it is a deliberate move to starve the press to death. Moreover, the media in Botswana generally struggles with access to official information. Despite these challenges, the print media seem resilient though with signs of saturation and reduced per capita circulation.

The broadcast sector in Botswana, similar to elsewhere in Africa, has its origins in the colonial period. Radio in Botswana dates back to pre-independence period where its origins are linked to colonialism. Broadcasting in Botswana, like in other African countries, may be traced to colonial times when the British Empire introduced radio stations to help communicate with the then Bechuanaland Protectorate (Zaffiro, 2000). In the early 1960s, the protectorate government also allowed its colonies to broadcast programmes that were aimed at the local people, and these included news of important and significant political meetings, policy statements and resolutions. Political parties were not allowed to buy airtime to broadcast political issues. At that time, the station focused on constitutional development and enhanced the spirit of African nationalism (Zaffiro, 2000). Post-independence radio has had a more developmental role in that in the context of Botswana, it was the dominant tool for communication between the government and the citizenry as the two parties forged to support each other in developing the livelihoods of Batswana. Radio in Botswana had also experienced significant changes in the 1990s, as there was a global move towards liberalisation of the communication industries. The outcome of shifts in ownership of the broadcast media saw an emergence of three commercial radio stations in Botswana, operating against two government owned radio channels. The state monopoly of the airwaves ended in 1999. The Botswana Telecommunications Authority (BTA), through the National Broadcasting Board, a regulatory body established under the Broadcasting Act (1998), licensed two private radio stations, *Yarona FM* and *Gabz FM*. They initially broadcast within Gaborone and surrounding areas, and were awarded national licenses in 2007 (www.bta.org.bw). A third station, *Duma FM*, was licensed in 2007, and is also national. The commercial channels have impacted the radio sector in that prior to their licensing in 1999s, the Botswana government made an unprecedented intervention by introducing a Broadcasting Act (1998). The act set the parameters for the broadcast media in Botswana, and provided for diversity of ownership structure of the broadcast media to include public service, commercial and community. But the Act, which was later repealed in 2012, had raised hopes in Batswana about the possibility of having community radio. This never occurred as there are currently no community radios, and the current legislation does not provide for either a public service or community broadcasting. The absence of these models of broadcasting raises questions on Botswana's liberal democracy, as diversity of media is assumed to be one of the cornerstones of democracy in Western societies.

Similar to radio, there are both public and private television stations in Botswana. Compared to radio and print media, the growth of television broadcasting in Botswana has been lethargic. A national television, Botswana Television, which is used as a case study in this chapter, was introduced in 2000. This is partly because at independence, the British colonial government had not set up a television service. Moreover, post-independence Botswana government did not prioritise the introduction of a television service citing various reasons such as financial constraints, insufficient power grid as well as unskilled manpower to run such a service. Nonetheless, since its introduction, the service has grown from five hours' to 24 hours' daily broadcasting. Similarly, the 'takeoff' of commercial television broadcasting in Botswana has been dismal, other than for subscription digital satellite television. The *eBotswana*, a version of the South African free-to-air channel *e.tv* Africa, was licensed in 2007, after taking over from the then Gaborone Broadcasting Corporation (GBC). While no empirical evidence exists to explain the poor performance of commercial television in Botswana, one can deduce that the inadequacy in local programming of *eBotswana* could be stifling the potential growth of the service. Local content is one of the key unique and selling points for most broadcast services. Both GBC and *eBotswana* have always been some form of relay channels for South African based programming. Nonetheless, these are also policy issues that need political commitment to nurture commercial broadcasting.

The Botswana media landscape has also witnessed significant strides in the use of modern information technologies, such as the Internet and social media platforms. As in other parts of the world, the Internet has taken off in Botswana with major urban centres enjoying access. However, access, uptake and usage of the internet in Botswana remains a challenge in most parts of the country. Among other factors, there is low broadband, high computer prices, power supply problems, and low IT literacy. The Internet has been used by the media with major newspapers (although just uploading content from the print copies) and private radio stations available online. Nonetheless, limited usage of the internet implies limited access to the media content through internet and social media platforms.The Botswana government has prioritised ICT as one of the ways to help the country to achieve economic prosperity. The government has also committed to improving Internet connectivity and promoting use of ICTs through a national ICT policy called *Maitlamo* (meaning commitment). Nonetheless, the limited access to the newer forms of communications imply that the traditional forms of communications such as the *kgotla*, and the 'old' channels such as radio, newspapers, and television, remain the dominant forms of development communication in Botswana.

GOVERNMENT MEDIA POLICIES AND PRACTICES BEFORE BTV (1966–2000)

The period 1966–2000 includes a period of media policy that preceded the commencement of a national television service in Botswana. During that period, broadcast media policy centred on radio broadcasting because it was the only existing state-owned broadcast media. Thus, the policy directives for radio broadcasting were to shape and influence television broadcasting. During this period, just before Botswana achieved independence (1963–1966) and soon after its independence (1963–1975), the media policy was framed by the government's Department of Information and Broadcasting's need to provide publicity about government activities, as well as mobilise the public to support the initiatives of the new government. It must be noted that, at this time, the crucial focus of the newly independent Botswana was establishing a united nation, avoiding racial rivalries, and safeguarding territorial and sovereign integrity (Republic of Botswana, 1966). In addition, the focus of the government was to introduce resource planning in lieu of the limited national financial resources (Republic of Botswana, 1968). In view of these national priorities and the socio-economic conditions at independence which included acute shortage of social services and non-existent infrastructure, the government-owned media were ascribed publicity and mobilization roles both within and outside Botswana. In this case, the media were to acquaint the Batswana about their country's problems and the objectives of their government, as well as publicise the government's development policies, consistent with the modernization theory of development communication, which ascribes the media a mobilization role in development.

Specifically, both film and radio were considered the most appropriate media for establishing communication between the government and the Batswana soon after independence. Moreover, radio was already an established medium that the colonial government had launched. Indeed, with illiteracy rates of approximately 75% (Republic of Botswana, 1968) at the time of independence, radio and visual media, such as film, were perhaps the most appropriate medium of communication. Furthermore, positive indications about the efficacy of radio were already evident as radio was said to have been instrumental in mobilising Batswana towards national objectives, such as national unity and self–reliance (Republic of Botswana, 1968).

The approach to broadcasting policy shortly after independence was focused on ensuring public support for government activities, mainly through exhortation, as well as mindset change. According to The Hughes Report, a consultancy report that has been widely used as a guiding document for government media in Botswana, the functions of the government media were the following:

1. Interpret the policy and actions of Government to the people:
    a. By a continuing service of information and public relations, and
    b. By campaign and concentrated publicity on particular subjects.

2. Encourage and assist the people of Botswana to take an increasing interest in and responsibility for economic, cultural, and political development of their country.
3. Advise government on public opinion and in the field of public relations.
4. Develop and exploit media, which will assist ministries and departments in performing their routine and extension duties.
5. Publicise Botswana outside its borders (Hughes Report, 1968).

Although there has been no explicit document in the form of a policy statement that articulates the role of television in national development in Botswana, these roles of radio continue to shape and influence the performance of a national television in Botswana.

The Hughes Report also established the expected relationship between the media and the government. The Report stated that state-owned radio was not to be adversarial towards government (Hughes, 1968). Furthermore, the Hughes report endorsed state control of the government-owned media and endorsed propaganda-based news reporting, dismissing impartial news broadcasting. Whereas Hughes' clarified the role of government media as a system that has to be supportive of the government as well as providing overt propagandist content, this is significant as it underscores the current perceptions of state control of the government-owned media in Botswana, including a national television.

The Hughes Report was prescriptive regarding programming. Hughes, consistent with the use of media for development as espoused by modernization (Lerner, 1958), advocated the use of radio to conduct adult education and use of foreign content that would be of interest to the Batswana. He also advocated cultural programming; using the national anthem at the beginning and at end of broadcasts; broadcasting Ministers' speeches; providing informal education-related programming, cultural magazine programs, and government spot announcements on topics such as elections, health, and animal husbandry. Such programs were intended to instill the spirit of nationalism. Additionally, the programs were to be packaged with the intention of praising the good work of any one community for others to follow. Hughes recommended that radio should cover government department programs, especially during off-peak hours. While this chapter does not focus on Btv programming output, a cursory analysis of Btv's program output reflects that radio programming requirements established by Hughes, have set a precedence for future state-owned broadcasters in Botswana, such as Btv. The channel's programming appears to be influenced by radio programming.

Another expert, Christie Lawrence, was sponsored by the Commonwealth Fund for Technical Co-operation, and was engaged 10 years after Hughes to advise the Botswana government on media policy options, and evaluate the operations and functions of the nation's government-owned media (Lawrence, 1978). Whereas the Lawrence Report endorsed the functions of the media as defined by Hughes, Lawrence's point of departure was how the functions should be implemented. Lawrence was of the view that the public relations role of state-owned media included the expectation that government should not suppress the truth, and that state-owned media should be allowed to publish material that is critical of the government or not to the government liking (1978). Commenting on Radio Botswana's programming, the Lawrence Report included the observation that "as long as the news and current affairs programs are regarded generally as providing solely a mouthpiece for Government, they will remain forever emasculate" (1978, p48). This signals a liberal orientation towards media's function, as opposed to the propaganda-oriented function that Hughes espoused. The liberal thinking that dominated Lawrence's report probably influenced the government to cling to Hughes' policy recommendations, because the

Botswana government did not adopt the Lawrence Report, unlike the Hughes Report, as guidelines for government media operations.

Overall, Botswana's policy options on state-owned media before the establishment of a national television broadcaster were consistent with the post-independence challenges facing the country's nation-building aspirations. As such, the media were directed to be supportive of the new state by providing public relations and publicity for the government's activities. Because Btv was launched almost 34 years after independence, and Botswana was, by then, a consolidated democratic state, the question arises as to whether the media policy is reflective of the democratic advancements in Botswana society. The next section explores the policy objectives during Btv's early years.

MEDIA POLICIES: THE ERA OF NATIONAL TELEVISION BROADCASTING

The data covering the early years of Btv (2000–2005) indicate that Btv's role was to emphasise the socio-cultural objectives of "reflecting Batswana to themselves", showcasing culture, and nation building (Botswana Television, 2004; Director Broadcasting Services, n.d).Thus, the role of television, as conceptualised by media professionals, in this case the Director of Broadcasting Services and the coordinator of the Btv project, reflected a shift from publicity and mobilization roles to a cultural identity role. A possible explanation of this shift might be the influence of the British nationals' understanding of the role of the media in a British context, where the media are cultural institutions. In this context, the influence of Kevin Hunt, a British expert engaged in the establishment of Botswana Television, might have nuances of what Servaes (1999) described as expatriate consultants mechanically transplanting their experiences of the West into non-Western contexts. Nonetheless, this cultural role of the media starkly contrasts the stated functions of government media, to which Btv belongs. For example, the National Development Plan 9 (NDP 9) enumerated the functions of government media, which were similar to the role of the media in the years before Btv's establishment, as follows:

- To interpret the policy and actions of Government to the people by continuing service of information and public relations and through campaigns and concentrated publicity on particular subjects.
- To encourage and assist the people of Botswana to take an increasing interest in and responsibility for the economic, cultural, and political development of their country.
- To develop and exploit media which will assist ministries and departments in performing their routine and extension duties.
- To publicise Botswana outside its borders (Ministry of Finance and Development Planning, 2003).

The role of television during the early years reflects two related but different policy goals: cultural identity and mobilisation of public support for government policies or national development. Although these are all social policy objectives, their practical implications in the Botswana context have varying levels of influence on content of the broadcaster. Whereas cultural goals reflect the need to convey distinctiveness and indigenisation, national development implied a focus on government activity, as national development in Botswana is a state-led and dominated practice. Thus, a national television through its programming has to grapple with these varied but realistic expectations.

Nonetheless, due to a lack of formal broadcasting policy, there appears to be limitations on local content to be able to project the above functions of Btv that relate to cultural identity and national development. The documents that were reviewed for this study revealed that government made no commitment to broadcast local content during the early period, although internally, media professionals opined that local content needed to increase (Botswana Television, 2003). For example, the then General Manager of Btv and a British national, Simon Higman, stated that local content would constitute approximately 60% of Btv's content (Gotlop, 2000). Moreover, although the Broadcasting Regulations recommended a minimum of 20% of local content on national television and 40% on radio (Broadcasting Regulations, 2004), these regulations could not be enforced because Btv is not licensed. The lack of enforcement impacted local content output on Btv, with the service continuing to be dominated by foreign content.

Regarding editorial independence, whereas the documents that were generated at national level were silent on this issue, those that were generated by media professionals at the departmental level included statements that Btv enjoyed editorial independence (Department of Broadcasting Services, n.d; Department of Information and Broadcasting n.d). The absence of articulation of editorial independence at a national level suggests an absence of commitment to this issue at government level. Moreover, such a decision could be an outcome of Hughes policy suggestions (1968), which gave less importance to government media's editorial independence. Guaranteeing broadcasters' editorial independence is intrinsic to media policy (Franklin, 2001); however, in the context of Botswana, the silence of a democratic government on such a significant policy issue might support concerns on Botswana's democracy, in that it displays signs of authoritarianism (Good, 1996). Conceivably, the reality on the ground at Btv in 2001, a year after the service's launch, reflects the dynamics of editorial independence of the national broadcaster. For example, the resignation of an expatriate Btv Head of News and Current Affairs, Chris Bishop, on grounds of government editorial interference suggests limited editorial independence at Btv.

During the mid-years, the Botswana government's thinking was motivated by the need to transform Botswana's broadcasting system to achieve a more diverse and socially responsible broadcasting agenda (Ministry of Communications Science and Technology, 2006). Two examples from the documents that were reviewed for this study illustrate this. First, in a speech at a consultative workshop on transforming the Department of Broadcasting Services into a "true" public service broadcasting outfit, a cabinet minister, Dorcus Makgato-Malesu, affirmed Botswana's commitment towards ensuring that both state radio and television were transformed into public service broadcasting output. Second, the then Minister of Communications Science and Technology, Pelonomi Venson Moitoi, presented the Draft Broadcasting Policy to parliament in 2006. The transformation of state broadcasting was explicitly stated in that document: "The Department of Broadcasting Services will be transformed into a public broadcasting services by an Act of Parliament and will fulfill the public mandate controlled by the public at large through an independent and representative board" (Ministry of Communications Science and Technology, 2006, p18). The document further enumerated various objectives for the public broadcaster as follows: to reflect issues of concern, provide diverse programs; provide a forum for democratic debate; showcase culture; act as a vehicle for development; provide coverage of significant events; provide quality programming; support local productions; be a reliable source of information; and provide a platform for innovation (Ministry of Communications Science and Technology, 2006). Of significance in the draft document was the introduction of three tiers of broadcasting in Botswana: public, commercial, and community broadcasting. The three-tier system of broadcasting would align with the Broadcasting Act, which provides for public service, commercial, and community media.

The draft broadcasting policy attracted resistance from members of parliament who opposed the introduction of community broadcasting. They feared that community broadcasting could cause harm by dividing the nation, because its content could be used to fuel tribal tensions (Botswana Press Agency,

2008). Indeed, these concerns could be valid in that community media was used to fuel ethnic tensions in countries such as Rwanda (Kellow and Steeves, 1998) and Croatia (Vigna, Enikolopov et al., 2011). In other contexts, it was deployed to fuel hate, as was the case with radio in Nazi Germany (Cambridge, 2007). Although this study does not include a focus on community broadcasting, community broadcasting can provide two essential lessons for the media policy debates in Botswana. First, although community broadcasting is considered as a threat to Botswana's national unity, the government must also recognise the potential of such a medium in giving voice to communities by allowing them to broadcast issues of immediate concern to them, as opposed to getting coverage of national issues as is the case with main-stream media. Second, the political elites in the Botswana Parliament's opposition to the draft broadcasting policy on the grounds that community radio could divide the nation elucidates the contextual ethnicity issues that Botswana must contend with as these issues permeate national policies, including the media. Currently, ethnic minority groups in Botswana are increasingly expressing disenfranchisement to public discourse in Botswana (Good, 1993, 1999a) with the exclusion of their languages. Led by educated members of the minority ethnic tribes, these groups have pressured government to amend Sections 77 to 79 of the Botswana Constitution because of their perceived tribal discrimination. The President established a commission of enquiry in 2000 to explore the issue. This culminated in a referendum and, finally, in the amendments of the sections, as well as in amended legislation, such as that relating to the House of Chiefs *(Ntlo ya Dikgosi)*. This reflects positive political reactions in a democratic society. In contrast to this amendment, the same decision was not extended to other areas, such as use of the ethnic minorities' language in schools and in the media. Inevitably, the government has shelved the broadcasting policy and the desire of the ethnic minorities to have their languages recognised continues. This is one of the issues that future media policy in Botswana must address.

Furthermore, the decision to shelve the broadcasting policy signals the possible absence of political will to transform the broadcasting sector as this sector is needed for political expediencies. Rather, the Draft Broadcasting Policy, while constituting a commendable effort, might have been a political gesture to quell the pressure that international organisations have exerted. Institutions such as UNESCO and the Southern African Broadcasting Association (SABA) advocate the reform of state broadcasting in Africa. However, internally, the government might have been unwilling to distance the media from government due to the dictates of the Draft Broadcasting Policy. Overall, similar to the media policy issues in the early years, Btv's attention during the mid-years of the service continued to be on improving transmission signal access and strengthening the commercial output of the service. In addition, the role of Btv in national development continued to be framed along the publicity role of the government media, as discussed in the preceding sections.

THE CURRENT AND FUTURE TELEVISION POLICIES

The current broadcasting policy environment (2012–2015) signals a complete departure from the reforms that were envisaged in the mid years of Btv. Although, as mentioned, a formal policy document does not exist, the documents that have been reviewed reveal two themes that highlight the policy direction of government broadcasting in Botswana. These are the dominance of the government-oriented communication role of Btv, and the establishment of legal recognition of the state broadcasting model. In this manner, policy has returned to the basic concept of the role of the media as prescribed by Hughes in 1968. Before the launch of Btv, the role of the media, as explained above, focused on the publicity role of government media. Unlike the early years of Btv (2000–2005) where the media professionals prioritised Btv's cultural role, in the current period, there appears to have been a shift among media professionals towards conceptualising the role of Btv as a tool for disseminating government information (Department of Broadcasting Services, 2012). The objectives of Btv, as stated in the current strategy document, include a focus on the state media's role in increasing awareness about government programs and policies. The expected outcome of information dissemination is an increased acceptance of Botswana government policies by its citizen. Thus, the focus of state media in Botswana is government communication. This can be explained by the fact that in a developing state such as Botswana, public institutions are tasked with supporting the developmental process of the country (Hillbom, 2014; Leftwich, 1995). The other peripheral reason for re-focusing government media into a government information tool can be explained contextually by the relationship between the state and commercial media. The need for the government to own, and, thus, be able to influence its own media, arouse out of the government officials' perception that the commercial media was adversarial.

Accordingly, the emerging regulatory framework in Botswana indicates a shift from envisaging the 'broadcaster' as a public service to legally recognising it as a state 'broadcaster'. In 2012, the Botswana parliament enacted a new law, the Communications Regulatory Authority Act (BOCRA Act), which replaced the Broadcasting Act (1998), as well as the regulatory authority, the Botswana Telecommunications Authority (BTA), with a new one, the Communications Regulatory Authority (BOCRA) (Botswana Government, 2012). Although the Act is an honourable intention consistent with contemporary trends of media convergence, it defies the traditional three-tier broadcasting system that is advocated for in Africa (Article 19, 2006) by legally endorsing the state broadcasting model, and abolishing public service and community broadcasting. Whereas the acknowledgement of state broadcasting is a reflection of a realistic regulatory framework in which the national context is considered, this act has considerable implications. The Communications Regulatory Act of 2012, exempts state-owned broadcasters from being regulated by the independent regulatory authority which the Act had established. The Act includes the statement that "a state broadcaster shall not require a license to operate" (Botswana Government, 2012). Moreover, the new Act does not clearly provide for public service or community broadcasting (see Section 34 on BOCRA Act) and is specific to commercial broadcasting. The decision to establish public service or community broadcasters is left to the discretion of the Minister. This is a major policy issue that should not be left to the discretion of the Minister. Because the Minister who is responsible for Communications has previously failed to push the National Broadcasting Policy through parliament, it remains to be seen whether the Minister will ever be able to ensure that the Botswana media system becomes diversified through public service and community broadcasting. This episode of legally recognising state broadcasting or "governmentisation" in Botswana can be explained by Botswana's political context. Political scientists argue that in Botswana, the state currently displays liberal authoritarian tendencies (Taylor, 2003; Botlhomilwe, Sebudubudu et al., 2011), and that civil society and opposition parties are weak (Molutsi and Holm, 1990). Hence, the state is likely to strengthen its control mechanisms in the absence of any strong institutions to counter such decisions.

Regarding local content, the current national development plan (NDP 10) includes the acknowledgement of the insufficiency of local content in the broadcasting sector, which is estimated at 20% (Ministry of Finance and Development Planning, 2009). Although in NDP10 the media is envisaged as a contributor towards Botswana becoming a knowledge-based society and as a source of entertainment, the suggested development strategies in the plan for this period focus on technology, in this instance, on the inevitable digital migration. However, NDP 10 is silent on increasing local content. Furthermore, the current NDP includes a focus on facilities, such as office accommodation, transmission, and distribution facilities, as well as on recruiting consultants to provide technical assistance to Btv (Ministry of Finance and Development Planning, 2009). Whereas these facilities are critical, the omission of any commitment to increase local content suggests that content production is relegated to a peripheral position, while transmission, access, and distribution are the core objectives of improving broadcasting coverage in Botswana.

Botswana as a developing nation like elsewhere in Africa is still to forge a media policy framework that can facilitate positive social change. It is evident there is a policy vacuum, and therefore, future research in Botswana and similar contexts can explore whether 'no policy' is actually a policy on itself. In such instances, a framework of how media is regulated and its performance is gauged, can be developed. This is because without a policy as it is currently, assessing the role of the media becomes a challenge, leaving the media to suspicions that it mainly serves the interests of the government of the day.

CONCLUSION

The chapter demonstrates that in Botswana, there has been focus on media access and infrastructure, at the detriment of local content. Furthermore, the chapter demonstrates that due to lack of a broadcasting policy, media content is dominated by government agenda, and limits access for minority groups. This chapter highlights that during the years soon after independence in 1966 to the 1990s, prior to the introduction of a national television, the role of the media in various documents reflected that the media had a publicity role. Nonetheless, clear statements about the role of this new medium in the country's socio-economic development did not accompany the introduction of a national television service.

In the absence of a broadcasting policy that stipulates the structure and role of Btv, the service remains susceptible to perceptions that certain sections of society use the station to pursue their own interests, as reflected by the framing of the role of the media as a tool for government communications in various documents reviewed in this study. To improve the performance of Btv as a modernisation agent, policy suggestions relating to three broad themes: the broadcasting model, objectives and, programming, should be addressed. The broadcasting model, in this instance the state broadcasting model if it is adopted for public television, should be clearly outlined, setting the parameters for national television's operations. This model will also address the specific mandate and objectives that should be pursued by national television, of which the goals will be nationally agreed to limit perceptions of abuse of state media by the ruling elites. Moreover, the policy should address the programming, as a way of ensuring local content that is commensurate with developmental and democratic ideals of Botswana society. Core to these ideals is citizen participation and empowerment in social change.

The chapter has outlined discernable policy changes in the communications sector in Botswana. Such include liberalization of the airwaves that saw the introduction of commercial radio stations in the 1990s. A tangible media policy intervention was the development of a Draft broadcasting policy in 2005, a document that was rejected by Botswana legislators on grounds that the introduction of community broadcasting was a potential threat to Botswana's democratic stability. The change in the regulatory framework to remove state-owned media from the authority of an independent regulator in 2012 is discussed as an example of how the government is responding to contextual pressures such as the growth of the opposition parties and dissenting voices in Botswana. The chapter concludes with an observation that in the absence of a documented media policy, the state owned Btv remains an elitist medium as the state strengthens its control of the service. In light of the "success" story of Botswana, it is rather contradictory, and even radical of the Botswana government to be strengthening its control of the public media. The theoretical implication of this is that the process of democratization and social change in some emerging democracies does not always translate into media policies that serve the interest of the general public, but rather the ruling elites in such societies.

REFERENCES

Amienyi. (2004). Broadcasting and national development in Africa. In C. C. Okigbo & F. Eribo (Eds.), *Development communication in Africa* (pp. 106-114). Lanham, MD: Rowman & Littlefield.

Botlhomilwe, M. Z., Sebudubudu, D., & Maripe, B. (2011). Limited freedom and intolerance in Botswana. *Journal of Contemporary African Studies*, *29*(3), 331–348. doi:10.1080/02589001.2011.581501

Broadcasting Act 1998. 72:04 (Botswana).

Broadcasting Regulations 2004. 72:04 (Botswana).

Cambridge, V. C. (2007). Milestones in communication and national development. In Y. R. Kamalipour (Ed.), *Global communication* (pp. 181–200). Belmont, CA: Thomson Wadsworth.

Communications and Regulatory Authority Act 2012. 72:03 (Botswana).

Department of Broadcasting Services. (2006). *Department of Broadcasting Services-Quality radio and television: PMS strategic plan 2006-2016*. Department of Information and Broadcasting. Retrieved from http://www.dib.gov.bw

Department of Broadcasting Services. (2012). *Department of Broadcasting Services Strategy Document 2012-2016*. Gaborone, Botswana: Department of Broadcasting Services. Retrieved from http://www.dib.gov.bw

Director Broadcasting Services. (n.d.). *From the Director*. Botswana Television. Retrieved from http://www.btv.gov.bw/from_the_director.html

Franklin, B. (Ed.). (2001). *British television policy: A reader*. London: Routledge.

Freedman, D. (2008). *The politics of media policy*. Cambridge, UK: Polity Press.

Good, K. (1993). At the end of the ladder: Radical inequalities in Botswana. *The Journal of Modern African Studies*, *31*(2), 203–230. doi:10.1017/S0022278X00011903

Good, K. (1996). Authoritarian liberalism: A defining characteristic of Botswana. *Journal of Contemporary African Studies*, *14*(1), 29–51. doi:10.1080/02589009608729580

Good, K. (2008). *Diamonds, dispossession and democracy in Botswana*. Retrieved from http://UQL.eblib.com.au/patron/FullRecord.aspx?p=1069026

Gotlop, K. (2000, July 14). Btv on July 31. *Daily News*. Retrieved from http://www.gov.bw

Hillbom, E. (2014). Cattle, diamonds and institutions: Main drivers of Botswanas economic development, 1850 to present. *Journal of International Development*, *26*(2), 155–176. doi:10.1002/jid.2957

Kellow, C. L., & Steeves, H. L. (1998). The role of radio in the Rwandan genocide. *Journal of Communication*, *48*(3), 107–128. doi:10.1111/j.1460-2466.1998.tb02762.x

Kijeski, T. (1995). Appropriate technology and the Botswana television experience: On the efficacy of waiting. *The Howard Journal of Communications*, *6*(4), 292–305. doi:10.1080/10646179509361704

Leftwich, A. (1995). Bringing politics back in: Towards a model of the developmental state. *The Journal of Development Studies*, *31*(3), 400–427. doi:10.1080/00220389508422370

Lerner, D. (1958). *The passing of traditional society: Modernisation in the Middle East*. Glencoe, IL: The Free Press.

Ministry of Communications Science and Technology. (2006). *Draft Botswana broadcasting policy*. Gaborone, Botswana: Author.

Mochaba, K., Raffinetti, C., & White, J. (2004). *SADC media law; A handbook for media practitioners; a comparative overview of media law and practice in Botswana, Swaziland and Zambia* (Vol. 2). Johannesburg, South Africa: Konrad Adenauer Foundation.

Molutsi, P. P., & Holm, J. D. (1990). Developing democracy when civil society is weak: The case of Botswana. *African Affairs*, *89*(356), 323–340.

Republic of Botswana. (1968). *National Development Plan (1968-1973)*. Gaberone, Botswana: Gaborone Government Printer.

Servaes, J. (1999). *Communication for development: One world multiple cultures*. Cresskill, NJ: Hampton Press.

Taylor, I. (2003). As good as it gets? Botswanas democratic development. *Journal of Contemporary African Studies*, *21*(2), 215–231. doi:10.1080/02589000305446

Vigna, S. D., Enikolopov, R., Mironova, V., Petrova, M., & Zhuravskaya, E. (2011). Cross-border media and nationalism: Evidence from Serbian radio in Croatia. Cambridge, MA: National Bureau of Economic Research Working Paper Series. National Bureau of Economic Research. Retrieved from http://www.nber.org/papers/w16989.pdf?new_window=1

Vilanilam, J. V. (2009). *Development communication in practice: India and the Millenium Development Goals*. Thousand Oaks, CA: Sage. doi:10.4135/9788132104148

Zaffiro, J. (2000). Broadcasting reform and democratization in Botswana. *Africa Today*, *47*(1), 87–102. doi:10.1353/at.2000.0023

KEY TERMS AND DEFINITIONS

Development Communication: A framework that emphasis the role of communication in facilitating positive social change in a society.

Govermentisation: A situation where decisions are made with the view to bolster the point of view of a current government.

Kgotla: A traditional public meeting place in Botswana, presided by a chief, and used for consultations, as well as for judicial matters such as arbitration and resolution of disputes.

Media Policy: Principles and decisions of both the government and media professionals that direct the actions of the media.

National Development: Processes and activities that are aimed at positive social change reflected through changes in economic, cultural, social, and political outlook of a country.

Policy Vacuum: An absence of a written policy statement on a specific issue.

State Media: A media system that it financially government funded, and its editorial independence guided by the government of the day and its key purpose is predominately to project a positive public image of a current government.

Section 4

Media, Health Behavior, and Development

The role of the media in promoting health issues in Africa.

Chapter 15
Examining Sexuality in the Context of HIV/AIDS Knowledge among Students of the North-West University Mafikeng Campus:
Does Gender Matter?

Paul Bigala
North-West University – Mafikeng, South Africa

ABSTRACT

The study investigated gender differences in attitudes towards sexuality in the context of HIV/AIDS among undergraduate students. The cross-sectional study was conducted among 1,498 randomly selected students. The majority of the students interviewed were females at 53.9% (M= 20.24, SD=2.98) compared to 46.1% males (M=21.67, SD=3.35). More Males (16%) responded in the affirmative with regard to having more than one sexual partner than female students (10%). Also, more male students (32.7%) than female students (18.6%) indicated that using condom reduces sexual pleasure and over 40% of both sets of respondents indicated that sexual intercourse should only be done between married couples. This study underscores the need to strengthen HIV/AIDS programmes to include issues on sexuality and encourage undergraduate students to attend these programmes.

INTRODUCTION

At the end of 2013, UNAIDS (2013) reported that about 95% of 35.2 million people living with HIV/AIDS worldwide were residing in developing nations and indicated that sub-Saharan Africa is affected the most by the HIV/AIDS pandemic. In sub-Saharan Africa, the southern African sub region remains the most affected with South Africa having the largest population living with HIV/AIDS in 2012. Despite the knowledge about HIV/AIDS pandemic the prevalence rate in South Africa continues to increase

DOI: 10.4018/978-1-5225-1859-4.ch015

(Makhoahle & Bagali, 2012; Reddy & Frantz, 2011; Republic of South Africa, 2012). This indicates that knowledge has not translated to safer sexual practices and prevention in South Africa (Makhoahle & Bagali, 2012).

Recent South African National HIV Prevalence, Incidence and Behaviour Survey showed that youth had the highest HIV prevalence (Shisana et al, 2014). The HIV prevalence among 20–24 years age group was estimated at 22.1% with women disproportionally affected (Republic of South Africa, 2012). In addition, studies in South African universities revealed that females aged 20–24 were at more risk of contracting sexually transmitted diseases compared to males (Akintola et al., 2012; Mutinta & Govender, 2012). The higher prevalence and risk of HIV infection among females compared to males may be attributed to the more HIV/AIDS preventive knowledge of the later (Marianne et al., 2013).

On the contrary, studies have indicated that female students were more knowledgeable because they participate more in HIV/AIDS programmes than males (Ebeniro, 2010; Maliki, 2010). The knowledge, attitudes, belief and practice study (KABP) in 21 South African higher educational institutions between 2008 and 2009 revealed that female students were more likely to have more knowledge on HIV/AIDS than their male counterparts. Similarly, higher knowledge of HIV/AIDS among female compared to male students have been reported in Ghana (Asante, 2013) and Nigeria (Maliki, 2010; Mbakwem-Aniebo et al., 2012).

Irrespective of differentials in HIV/AIDS knowledge by gender, university environment provides students with unlimited opportunities for sexual exploration (Reddy & Frantz, 2011, HEAIDS, 2010). Several studies have shown that young South Africans continue to engage in risky behaviour as they perceive themselves to be at low risk of HIV infection (Maswanya et al., 1999; Pettifor et al., 2009; Macintyre et al., 2004). This suggests that knowledge and attitude which would culminate to HIV preventative measures remain inadequate among university students (Akintola et al., 2012; Petiffor et al., 2005, Tagoe & Aggor, 2009; Oppong & Oti-Boadi, 2013).

Given that university students are sexually active, engaging in risky sexual behaviour, the North-West University responded to the Government's call to mitigate the spread of the pandemic through the university's core functions of teaching/learning and research. North-West University has continued implementation of HIV/AIDS interventions such as prevention, treatment, care and support programme. However, the dividends of this programme have not been assessed. Thus, this study examined the gender differences in HIV/AIDS knowledge and beliefs among undergraduate students of Mafikeng Campus, North-West University. It provided an empirical assessment of the University's HIV/AIDS prevention programmes. Understanding the pattern in HIV/AIDS knowledge and beliefs by gender and the mitigating impact of the current programme on HIV/AIDS among students would guide the thrust of future intervention strategy. In line with the objective of the programme, it was hypothesized that knowledge of HIV/AIDS prevention is significantly higher among female undergraduates compared to their male counterparts. Secondly, female undergraduates had more access to HIV/AIDS programmes than their male counterparts.

STUDY METHODS

Study Setting

The study was conducted among undergraduate students aged 16–24 years in North-West University, Mafikeng Campus. The Mafikeng Campus is the second-largest of the University's three campuses and is situated in Mafikeng, the capital of the North-West Province. The student population of the five faculties of Education; Law; Agriculture, Science and Technology; Human and Social Sciences; Commerce and Administration from over 22 countries is about 10,114. As a result of the awareness of high prevalence of HIV rates in the country which is affecting mainly the youth, the University Management created a HIV/AIDS Coordinating office to sensitize students and all the members of the university community on HIV/AIDS information and programmes.

With its strong passion for community service and skilful blend of academic and research programmes, the Mafikeng Campus embodies the North-West University's commitment to being locally engaged, nationally relevant and internationally recognized. The North-West University has recognised the devastating consequences of the HIV/AIDS pandemic to students and staff alike. As a result, it has worked together with its sister campuses in mitigating the impact of HIV/AIDS. The HIV/AIDS co-ordinating office is not only interested in reducing the impact of HIV/AIDS but also to protect and up-hold human rights, ethical principles and gender equity; provide a comprehensive leadership response including staff and student structures; minimize short and long-term impact on the risk on institution/campus; and de-stigmatize HIV/AIDS. The services provided by this office, in collaboration with the health centre, include prevention and peer education programmes, awareness campaigns, voluntary counselling and testing/HIV counselling and testing.

Study Design

The design for the study was cross-sectional where the participants were examined based on prepared questions on demographic, HIV/AIDS knowledge, HIV/AIDS attitudes related to sexuality, awareness and access to HIV/AIDS prevention programme on campus, and in addition a section on general self-esteem of students.

Training of the Research Assistants and Data Collection

The data were collected by 10 trained peer educators who are students from the five faculties and also work in collaboration with the University HIV/AIDS unit. They were trained for two days on the contents of the questionnaire designed for this study. The questionnaire was pilot-tested for internal consistency, reliability and validity of its contents.

Sample Method

The data collection involved a multi-stage sampling approach where the selected participants were taken from each of the five faculties in the University. At the first stage departments were randomly selected from the faculties and thereafter, students were randomly selected from each of the departments. However, participants were selected from each year of study to ensure that the data composition includes students from each level. At the end of the selection, 1,498 students were selected from Faculty of Commerce, Humanities and Social Sciences, Law, Agriculture and Science, and Education. The number allotted to each faculty was obtained using the proportionate sample approach.

Study Approval

Permission to conduct this study was sought from the Ethics Committee of the University. An ethics number was granted by the Institutional office of the University. The Dean of Student Affairs who is in charge of students on campus also approved this study.

Data Analysis

The Statistical Package for the Social Sciences (SPSS) version 21.0 was used for data analysis where standard descriptive statistics were used to describe some of the findings. Chi-square was used to explore the associations between the various variables under the themes of students' HIV/AIDS knowledge, and attitudes related to sexuality by gender. A composite variable was created to measure students' attitudes towards their own sexuality. The scores were categorised into two segments based on the mean and median score: those less than mean scores were classified as negative and those above were classified as positive attitudes.

Limitations

This study did not look at whether students had ever had sex as it was not asked of them. This would have given an indication of their sexuality in the context of HIV/AIDS knowledge. As a result, aspects such as number of sexual partners were also not included or asked. A qualitative research would need to be done to ascertain their sexual attitudes especially in relation to HIV/AIDS knowledge.

Results

Table 1 shows the characteristics of the sample in the study. Of the 1,498 participants, slightly over half of the respondents were females (n=807, 53.9%). The majority of the undergraduate students sampled were between the ages of 16–20 years with the mean age of 21.67 years (SD=3.35). Most of the undergraduate students were in the Commerce, Education, Science and Social Science Faculties respectively. Approximately 83% of the undergraduates were Christians. The majority of the interviewed students were undergraduates, in the first and second year of study.

Students' response to their sources of HIV/AIDS information is shown in the Table 2. Majority of the students had some basic knowledge about HIV/AIDS such as whether they had heard of HIV/AIDS, whether an infected person can look healthy and if there was a cure for AIDS. Furthermore, over 80.0% of the undergraduate students had ever tested for HIV. In regards to sources, the campus clinic (25.7%) and peer educators (21.5%) respectively were the main sources of HIV/AIDS information. Conversely, peer educators provided the most effective sources of HIV/AIDS information followed by the campus clinic at 24.2% and 19.4% respectively.

The distribution of HIV knowledge by gender is presented in Table 3. The above responses are in the affirmative. The levels of knowledge on transmission of HIV between gender was similar for some responses except for the question on whether HIV can be transmitted through sharing a glass of water with an HIV+ person (15.6% males, 11.9% females $p<0.01$), whether HIV+ women will have babies born with HIV (27.9% males, 26.4% females; $p<0.001$), whether a flavoured condom works better than a latex condom (17.9% males, 10% females; $p<0.001$),whether women are always tested for HIV during

Table 1. Socio-demographic information of the respondents (N=1498)

Characteristics	N	%
Gender		
Male (M=21.84,SD=3.1)	691	46.1
Female (M=21.5,SD=3.6)	807	53.9
Age (M=21.67,SD=3.35)		
16–20	563	37.6
21–22	543	36.2
23+	392	26.2
Faculty		
Commerce	352	23.5
Law	184	12.3
Science	320	21.4
Social Science	317	21.2
Education	325	21.7
Religious Affiliation		
Christianity	1249	83.4
Islam	52	3.5
African Traditional	145	9.7
Others	52	3.5
Level of Study		
First	489	32.6
Second	471	31.4
Third	370	24.7
Fourth	168	11.2

Table 2. Students response to sources of HIV/AIDS information

Characteristics	Yes (%)	No (%)
Heard of HIV/AIDS?	97.5	2.5
Can a person infected with HIV look healthy?	94.0	6.0
Can someone die from AIDS?	86.3	12.7
Is there a cure for AIDS?	19.4	80.4
Did you ever test for HIV?	81.7	18.3
	Your Source about HIV/AIDS	**Most Effective Source**
University programmes	14.6	17.6
Friends on campus	13.4	14.6
Campus radio	10.9	17.7
Posters on campus	13.9	6.6
Campus clinic	25.7	19.4
Peer educators	21.5	24.2

pap smears (27.1% males, 40.1% females; p<0.01) and whether it was possible to get HIV when a person is tattooed (50.2% males,43.3% females; p<0.01). Overall there was no significant difference in HIV/AIDS knowledge between males and females (M=25.3, SD=3.9; F=25.6, SD=3.8; T=-1.73, p<.05). This rejects the hypothesis that female undergraduates had significantly higher HIV/AIDS knowledge than their male counterparts.

Gender differences on attitudes related to sexuality amongst students is shown in the Table 4. The above responses are in the affirmative. HIV/AIDS related behaviour between the sexes was different for most of the related responses. For instance there were significant differences in regards to whether using a male condom reduced sexual pleasure with 32.7% males saying yes compared to just 18.6% females (p<0.0001). Also, further differences can be seen in regards to whether a man is offended if a woman insisted that he used a condom (26.9% males compared to 24.7% females, p<0.0001). However, similar responses were got in relation to accepting of blood after it has been screened for HIV and sharing of equipment used to perform circumcision can lead to HIV infection. Overall, there are significant

Table 3. Association between gender and HIV/AIDS knowledge

Characteristics	Overall (%)	Males (%)	Females (%)	Chi-Square (X^2)
Can get HIV by sharing a glass of water with an HIV+ person?	13.6	15.6	11.9	14.075***
HIV+ women will have babies born with HIV?	27.1	27.9	26.4	15.539****
All infected with HIV show serious signs of being infected?	14.3	13.3	14.7	3.476
Is there a vaccine that can stop people from getting HIV?	17.6	18.1	17.2	3.352
People likely to get HIV by deep kissing if partner is HIV+?	18.8	21.0	17.0	3.974
A woman get HIV during her periods when using a condom?	18.3	19.5	17.2	1.375
Female condom to help stop spread of HIV?	55.3	56.7	54.2	3.346
A flavoured condom works better than a latex condom?	13.7	17.9	10.0	21.583****
Can a person on antibiotics get HIV?	34.8	35.8	34.0	0.689
Take a test for HIV a week after having sex can tell if HIV+?	16.4	18.5	14.6	4.259
Using baby oil with condoms increase chance of getting HIV?	10.6	10.9	10.4	2.028
Women always tested for HIV during their pap smears?	33.6	27.1	40.1	9.547**
Is it possible to get HIV when a person gets a tattoo?	46.5	50.2	43.3	9.837**

differences between male and female undergraduate students in regards to attitudes related to sexuality (*M*=13.5; *SD*=2.23; *F*=14; *SD*=2.27; t=-3.590, p<0.0001) with males more likely to engage in more risky behaviour than female students.

Table 5 presents the two models of determinants of students' attitudes towards their sexuality by gender. Both models show that older age, religious affiliation, faculty and HIV/AIDS knowledge were the main determinants of students' attitudes towards sexuality. In addition, level of study was a determinant of attitudes towards sexuality among male students.

Model 1 shows that female students 25 years and over were 1.71 (p<0.05) times more likely to have a positive attitude regarding their sexuality compared to those aged 15-19. Female students who were affiliated to Islam and other religions were 1.76 (p<0.01) times more likely to have a positive attitude to their sexuality compared to females who were affiliated to the Christian faith. Conversely, female

Table 4. Association between gender and attitudes related to sexuality

Characteristics	Overall (%)	Males (%)	Females (%)	Chi-Square (X^2)
Having a male condom, implies he/she is planning to have sex	35.0	37.6	32.8	6.739*
Using a male condom reduces sexual pleasure	25.1	32.7	18.6	51.335****
A man is offended if a woman insists he use a condom	25.7	26.9	24.7	30.747****
It is embarrassing to buy or ask for condoms	23.2	22.0	24.2	18.742****
men and women need to have more than one sexual partner	12.6	15.6	9.9	17.983****
sexual intercourse should only be between married couples	45.0	40.1	49.2	12.550**
A person can safely accept screened blood for transfusion	61.9	61.9	62.0	1.241
Sharing of equipment for circumcision can lead to infection	73.8	74.8	73.0	1.196

Table 5. Logistic regression showing attitudes towards sexuality by gender among University students

Characteristics	Model 1: Females		Model 2: Males	
	OR	IC	OR	IC
Age				
15-19	1.00		1.00	
20-24	1.34	0.72-2.51	1.56	0.76-3.19
25+	1.71*	1.02-2.87	1.95*	1.11-3.45
Level of Study				
first	1.00		1.00	
second	1.38	0.77-2.49	1.76*	1.01-3.16
third	0.76	0.44-1.34	1.09	0.63-1.92
fourth	0.87	0.51-1.52	1.06	0.61-1.87
Religion				
Christianity	1.00		1.00	
Islam and others	1.76**	1.11-2.79	1.50*	1.03-2.19
Faculty				
Commerce	1.00		1.00	
Law	0.51***	0.33-0.79	0.40****	0.25-0.65
FAST	0.59	0.35-1.004	0.32****	0.18-0.56
HSS	0.57*	0.36-0.91	0.56*	0.35-0.91
Education	0.78	0.51-1.201	0.81	0.47-1.39
Source of Information				
University programme	1.00		1.00	
Friends on campus	1.11	0.77-1.59	1.21	0.79-1.83
Campus clinic	1.16	0.80-1.69	1.29	0.84-1.97
HIV/Knowledge				
High	1.00		1.00	
Low	0.66**	0.48-0.89	0.67*	0.49-0.93

Level of confidence * =0.05, **=p<0.01; ***=p<0.001; p<0.0001 OR=Odds Ratio; CI=confidence Interval

students from the faculties of law (OR=0.51, p<0.001) and Humanities and Social Sciences (OR=0.57, p<0.05) were less likely to have positive attitudes to their sexuality. Additionally, female students with low knowledge of HIV/AIDS prevention were 0.34 times (p<0.01) less likely to have a positive attitude to their sexuality. Model 2 shows that older male students (25 years and over) were more likely (OR=1.95, p<0.05) than their younger counterparts to have positive attitudes to their sexuality. In addition, second year male students and those affiliated to the Islamic faith and other religions were 1.76 times (p<0.05) and 1.5 times (p<0.05) respectively more likely to have positive attitudes regarding their sexuality. Furthermore, male students who belonged to the Faculties of Law (OR=0.40, p<0.0001), compared to males in the Faculty of Commerce, those in Law, Agriculture Science and Technology, Humanities and Social

Sciences faculties were less likely to have a positive attitude towards their sexuality. Male students with low knowledge of HIV/AIDS prevention were less likely to have a positive attitude to their sexuality.

DISCUSSION

The study was conducted to examine whether there is gender difference in HIV/AIDS knowledge and attitudes towards sexuality amongst undergraduate students of the North-West University, Mafikeng Campus. This study has shown that majority of the students had general information about HIV/AIDS. This is in line with other studies done in South African Universities (Reddy & Frantz, 2011; HEAIDS, 2010, Mbelle et al, 2014). The finding that peer educators had influence in mediating HIV/AIDS related information is consistent with previous report in Zimbabwe (Terry et al, 2006). The effectiveness of peer educators in communicating sensitive issues about HIV/AIDS is a positive aspect of peer influence. The use of posters was the least effective source of disseminating HIV/AIDS information, this contradicting the findings in Ghana (Nunyenge, 2013).

Students in general provided the correct responses with regards to HIV/AIDS related knowledge with a few exceptions such as whether a person can get HIV through a tattoo and whether women always tested for HIV during their pap smears. These differences in responses by gender highlight the need to expand range of information in HIV/AIDS awareness programme.

This result supports similar studies where there were no gender differences on HIV knowledge and attitudes amongst students (Durojaiye, 2011, Ahmed et al., 2009), but contradicts previous findings in Ghana (Oppong et al., 2014) and Nigeria (Chng et al., 2005) where gender differences were observed. It would appear that students have taken the impact of HIV/AIDS pandemic in South Africa more seriously and were receptive to HIV/AIDS information from the university campus. In addition, the presence of peer educators who were undergraduate students themselves provided an excellent platform to easily collect detailed HIV/AIDS information.

Our findings suggested significant differences with regards to attitudes related to sexuality of students by gender. The findings that males were more likely to engage in risky behaviour than their females is similar to studies in Uganda and Zimbabwe that reported multiple sexual partners among male students (Agardh et al., 2012, Shumba et al., 2011, Cederbaum et al., 2006). This may be attributed to males' attempt to prove their masculinity coupled with negative attitudes towards condom use. Furthermore, males with more traditional ideologies were significantly more likely to support amongst others having more casual sex partners (Shearer et al., 2005). This suggests that there is a need for greater sexual education related to sexual matters to be included in the university awareness programmes to mitigate males' risky sexual behaviour.

Consistent with studies done in Ghana and Madagascar (Rahamefy et al., 2008, Fiaveh, 2012, Calsyn et al, 2013), male students were reluctant to use condoms regularly because condom reduced sexual pleasure. Some of the reasons given for against condom use among men were that it changed the mood, did not fit well and that condoms did not feel good or were unnatural (Calysn et al., 2013). This suggests that either some male undergraduate students may have misinformation about the male condom or the masculine attitudes of male undergraduates instigate this wrong perception (Szabo & Short, 2000; Fiaveh, 2011).

Further misconceptions about condom use were seen with regards to buying or asking for condoms between male and female undergraduates. It is not unexpected that female undergraduates found it embarrassing to buy or ask for condoms compared to their male counterparts (24.2% females versus 22% males, $p<0.0001$). This is due to the perception that purchasing a condom by females depicts promiscuous behaviour. In addition, it is not females' responsibility to have condoms, but that of the males. Male dominance in the society may also play a role in the behaviour of the females, in which case they were expected to be submissive to their male partners (Santana et al., 2006, Van Staden & Badenhorst, 2009; Pelzer & Promtussananon, 2005).

The multivariate results show that older students were more likely to have a positive attitude to their sexuality. This is to be expected as they have more information regarding issues related to their sexuality especially in the context of HIV/AIDS prevention and awareness. Students affiliated to Islam and other religions were more receptive on issues related their sexuality. It is plausible that Islam and other religious faiths are more open on issues related to sexuality. Some studies have postulated that the Islamic faith and other religions excluding Christianity encourage open discussions on issue related to sexuality (Aguwa, 2006, Neal et al, 2015). Students who belonged to the Faculties of Law and Education, and male students who belonged to the Faculty of Agriculture Science and Technology (FAST) were less likely to have positive attitudes regarding their sexuality. A reason for this may be related to the differences in curricula especially in the social sciences where modules related to HIV/AIDS prevention are taught. HIV/AIDS programmes are yet to be made mandatory throughout the University. Students who had low levels of knowledge regarding HIV/AIDS were also less likely to have a positive attitude regarding their sexuality. The reason for this finding is that these students do not have adequate information on sexual issues especially in the context of HIV/AIDS prevention.

FUTURE RESEARCH

This Research was mainly quantitative in nature, undertaken in all three main Campuses of the North West University. Efforts are under way to undertake this study from a qualitative perspective to fully understand the extent of HIV/AIDS Knowledge among undergraduate students and their awareness of campus programmes. It is hoped that the results would aid the quantitative results.

CONCLUSION

In conclusion, this study has shown that students had general knowledge of HIV/AIDS. However, significant gender differences in relation to attitudes towards sex were observed, with more male students susceptible to risky behaviour than female students. Given the availability of HIV/AIDS information which includes reproductive health care, there is a need for programmes to emphasize condom use and promote one partner as the most effective ways of preventing the spread of HIV/AIDS. Such programmes must be geared towards males with traditional ideologies over sexual behaviour.

This study recommends that the HIV/AIDS office at the Mafikeng Campus should put in place strategies to include more contents on sexual education especially in relation to HIV/AIDS prevention. A more qualitative approach to explore the motivation for some of the sexual behaviours that were investigated in this study is needed.

ACKNOWLEDGMENT

The authors thank the Niche area in the Faculty of Humanities and Social Sciences, North-West University, Mafikeng Campus, for funding this study.

REFERENCES

Agardh, A., Cantor-Graae, E., & Ostergren, P. (2012). Youth, sexual risk-takingbehaviour, and mental health: A study of University students in Uganda. *International Journal of Behavioral Medicine*, *19*(2), 208–216. doi:10.1007/s12529-011-9159-4 PMID:21590465

Ahmed, S., Hassali, M., & Abdul-Aziz, N. (2009). An assessment of the knowledge, attitudes and risk Perceptions of Pharmacy Students regarding HIV/AIDS. *American Journal of Pharmaceutical Education*, *73*(15), 1–7. PMID:19513153

Akintola, O., Ngubane, L., & Makhaba, L. (2012). I did it for him, not for me An exploratory study of factors influencing sexual debut among female university student in Durban, South Africa. *Journal of Health Psychology*, *17*(1), 143–153. doi:10.1177/1359105311410512 PMID:21672963

Asante, K. (2013). HIV/AIDS knowledge and uptake of HIV counseling and testing among undergraduate private University Students in Accra, Ghana. *Reproductive Health Journal*, *10*(17), 1–8.

Badenhorst, G., Van Staden, A., & Coetsee, E. (2008). HIV/AIDS risk factors among residence students at the University of the Free State. *Curationis*, *31*(3), 27–35. doi:10.4102/curationis.v31i3.1028 PMID:19177968

Calsyn, D., Peavy, M., Wells, E., Campbell, A. et al, 2013. Differences between men and women in Condom use, attitudes and skills in substance abuse treatment seekers. *American Journal of Addicts, 22*(2), 150-157. Doi:101111/j.1521-0391.000312

Cederbaum, J., Coleman, L., Goller, G., & Jemmott, S. (2006). Understanding the HIV risk reduction needs of heterosexual African American substance-abusing men. *The Journal of the Association of Nurses in AIDS Care*, *17*(6), 28–37. doi:10.1016/j.jana.2006.09.001 PMID:17113481

Chng, L., Eke-Huber, E., Eaddy, S., & Collins, J. (2005). Nigerian college Students: HIV Knowledge, perceived susceptibility for HIV and sexual behaviours. *College Student Journal*, *39*, 60–71.

Durojaiye, O. C. (2011). Knowledge, attitude and practice of HIV/AIDS: Behaviour change among tertiary education students in Lagos, Nigeria. *Annals of Tropical Medicine and Public Health*, *4*(1), 18–24. doi:10.4103/1755-6783.80516

Ebeniro, C. (2010). Knowledge and beliefs about HIV/AIDS amongst male and female students of Nigerian Universities. *Journal of Comparative Research in Anthropology & Sociology*, *1*(1), 121–131.

Fiaveh, D. Y. (2011). *Male Perspective on condom use: implications for STI/HIV*. Saarbrucken, Germany: Lambert Academic Publishing.

Fiaveh, D. Y. (2012). Condom Myths and Misconceptions: The Male perspective. *Global Journal of Medical Research*, *12*(5), 43–50.

HEAIDS. (2010). *HIV prevalence and Related factors – Higher Education Sector Study, South Africa, 2008–2009*. Pretoria: Higher Education South Africa.

Macintyre, K., Rutenberg, N., Brown, L., & Karim, A. (2004). Understanding perceptions of HIV risk among adolescents in KwaZulu-Natal. *AIDS and Behavior*, *8*(3), 237–250. doi:10.1023/B:AIBE.0000044072.71361.b3 PMID:15475673

Makhoahle, P., & Bagali, T. M. (2012). The Impact of HIV/AIDS awareness campaigns on students who enrolled from 2009-2011 at central University of Technology, Free State. *Interim: Interdisciplinary Journal*, *12*(1), 15–25.

Maliki, A. (2010). Attitude of Niger Delta University undergraduates towards HIV/AIDS Campaigns. *Edo Journal of Counselling*, *3*(2), 210–219.

Marianne, P., Majumdar, B., Akhtar-Danesh, N., Boblin, S., & Sherifali, D. 2013. The gendered Knowledge of HIV/AIDS among urban University Students in Southern Ethiopia.*Proceedings of 1st annual International Interdisciplinary Conference*.

Maswanya, E., Moji, K., Horiguchi, I., Nagata, K., Aoyagi, K., Honda, S., & Takemoto, T. (1999). Knowledge, risk perception of AIDS and reported sexual behaviour among students in secondary schools and colleges in Tanzania. *Health Education Research: Theory and Practise*, *14*(2), 185–196. doi:10.1093/her/14.2.185 PMID:10387499

Mbakwem-Aniebo, C., Ezekoye, C., & Okonko, I. O. (2012). Knowledge about HIV/AIDS and reported risk behaviours among freshmen of the University of Port Harcourt, Southern Nigeria. *World Applied Sciences Journal*, *16*(8), 1093–1103.

Mbelle, N., Setswe, G., Sifunda, S., Mabaso, M., & Maduna, V. (2014). *HIV and AIDS related Knowledge,attitudes and behaviours of students and staff at South Africa's technical and vocational education and training colleges*. Pretoria: HSRC.

Mutinta, G., & Govender, K. (2012). The social-environmental determinants of students' sexual risk behavior and HIV prevention at the University of Kwazulu-Natal. *Journal of Human Ecology (Delhi, India)*, *38*, 17–29.

Nunyenge, R. (2013). The meaning people make of HIV Posters: A study on health improvement at Jirapa District in the upper West Region of Ghana. *Journal of Biology. Agriculture and Healthcare*, *12*, 23–34.

Oppong, A. K., MensahBoafo, I., &Nyamekye, M. (2014). Identifying gender differences in Ghanian university Students' Sexual Practices, Attitudes and Knowledge Regarding HIV. *The Journal of Psychology*, *5*(1), 9–18.

Oppong, A. K., & Oti-Boadi, M. (2013). HIV/AIDS knowledge among undergraduate university students: Implication for health education programmes in Ghana. *African Health Sciences*, *13*, 270–277. PMID:24235924

Peltzer, K., & Promtussananon, S. (2005). HIV/AIDS Knowledge and Sexual Behaviour among Junior Secondary school students in South Africa. *Journal of Social Sciences*, *1*(1), 1–8. doi:10.3844/jssp.2005.1.8

Pettifor, A., Kleinschmidt, I., Levin, J., Ries, H., Macphail, C., & Hlongwa-Madikizela, L. (2005). A community-based study to examine the effect of a youth prevention intervention on young people age 15–24 in South Africa: Results of the baseline study. *Tropical Medicine & International Health*, *10*(10), 971–980. doi:10.1111/j.1365-3156.2005.01483.x PMID:16185231

Pettifor, A., OBrien, K., MacPhail, C., Miller, W. C., & Rees, H. (2009). Early coital debut and associated HIV risk factors among young women and men in South Africa. *International Perspectives on Sexual and Reproductive Health*, *35*(02), 74–82. doi:10.1363/3508209 PMID:19620092

Rahamefy, O., Rivard, M. &Ravaoarinoro, T. 2008. Sexual behaviour and condom use among University students in Madagascar. *Journal of Social Aspects of HIV/AIDS, 5*(1), 28–35.

Reddy, P. & Frantz, J. 2011. HIV/AIDS Knowledge, behaviour and beliefs among South African University students. *Journal of Social Aspects of HIV/AIDS, 8*(4), 166–170.

Republic of South Africa. (2012). *Global AIDS Progress Report*. Pretoria, South Africa: Department of Health.

Santana, C., Raj, A., Decker, S., La Marche, A., & Silverman, J. (2006). Masculine Gender Roles Associated with increased sexual Risk and Intimate partner Violence Perpetration among Young Adult men. *Journal of Urban Health: Bulletin of the New York Academy of Medicine*, *83*(4), 575–585. doi:10.1007/s11524-006-9061-6 PMID:16845496

Shearer, C. L., Hosterman, S. J., Gillen, M. M., & Lefkowitz, E. S. (2005). Are traditional gender role attitudes associated with risky sexual behavior and condom-related beliefs. *Sex Roles*, *52*(5), 311–324. doi:10.1007/s11199-005-2675-4

Shisana, O., Rehle, T., Simbayi, L. C., Zuma, K., Jooste, S., & Zungu, N. et al.. (2014). *South African National HIV Prevalence, Incidence and Behaviour Survey, 2012*. Cape Town: HSRC Press.

Shumba, A., Mapfumo, J., & Chademana, J. (2011). The prevalence of concurrent sexual partnerships among students in Institutions of higher Education in Zimbabwe. *Journal of Human Ecology (Delhi, India)*, *34*(1), 53–65.

Szabo, R., & Short, R. V. (2000). How does male circumcision protect against HIV infection? *British Medical Journal*, *320*(7249), 1592–1594. doi:10.1136/bmj.320.7249.1592 PMID:10845974

Tagoe, M., & Aggor, R. A. (2009). Knowledge, behaviour, perceptions and attitudes of University of Ghana students towards HIV/AIDS: What does behavioural surveillance survey tell us? *Journal of Health and Human Services Administration*, *32*(1), 51–84. PMID:19558033

Terry, P. E., Mhloyi, M., Masvaure, T., & Adlis, S. (2006). An examination of knowledge, attitudes and practises related to HIV/AIDS prevention in Zimbabwe university students: Comparing intervention programme participants and non-participants. *International Journal of Infectious Diseases*, *10*(1), 38–46. doi:10.1016/j.ijid.2004.10.007 PMID:16183316

UNAIDS. (2013). UNAIDS Report on the Global AIDS epidemic update, 2013. Geneva: Switzerland: UNAIDS.

Van Staden, A., & Badenhorst, G. (2009). Reviewing gender and cultural factors associated with HIV/AIDS among university students in the South African context. *Curationais*, *32*(4), 19–28.

KEY TERMS AND DEFINITIONS

HIV/AIDS: Acquired immune deficiency syndrome (AIDS) is an infectious disease caused by the human immunodeficiency virus (HIV). There are two variants of the HIV virus, HIV-1 and HIV-2, both of which ultimately cause AIDS.

HIV/AIDS Knowledge: Facts gained about relevant information on prevention of the pandemic, where one can be tested and the misconceptions about how its spread.

HIV/AIDS Programme: The HIV and AIDS Programme has developed a model integrated response to the epidemic that incorporates teaching, research, care and support, community outreach, advocacy, prevention and management of the epidemic.

Chapter 16
Awareness and Education on Viral Infections in Nigeria Using Edutainment

Suleimanu Usaini
Covenant University, Nigeria

Olufunke Omole
Covenant University, Nigeria

Tolulope Kayode-Adedeji
Covenant University, Nigeria

Tunji Oyedepo
Covenant University, Nigeria

ABSTRACT

Researches in clinical and medical science have shown that Human Immunodeficiency Virus (HIV), Hepatitis B, and to some extent Hepatitis C viruses, constitute a major public health challenge in the Sub-Saharan Africa. This is without prejudice to the Ebola virus disease (EVD) that is more deadly than the other three put together, and had affected some countries in West Africa- Liberia, Sierra Leone and Guinea. There is little awareness and education via the mass media on some common viral infections in Nigeria like Hepatitis compared to HIV. Therefore, there is dire need for information, sensitisation and education on the viruses, means of transmission, preventive measures and also therapy. Edutainment is the fusion of education into entertainment programming which can come in form of drama, music, poetry and lots more, and it had been used as a platform to create awareness for positive reproductive health and HIV/AIDS by leading health organisations like WHO, USAID, DFID, SFH, and UNFPA. This paper therefore, examines why edutainment should be used and how it can be used to educate media audience in Nigeria on some viral infections that pose serious health risks and how they can live healthy lives.

INTRODUCTION

Communication is one discipline that is dynamic and evolving over the years as different aspects have emerged due to the dynamic nature of man and his environment. There is no doubt that communication plays a significant role in our lives, and it forms the crucible of human existence. In every foundation class of communication, it is rightly mentioned that the types of communication are: Intrapersonal, Interpersonal, Group, Organisational and Mass communication. An addition to this list is health com-

DOI: 10.4018/978-1-5225-1859-4.ch016

munication; communication scholars like Corcoran (2007) otherwise call it communication in health. She is of the opinion that communication in health can be defined generally the same way communication has been defined as a transactional process. Kreps (2003) as cited in Corcoran (2007, p. 8) "summarises the addition of 'health' to the definition of communication as a 'resource' that allows health messages (for example prevention, risk or awareness) to be used in the education and avoidance of ill health. This broad definition incorporates the fact that health communication can take place at many levels and embodies a holistic approach to health promotion."

It will not be out of place to state that health communication emphasises health promotion and education, that aims at reducing unhealthy practices and improving public health. This is very vital, especially in developing countries as they strived to achieve some of the Millennium Development Goals (MDGs) by 2015 which were health related: reducing child mortality, improving maternal health, combating HIV/AIDS, malaria and other diseases.

For health communication to be effective, the mass media are indispensable because they serve as tools for the 'transactional processes'. The mass media play important roles in our lives as they perform their functions: sources of information, education and entertainment. Ebo (1999) acknowledges that channels of mass communication perform peculiar functions as they educate, entertain, inform, mould opinion, and socialise thereby satisfying the needs and aspirations of the audience. Assay and Adigu (2008) note that in both long-and-short term bases, the media can make substantial contributions to the process of creating public awareness whether the aims of such efforts are to: Bring about an increase in information or knowledge of the public or target group about a specific issue; generate a change in their opinions or attitude; or motivate them to action that is to change their behaviour and habits.

This presupposes from the above mentioned the need to employ the mass media wisely and effectively in health education interventions, and public health campaigns in the areas of tackling viral infections and promoting healthy living. Therefore, this paper explores how edutainment as a media campaign strategy can be employed by media organisations, governmental and non-governmental organisations, community stakeholders, corporate organisations, etc. to promote awareness and education on viral infections prevalent in Nigeria, their modes of transmission, prevention and possible cure.

THEORETICAL UNDERPINNING

British psychologist, Stephenson (1967) as cited by Folarin (2005, p. 170), divides man's activities into work and play; the former meaning reality and production, while the latter deals with entertainment, relaxation or self-satisfaction. He further says that the audience use the mass media (both old and new) for play- pleasure and entertainment than for information and serious work. This has led to a constant criticism of television and radio in Nigeria, on its focus on entertainment more rather than on development purposes (Folarin, 2005). Rather than just criticise the media, it will be profitable to maximise its emphasis on entertainment by juxtaposing it with the educational function to promote positive social behaviour that will lead to positive social change.

In this section therefore, two theories that capture the two functions will be discussed as it relates to health communication. The theories are Social Learning Theory and Mass Entertainment Theory.

Social Learning Theory

The Social Learning Theory emerged from an extensive work of Albert Bandura, who was a psychologist at Stanford University. The theory posits that much learning takes place through observing the behaviour of others (Anaeto, Onabanjo & Osifeso, 2008). Bandura (1986) posits that people learn behaviours, emotional reactions, and attitudes from role models whom they wish to emulate. This he arrived at from his earliest studies to support the theory, fondly called the "Bobo Doll Studies". Defleur and Rokeach (1989) cited in Asemah (2011, p. 190) opine thus:

It is a general explanation of how people acquire new forms of behaviours. It is called "social" because it attempts to explain how individuals observe other people's actions and how they come to adopt those patterns of actions as personal modes of responses to problems, conditions or events in their own lives.

The theory applies basically to television and other forms of televised media such as film and cinema, and internet TV. Wirtz (2008) citing Albert Bandura, states that children and adults acquire attitudes, emotional responses, and new styles of conduct through filmed and televised modelling. Furthermore, it has a general application to socialising effects of the mass media, most especially television and adoption of various models of action that applies to many everyday matters such as clothing, appearance, style, eating and drinking, modes of interaction, and personal consumption (McQuail, 2010). It is important to add that the theory can also be adopted and applied to information seeking, social mobilisation, public enlightenment and attitude change. These are vital ingredients for the adoption and use of edutainment in health campaigns.

According to Wirtz (2008) citing Lefkowitz, Eron, Walder and Huesmann (1977), there are three stages that serve as a link between the audience watching or being exposed to the media content, and replicating what was learnt or something similar.

- **Attention:** The first step is to grab the audiences' attention by producing captivating programmes. There is no doubt that it is easier to catch the attention of the audience with edutainment programmes using television than radio because of the advantages of sight and sound. Anyway no matter the medium to be selected it is important to have in mind that the uttermost attention of the audience is a priority.
- **Retention:** Social learning theorists agree that mass media audience mostly learn things by vicariously experiencing them. The TV audience can watch the most graphic and explicit acts and experience the thrills, the fear, the strength, etc, in the safety and comfort of their rooms, and houses before their TV screens. Therefore, they interpret these experiences according to their cognitive and emotional levels, and then store them in their memories. The processed experiences as stored in the audiences' memories may remain unused and untapped for years, but they may contribute towards shaping future active or passive experiences.
- **Motivation:** This is the stage where the audience will want to try out lessons learnt from the media experiences. Some media audience learn how to handle life situations and circumstances by applying the techniques or strategies the on-screen character who was likely in similar situations and obtained positive results. This supposes, that edutainment programmes in whatever format they are produced and aired on TV and radio or via internet, have great potentials to inform and educate the audience, nevertheless influence their attitudes towards viral infections in their environments.

The emphasis, therefore, is that edutainment programmes in the electronic and new media have potency for serving as sources of information for the audience on positive reproductive health such as sexual attitudes, family planning, and issues relating to Sexually Transmitted Infections (STIs) such as HIV/AIDS, Syphilis, and Gonorrhea. They have also being used to educate the audience on positive and beneficial health practices such as abstinence, prevention of STIs through the use of condoms, regular tests and medical check-ups, prevention of mother to child transmission (PMTCT), and showing love to people living with HIV/AIDS (PLWHA). It therefore, suffices to state that edutainment programmes via the same media can be used to educate the audience on possible viral infections in their environment and precautionary measures to adopt in order to avoid being infected and live healthy lives.

Mass Entertainment Theory

Harold Mendelsohn, who was a social psychologist at University of Denver, in his book *Mass Entertainment*, published in 1966, did a discourse on the need for entertainment in the society via the mass media. He proposed the Mass Entertainment Theory which asserts that television and other mass media perform vital social functions because they relax or otherwise entertain the average people in the society. Guback (1968) who did a review on Mendelsohn's book pointed out that mass entertainment is not only good for us but a very necessary part of our lives. He said the mass media, especially television, have now made it possible to cater effectively to the "crudest" wants of tremendous numbers of people by providing entertainment. Mendelsohn further, argued that average people needed the relaxation and harmless escapism that television entertainment offer and if television entertainment was not available, people would find other avenues for easing the strain and stress of daily life. Moreover, television simply served these needs more easily, powerfully, and efficiently than other alternatives (Baran & Davis, 2003).

According to Baran & Davis (2003, p. 174), Mendelsohn emphasised that "Television entertainment did not disrupt or debase high culture, it merely gave average people a more attractive alternative to operas or symphony concerts. It did not distract people from important activities like religion, politics, or family life; rather, it helped them relax so that they could later engage in these activities with renewed interest and energy." The Mass Entertainment Theory therefore, posits the importance of entertainment to media audience and the viability of using entertainment contents to educate the audience concerning health matters.

If entertainment is this important to the audience, then it is wisdom to utilise it as a means of education, especially health education. Entertainment with a mix of education is what gives edutainment, and it is without doubt a very relaxed way of creating awareness and education concerning viral infections in Nigeria.

The Social Learning Theory describes how role models can be used to produce entertainment programmes for television, radio and the internet in order to meet the entertainment needs of the audience as highlighted by the Mass Entertainment Theory. The contents of the entertainment programmes should focus on how the talents or models adopt healthy practices and lifestyles to avoid viral infections. It is believed that the media audience will learn such behaviours and attitudes from the models, and will also strive to live healthy lifestyles in order to avoid viral infections.

Edutainment in the Nigerian Media

Education in the media has always been connected to a major responsibility of the media with respect to its heterogeneous audience. According to Moemeka (2000), the major functions of the mass media are for the purpose of information, education and entertainment among others. Media organisations in time past have continued to strive to achieve the aim of ensuring that these functions are achieved through the diverse mass media forms such as print, electronic or broadcast, and internet. For example, Umeh (1989, p.56) opines that "leaders in the past, present the educational factor to the people in order to secure approval and support for the introduction and expansion of television". This he explained was always a vote catcher for politicians who were interested in using the media for propaganda. "Observers at the time believed that although the idea was welcomed and useful in many respects, they nevertheless served partisan political purposes for their various governments" (Umeh, 1989, p. 57).

It was further gathered that the media were used as ideological/political apparatuses in the 1970s and 1980s because of high government ownership (Oso, 2012, p. 8-9). Today, the media are now market driven as commercialisation of this various media outlets continue to spread (Akeem, Oyeyinka, Qasim, Lateef, Omolayo & Onyinyechi, 2013, p. 10). The drive for media commercialisation continues to increase as sponsors of entertainment programmes use the entertainment function of the media to promote their brands, products and services. According to Sanusi, Omowale, and Myke (2015, p. 9), entertainment function is "even more relevant in the Nigerian media industry as more people rely on the media for entertainment." Bosshart, and Hellmular (2009, p. 3) argue that "media entertainment, now decentralised and omni-present in our lives, has transformed our society into a hedonist one. We have more technical opportunities to enjoy entertainment, but we also see that entertainment has grown and affects more and more areas such as sports, politics, information and education."

Although new media have emerged, Nwulu, Adekunle, Oranugo, and Adewale (2010, p. 1) stress that "television remains one of the most important media worldwide for disseminating information and providing entertainment." Brown (1992, p. 253) affirms this when he notes that "audience exposure to entertainment television programmes worldwide is radically increasing". Consequently, "several nations are infusing pro-social messages within the content of entertainment television programmes to promote development" (Brown, 1992, p. 253). While stating that edutainment is a better term to characterise socialist regimes, Imre (2012, p. 120) affirms that television as a home-based medium has a strong appeal that could form intimacy with its audience. This could account for why most non-governmental organisations have used this medium for the purpose of educating the public about important societal issues like public health and procedures. These organisations identify entertainment programming such as drama, TV series, TV soap opera and film among others as the most effective way of reaching the public with campaign contents for behavioural change.

Moyer-Guse, Jain, and Chung (2012, p. 1010) posit that "A growing body of research has examined how narrative-based entertainment programmes influence viewers' attitudes and behaviours across a variety of health or social issues." However, Nwadigwe (2012, p. 105) complains that their efforts have not been effective as the process of educating the public about health issues is slow and not encouraging. Stating examples, he notes further that "observations indicate that despite the mass media campaign, female circumcision is still practised even by urban dwellers, while STIs and abortions are on the increase especially among young people who often patronise quacks for solutions" (Nwadigwe, 2012, p. 109).

Concept of Edutainment

Edutainment is derived from the "combination of two words which are brought together to form a new word" (Adedimeji & Abdulbaqi, 2007, p. 41). Edutainment is made up of education and entertainment resources. According to the Government and Social Development Resource Centre, (2011, p.1), entertainment education or edutainment is a "contemporary form of behaviour change communication" and a "research- based communication process that aims to both entertain and educate, in order to increase audience knowledge; and change attitudes, social norms, and behaviour" (UNICEF, 2005, p. 23).

Edutainment is seen as a strategy used to drive societal and behavioural change. Entertainment-education as described by Hether, Huang, Beck, Murphy and Valente (2008, p. 809) "is a communication strategy that has been popular among international health promotion programme planners since the 1950s which have been integrated into dramatic serial TV programmes, and other forms of entertainment, in countries." The combination of both terms to effect societal change "can be traced historically to the timeless art of storytelling" (Brown & Singhal, 1999, p. 263-264). Edutainment is said to have emerged from "marketing, persuasion theory and diffusion of innovation theory" (Tufte, 2001, p. 29).

According to UNICEF (2005, p. 22-23) the concept of edutainment was introduced by the BBC in 1951 in Peru. It was done in order to provide information on agricultural techniques. The edutainment programme wass a "commercial soap opera that spun off unanticipated pro-social behaviours among its audience members. The programme 'PANTEL' was produced by Simplemente Maria and went on air in 1969 in Peru (UNICEF 2005, p. 22). The use of edutainment concept of programming was initiated to stop the degrading nature of entertainment and boring nature of educative programme. Brown (1992) notes that the use of edutainment has proved successful and effective in most developing continents such as Latin America, Asia and Africa.

Viral Infections in Nigeria

A viral infection is a proliferation of a harmful virus inside the body (Lodish, Berk, Zipursky, Matsudaira, Baltimore & Darnell, 2000). Viruses reproduce by infecting a host; introducing their genetic material into the cells and hijacking the cell's internal machinery to make more virus particles. Viral illness occurs as a result of cell damage, tissue destruction, and the associated immune response.

Alberts, Alexander, Lewis, Raff, Roberts and Walter (2002) explain that certain viruses may be inactive after the initial infection. For example, a cold sore can erupt and later heal. The cold sore virus remains in the victim's cells in a dormant state. At a later date, a trigger, such as stress, sunlight, or something else, may reactivate the virus and lead to new symptoms. In the process, the virus may reproduce itself, release new virus particles, and kills more host cells.

Viral infections according to Centres for Disease Control and Prevention (2015) are contagious for varying periods of time, depending on the virus. The contagious period of a virus is not necessarily the same as the incubation period. It adds that viruses and bacteria are identified as two types of potentially disease-causing particles, but viruses are much smaller than bacteria. It is, however, noted that while the symptoms of viral and bacterial illnesses are sometimes similar, laboratory tests may help clarify whether an illness is due to a virus, bacteria, or other infectious agent or disease process.

Roossinck (2011), states that virus transmission can occur in different ways. Some viruses can spread through touch, saliva, or even the air. Other viruses can be transmitted through sexual contact or by sharing contaminated needles and sharp objects. Insects including ticks and mosquitoes can act as vectors or carriers, transmitting a virus from one host to another. Contaminated food and water are other potential sources of viral infection. The University of Rochester Medical Center (2015) identifies categories of viral infections as follows:

Respiratory Viral Infections

These infections affect the lungs, nose, and throat. The viruses are most commonly spread by inhaling droplets containing virus particles. Examples include:

- **Rhinovirus:** The virus that most often causes the common cold. However, there are more than 200 different viruses that can cause colds.
- **Seasonal Influenza:** Flu symptoms are more severe than cold symptoms and often include body aches and severe fatigue. The flu also tends to come on more suddenly than a cold.
- **Respiratory Syncytial Virus (RSV):** An infection that can cause both upper respiratory infections (like colds) and lower respiratory infections (like pneumonia and bronchiolitis). It can be very severe in infants, small children, and elderly adults.

This group of infections can be prevented or reduced by frequent hand-washing, covering the nose and mouth when coughing or sneezing, and avoiding contact with infected individuals. Disinfecting hard surfaces and not touching the eyes, nose, and mouth can help reduce transmission as well.

Viral Skin Infections

This category of infections can range from mild to severe and often produce a rash. Examples of viral skin infections include:

- **Molluscum Contagiosum:** It causes small, flesh-colored bumps most often in children ages one to ten years old. However, people of any age can acquire the virus. The bumps usually disappear without treatment, usually in 6 to 12 months.
- **Herpes Simplex Virus-1 (HSV-1):** The common virus that causes cold sores. It is transmitted through saliva by kissing or sharing food or drink with an infected individual. Sometimes, HSV-1 causes genital herpes – a genital infection.
- **Varicella-Zoster Virus (Vzv):** Causes itchy, oozing blisters, fatigue, and high fever characteristic of chickenpox.

One major way to prevent viral skin infections is to avoid skin-to-skin contact (especially areas that have a rash or sores) with an infected individual. Some viral skin infections, such as varicella-zoster virus, are also transmitted by an airborne route. Communal showers, swimming pools, and contaminated towels can also potentially harbor certain viruses.

Foodborne Viral Infections

These are viruses that are the most common causes of food poisoning. The symptoms of these infections vary depending on the virus involved.

- **Hepatitis A:** A virus that affects the liver for a few weeks up to several months. Symptoms may include yellow skin, nausea, diarrhea, and vomiting. Up to 15% of infected individuals experience recurrent illness within six months of infection.
- **Norovirus:** This has been reported to be responsible for outbreaks of severe gastrointestinal illness that happen on cruise ships, but it causes disease in many situations and locations.
- **Rotavirus:** Causes severe, watery diarrhea that can lead to dehydration. Anyone can get rotavirus, but the illness occurs most often in babies and young children.

Rotaviruses and noroviruses are responsible for many (but not all) cases of viral gastroenteritis, which causes inflammation of the stomach and intestines. Foodborne viral illnesses are transmitted via the fecal-oral route. This means that a person gets the virus by ingesting virus particles that were shed through the faeces of an infected person. Contaminated water is another potential source of infection.

Sexually Transmitted Viral Infections

These are spread through contact with bodily fluids. Some sexually transmitted infections can also be transmitted via the blood (blood-borne transmission). They include:

- **Human Papillomavirus (HPV):** There are many different types of HPV. Some cause genital warts, while others increase the risk of cervical cancer. Vaccination can protect against cancer-causing strains of HPV.
- **Hepatitis B:** A virus that causes inflammation in the liver. It is transmitted through contaminated blood and bodily fluids. Some people with the virus do not have any symptoms while others feel like they have the flu. The hepatitis B vaccine is more than 90% effective at preventing infection.
- **Genital Herpes:** A common sexually-transmitted infection caused by herpes simplex virus-2 (HSV-2). Herpes simplex virus-1 (HSV-1), the virus responsible for cold sores, can also sometimes cause genital herpes. There is no cure for genital herpes. Painful sores often recur during outbreaks. However, antiviral medications can decrease both the number and length of outbreaks.
- **Human Immunodeficiency Virus (HIV):** A virus that affects certain types of T cells of the immune system. Progression of the infection decreases the body's ability to fight disease and infection, leading to acquired immune deficiency syndrome (AIDS). HIV is transmitted by coming into contact with blood or bodily fluids of an infected person.

The risk of getting a sexually-transmitted viral infections can be reduced by abstaining from sex or only having sex while in a monogamous relationship with someone who does not have a sexually-transmitted infection. Using a condom decreases, but does not entirely eliminate, the risk of acquiring a sexually-transmitted infection. Minimising the number of sexual partners and avoiding intravenous drug use are other ways to reduce the risk of acquiring sexually-transmitted and blood borne viral infections.

Other Viral Infections

There are many other viruses that cause many other infections ranging from mild to life-threatening. They include:

- **Epstein-Barr Virus (EBV):** A type of herpes virus that is associated with fever, fatigue, swollen lymph nodes, and an enlarged spleen. EBV is a very common virus that causes mononucleosis ("mono"). More than 90% of adults have been infected with this "kissing disease" that is spread primarily through saliva.
- **West Nile Virus (WNV):** A virus that is most commonly transmitted by infected mosquitos. Most people with WNV do not have any symptoms while others develop a fever, headache, and other symptoms. Less than 1% of people with WNV develop inflammation of the brain (encephalitis) or inflammation of the tissue surrounding the brain and spinal cord (meningitis).
- **Viral Meningitis:** An inflammation of the lining of the brain and spinal cord that causes headache, fever, stiff neck, and other symptoms. Many viruses can cause viral meningitis, but a group of viruses called enter viruses are most often responsible.
- **Ebola Virus Disease (EVD):** Formerly known as Ebola haemorrhagic fever, is a severe, often fatal illness in humans. The virus is transmitted to people from wild animals and spreads in the human population through human-to-human transmission. According to the WHO, Good outbreak control relies on applying a package of interventions, namely case management, surveillance and contact tracing, a good laboratory service, safe burials and social mobilisation.

On antiviral medication and treatment, Janniger, Eastern, and Hospenthal (2014) state that many viral infections resolve on their own without treatment. Other times, treatment of viral infections focuses on symptom relief, not fighting the virus. For example, cold medicine helps alleviate the pain and congestion associated with the cold, but it does not act directly on the cold virus.

There are some medications that work directly on viruses Pyrsopoulos and Reddy (2013). These are called antiviral medications. They work by inhibiting the production of virus particles. Some interfere with the production of viral DNA. Others prevent viruses from entering host cells. There are other ways in which these medications work. They further add, in general, antiviral medications are most effective when they are taken early on in the course of an initial viral infection or a current outbreak.

Using Edutainment to Create Awareness on Viral Infections in Nigeria

Nina Devries (2015), a correspondent with Al Jazeera news network reports that recently, there have been new cases of Ebola virus disease (EVD) in Port Loko, Sierra Leone, and this has prompted a renewed emphasis on using creative ways to increase awareness of the virus in the country through drama and comedy. Entertainment-education is being utilised by a popular local comedian, Amadu Kalakoh and a drama group to educate as they entertain the locals in the community on how to stay safe and help fight the spread of Ebola. Entertainment programmes remain the most viable medium of consciously reaching the public with campaign messages (Brown, Steele & Walsh-Childers, 2002). This could be because of the element of entertainment that exist in such content as Bosshart, and Hellmullar (2009, p. 4) are of the opinion that "Entertainment serves as a survival kit for daily life that makes it liveable; it serves as a vehicle in finding a fit with the environment. Entertainment reduces the gap between reality and utopia (in our minds); it allows us to live with contradictions, inconsistencies, and inadequacies; and it offers venues for self-directed self-experiences, self-enhancement as well as self-fulfilment or self-realisation. From this point of view, entertainment sustains humans".

Most pervious entertainment-education campaigns (EEC) were implemented mainly in developing countries for improvement of public's knowledge, their behaviour changes, and ultimately social change. The effort to use awareness to cause social and behavioural change has long been on. One of the promising ways to cause this change according to Haider (2012, p. 3) is through edutainment programmes such as radio drama series, TV Soap operas, film, and theatre; by focusing "especially on family planning, HIV/AIDS prevention, gender equality, and adult literacy".

According to Alawode, Atofojowo, and Fatonji (2015, p. 294) "Nigeria, Sub-Saharan Africa is currently faced with numerous challenges ranging from security, poverty, mass unemployment, corruption to health challenges." Consequently, it is necessary to initiate edutainment contents using the mass media for positive societal education and change. They, further point out that most of these edutainment contents in Nigerian television and radio stations have addressed issues boiling on malaria, cancer, use of contraception, HIV, family planning, child abuse, violence against women, and child trafficking (Alawode, Atofojowo, & Fatonji, 2015, p. 297). These issues are those that have been identified with most developing countries especially in Africa. On the other hand, there is need to identify with the present epidemic that now stands as a present societal problem which affects human existence irrespective of the population affected in the society.

Edutainment contents tend to focus on important health issues, especially health issues that have long been in existence. These edutainment campaigns seem to address old problems in new ways. Over the years, the mass media through edutainment programming and community outreach have been the major media used to reach out to people on societal health change. Other forms of innovative approaches suggested by United Nations (2012) include but not restricted to are: games, electronic technologies (mobile phones/computers), street theatre, art, music and cultural activities.

Although according to Zonta Organisation (n.d, p.1) "Edutainment approach, which combines traditional and contemporary media, is proven to be both cost effective and efficient in reaching women and men on a much wider scale." New health issues are beginning to emerge which need urgent solutions. The European food information council (2015, p.1) affirms that "chronic diseases- heart and lung disease, cancer and diabetes- are having a negative economic impact on both the developed and developing world and should thus be adequately addressed by domestic and international policy makers." On the other hand, viral infectious diseases are common within the Nigerian environment. These viral infections include Hepatitis, Yellow fever, Typhoid fever. According to Hwang, and Cheung, (2011, p.11) "Africa has the second largest number of chronic carriers after Asia and is considered a region of high endemicity."

According to Emechebe, Emodi, Ikefuna, Ilechukwu, Igwe, Ejiofor, and Ilechukwu (2009, p.16) "Since over 30years when pioneer works were done in Nigeria to the recent times the prevalence of hepatitis B virus (HBV) infection has remained very high. In Nigeria, the transmission of hepatitis B virus occurs mainly during childhood and all the risk factors like blood transfusion, sexual promiscuity, and lower socio-economic status" are feasible in adults. Musa, Bussell, Borodo, Samaila, and Femi (2015, p. 2) state thus, "results on the increase and spread of HBV suggests that "large numbers of pregnant women and children were exposed to HBV from 2000 to 2013. Increased efforts to prevent new HBV infections are urgently needed in Nigeria."

Emechebe, Emodi, Ikefuna, Ilechukwu, Igwe, Ejiofor and Ilechuku (2010, p. 3) suggest that "HBV infection in Nigeria can be prevented or drastically reduced through health education of the general population on the various mode of transmission of HBV and preventive measures". But over the years, edutainment programmes have focused so much attention on addressing issues dealing with HIV/AIDS in Nigeria and across affected countries. Lots of authors have also focused on the use of edutainment programmes mainly for educating the public on issues relating to transmission and prevention of sexually transmitted diseases such as HIV/AID and care for people living with the virus (PLWHA). This is the reason studies have been carried out by different authors to encourage this method of campaign on health issues and to find out its effectiveness through empirical studies (Farhana & Ahmed, 2008; Kafewo, 2008; Literat & Chan, 2014).

Consequently, other pressing health issues in Nigeria that are endangering people's lives receive little or no attention. There is a dire need for contents to be created in edutainment programmes to not only create awareness but also mobilise people to put on their guards to avoid being infected. Sometime ago, the Ebola virus disease crept into Nigeria, but due to timely intervention, its spread was halted. It is advisable to use edutainment programmes to create and maintain awareness among media audience about the Ebola virus to prevent any more epidemic in the country. This position was affirmed by Farhana and Ahmed (2008), stating that entertainment-education approaches to promoting healthy living and disease prevention are popular and viable methods for health intervention programmes especially that target adolescents and young adults. Therefore, the same avenues can be maximised for viral infections intervention programmes affecting Nigerians.

According to Barz and Cohen (2011) other forms that could be used in disseminating health messages include central African pop, carribian dance music, concert performance, documentaries, posters and graphic designs on T-shirts. Entertainment especially in music is a viable way to communicate these health issues in Nigeria. There is need to begin to use edutainment and other forms that attract people affected to disseminate information in order to create awareness on these viral diseases in Nigeria. This will ensure well-being of the people which results in societal development.

There is need to encourage future research endeavours in the following areas: examining the viable sources of information on viral infections in Nigeria; investigating how reliable and effective some edutainment programmes produced and aired via some media platforms have been used to communicate the incidences of some viral infections and the strategies to avoid contracting such diseases, including ways to get treatment if infected.

CONCLUSION

Entertainment-education otherwise known as edutainment has proven to be a viable programme format on media platforms such as radio and television for creating awareness and providing education on viral infections in Nigeria. This is because edutainment aims at providing education to the audience in a relaxed manner, using a 'not to serious' approach. There are persistent and emerging health issues and therefore, requires urgent attention from all and sundry in the society.

The position of this paper is that media and health organisations, independent producers and content providers should first take up the challenge to educate Nigerians on the health issues associated with viral infections, and possible ways to stay safe without being infected. Secondly, in cases whereby there are persons infected with any of the viruses discussed in this paper, they should be educated on the viable and possible therapies. Programmes like drama, soap operas, music, comedy skits, etc., can be employed to carryout education on viral infections. It is possible to help people live better healthy lives, but it begins with information and education on the health issues and how to avoid being infected.

REFERENCES

Adedimeji, M. A., & Abdulbaqi, S. S. (2007). *A communicative approach to the prevention and control of HIV/AIIDS pandemic in Nigeria*. Paper presented at the Faculty of Business and Social Sciences, University of Ilorin Second National Conference themed "Nigeria Beyond 2007: Issues, Perspectives and Challenges" held at the FBSS Lecture Theatre, University of Ilorin, Ilorin, Nigeria.

Akeem, R. A., Oyeyinka, O. T., Qasim, A. O., Lateef, O., Omolayo, J., & Onyinyechi, N. (2013). Deregulation of broadcast media in Nigeria: An appraisal. *Kuwait Chapter of Arabian Journal of Business and Management Review*, *3*(3), 6–13. doi:10.12816/0017469

Alawode, S. O., Atofojomo, O., & Fatonji, S. (2015). Home video depictions of health and moral decadences as correlates of entertainment-education in Nigeria. *Mediterranean Journal of Social Sciences.*, *6*(2), 293–307.

Alberts, B., Alexander, J., Lewis, J., Raff, M., Roberts, K., & Walter, P. (2002). *Molecular biology of the cell* (4th ed.). New York: Garland Science.

Anaeto, S. G., Onabanjo, S. O., & Osifeso, B. J. (2008). *Models and theories of communication*. African Renaissance Books Incorporated.

Assay, B. E., & Adigu, R. O. (2008). The impact of mass media sensitisation on the control of tobacoo abuse in selected local government of Delta State. In E. M. Mojaye, O. O. Oyewo, R. M'Bayo, & I. A. Sobowale (Eds.), *Health communication, gender violence and ICTs in Nigeria* (pp. 65–76). Ibadan: Ibadan University Press.

Bandura, A. (1986). *Social foundations of thought and action*. Englewood Cliffs, NJ: Prentice-Hall.

Baran, J. S., & Davis, K. D. (2003). *Mass communication theory: foundations, ferment, and future*. Belmont, CA: Thomson Wadsworth.

Barz, G., & Cohen, J. (2011). *The culture of AIDS in Africa: Hope and healing through music and arts*. Oxford, UK: Oxford University. doi:10.1093/acprof:osobl/9780199744473.001.0001

Bosshart, L., & Hellmuller, L. (2009). Pervasive entertainment ubiquitous entertainment. *Communication Research Trend*, *28*(2), 1–18.

Brown, J. D., Steele, J. R., & Walsh-Childers, K. (2002). Sexual teens, sexual media: Investigating media's influence on adolescent sexuality. London: Lawrence Erlbaum.

Brown, W. (1992). The use of entertainment television programs for promoting prosocial messages. *The Howard Journal of Communications*, *3*(3&4), 253–266. doi:10.1080/10646179209359754

Brown, W. J., & Singhal, A. (1999). Entertainment-education media strategies for social change. In D. P. Demers & K. Viswanath (Eds.), *Mass media, social control and social change* (pp. 263–280). Ames, IA: Iowa State University Press.

Centers for Disease Control and Prevention. (2015). *Sexually transmitted diseases (STDs).* Retrieved from http://cdc.gov/std/

Corcoran, N. (2007). Communicating health: Strategies for health promotion Thousand Oaks, CA: Sage.

Devries, N. (2015). *Sierra Leone faces setback in Ebola fight.* Retrieved from http://aljazeera.com/news/2015/09/sierra-leone-faces-setback-ebola-fight-150920094802322.html

Ebo, S. J. (1999). *African mass media systems*. Enugu: Computer Edge.

Emechebe, G. O., Emodi, I. J., Ikefuna, A. N., Ilechukwu, A. C., Igwe, W.C., Ejiofor, O. S., & Ilechukwu, C. A. (2009). Hepatitis B virus infection in Nigeria – a review. *African Journal online, 50*(1), 8-22.

Emechebe, G. O., Emodi, I. J., Ikefuna, A. N., Ilechukwu, G. C., Igwe, W. C., Ejiofor, O. S., & Ilechukwu, C. A. (2010). Hepatitis B virus infection in Nigeria– a review. *Nigerian Medical Journal, 50*(1), 18–22.

European Food Information Council. (2015). Retrieved from http://www.eufic.org/page/en/show/latest-science-news/fftid/chronic-disease/

Farhana, S., & Ahmed, J. U. (2008). Entertainment-education to promote safe motherhood: In the context of Bangladesh. *Canadian Social Science*, *4*(4), 16–26.

Governance and Social Development Resources Centre. (2011). *Helpdesk research report: Communication initiatives to change attitude and behaviours.* Retrieved from http://gsdrc.org/doc/open/pdf

Guback, T. H. (1968). Review of mass entertainment by Harold Mendelsohn. *Journal of Aesthetic Education, 2*(3), 147-149.

Haider, H. (2012). *Helpdesk research report: Changing attitudes and behaviours in relation to gender equality.* Governance and Social Development Resource Centre (GSDRC), Birmingham. Retrieved from http://www.gsdrc.org/docs/open/HDQ817.pdf

Hether, H. J., Huang, G. C., Beck, V., Murphy, S., & Valente, T. W. (2008). Entertainment –education in a media-saturated environment. Examining the impact of single and multiple exposures to breast cancer storylines on two popular medical dramas. *Journal of Health Communication*, *13*(8), 808–823. doi:10.1080/10810730802487471 PMID:19051115

Hwang, E. W., & Cheung, R. (2011). Global epidemiology of hepatitis B virus (HBV) infection. *The Northern. The American Journal of the Medical Sciences*, *4*(1), 7–13.

Imre, A. (2012). Adventures in early socialist television edutainment. *Journal of Popular Film and Television*, *40*(3), 119–130. doi:10.1080/01956051.2012.697790

Janniger, K. C., Eastern, S. J., & Hospenthal, R. D. (2014). *Herps Zoster treatment and management.* Retrieved from http://emedicine.medscape.com/article/1132465-treatment

Kafewo, S. A. (2008). Using drama for school-based adolescent sexuality education in Zaria, Nigeria. *Reproductive Health Matters*, *16*(31), 203–210. doi:10.1016/S0968-8080(08)31356-1 PMID:18513621

Literat, I., & Chen, N. (2014). Communication infrastructure theory and entertainment-education: An integrative model for health communication. *Communication Theory*, *24*(1), 83–103. doi:10.1111/comt.12011

Lodish, H., Berk, A., Zipursky, L., Matsudaira, P., Baltimore, D., & Darnell, J. (2000). *Molecular cell biology* (4th ed.). New York: W. H. Freeman.

McQuail, D. (2010). *Mass communication theory* (5th ed.). London: Sage Publications.

Mendelsohn, H. (1966). *Mass entertainment.* New Haven, CT: College and University Press.

Moemeka, A. (2000). *Report handbook: An introduction to effective journalism.* Morris Publisher.

Moyer-Guse, E., Jain, P., & Chung, A. H. (2012). Reinforcement or reactance? Examining the effect of an explicit persuasive appeal following an entertainment-education narrative. *Journal of Communication*, *62*(6), 1010–1027. doi:10.1111/j.1460-2466.2012.01680.x

Musa, B., Bussell, S., Borodo, M. M., Samaila, A. A., & Femi, O. L. (2015). Prevalence of hepatitis B virus infection in Nigeria, 20002013: A systematic review and meta-analysis. *Nigerian Journal of Clinical Practice*, *18*(2), 163–172. doi:10.4103/1119-3077.151035 PMID:25665986

Nwadigwe, C. E. (2012). Theatre for development: An alternative programme for reproductive health communication in urban Nigeria. *African Sociological Review*, *16*(2), 102–118.

Nwulu, N., Adekanbi, A., Oranugo, T., & Adewale, Y. (2010). Television broadcasting in Africa. *Telecommunications Conference (HISTELCON), 2010 Second IEEE Region 8 Conference.*

Oso, L. (2012). The burden of history: The press and the 2003 general elections in Nigeria. In S. T. Babalola & T. Azeez (Eds.), *Critical perspectives in language, literature and communication studies.* Obafemi Awolowo University Press.

Pyrsopoulous, T. N., & Reddy, R. K. (2015). *Hepatitis B treatment and management.* Retrieved from http://emedicine.mediascape.com/article/177632-treatment

Roossinck, M. J. (2011). The good virus: Viral mutualistic symbioses. *Nature Reviews. Microbiology*, *9*(2), 99–108. doi:10.1038/nrmicro2491 PMID:21200397

Sanusi, B. O., Omowale, A., & Myke, E. (2015). Changing roles in the Nigerian media industry: A study of multimedia journalism. *Kuwait Chapter of Arabian Journal of Business and Management Review*, *4*(12), 9–17. doi:10.12816/0019000

Tufte, E. R. (2001). *The visual display of quantitative information* (2nd ed.). Cheshire, CT: Graphics Press.

Umeh, C. C. (1989). The advent and growth of television broadcasting in Nigeria: Its political and educational overtone. *Africa Media Review*, *3*(2), 54–66. PMID:12342789

UNICEF. (2015). *Strategic communication for behaviour and social change in South Asia.* Retrieved from http://www.unicef.org/rosa/Strategic_Communication_for_Behaviour_and_Social_Change.pdf

United Nations. (2012). *Community mobilization, outreach and mass media*. Retrieved from http://www.endvawnow.org/en/modules/view/3-campaigns.html

University of Rochester Medical Center. (2015). *Is it a virus or a bacterium? Know the difference*. Retrieved from http://www.urmc.rochester.edu./encyclopedia/content.aspx

Wirtz, B. (2008). *What is social learning theory?* Retrieved 30 April, 2010 from http://www,xyhd.tv/2008/uncategorised/what-is-social-learning-theory

World Health Organisation. (2015). *Ebola virus disease: Fact sheet*. Retrieved from http://who.int/mediacentre/factsheets/fs103/en/

Zonta Organisation. (n.d.). *Mass communication with a purpose: Global partnership on edutainment for social change*. Retrieved from https://www.zonta.org/Portals/0/PDFs/04c_Edutainment%20Description.pdf

KEY TERMS AND DEFINITIONS

Ebola: Also known as Ebola Virus Disease (EVD), is a type of hemorrhagic (blood) fever that is caused by the Ebola virus. It is a fatal disease that affects humans and non-humans such as monkeys, chimpanzees and bats. Some of its symptoms are high fever, severe internal bleeding, and gastrointestinal distress. It is spread through contact with body fluid from infected persons.

Education: The process of facilitating the learning experience through the act of delivering and acquiring general and specialized knowledge, developing the power of reasoning, judgment, skills, beliefs, values, character and discipline within and without the walls of a school system.

Edutainment: A concept that is finding wide acceptance in terms of content development and production for film/television, radio and even the internet. It is the fusion of education and entertainment, designed to educate and entertain the audience, because the content contains educational value.

Entertainment: The act of giving pleasure, delight and amusement to an audience. Entertainment is a way of rendering pleasurable diversion and recreation away from the daily chores, hard work and, hustle and bustle of life. It comes in form of stage shows and performances, music, drama, films, radio and television shows.

Health Communication: A type of communication that comprises of the study, practice and use of communication strategies, in promoting health information through public health campaigns and health education, with the aim of improving personal and community health literacy, and health decisions that will lead to healthy lifestyles.

Hepatitis: Hepatitis simply is inflammation of the liver, caused by the hepatitis virus. The hepatitis virus primarily attacks the liver causing damages that can result in death of an infected person. Viral hepatitis includes hepatitis A, B, C, D, E and the most common is hepatitis B which can be transmitted through bodily fluids.

HIV/AIDS: The human immunodeficiency virus (HIV) attacks and damages the human immune system, thereby affecting the potentials and ability of the immune system to fight organisms that cause sicknesses and diseases. The virus is spread through sexual intercourse, nevertheless, it can be transmitted through other means of contact with infected blood. Acquired immunodeficiency syndrome (AIDS) is the last or advanced stage of HIV; a stage whereby the body's immune system is too weak to fight off infections.

MDGs: The Millenium Development Goals (MDGs) were eight goals set by the United Nations in 2000 during its Millenium Summit. At the summit, the United Nations Millenium Declaration was adopted, which comprised of international set plans that will be achieved within the space of 15 years (2000-2015). Therefore, all countries and nations that were members of the United Nations subscribed to the demands and all were expected to work towards the fulfilment of the goals by 2015.

STIs: According to the World Health Organization (WHO), sexually transmitted infections (STIs), also known as sexually transmitted diseases (STDs) or venereal diseases (VD), are infections that are spread from person to person through sexual intercourse. Examples of STIs are: syphilis, gonorrhea, chlamydia, hepatitis B, HIV, just to mention a few.

Viral Infection: Simply put, it is any infection on the human body caused by a virus. Some diseases caused by viral infections include chicken pox, AIDS, herpes, hepatitis, influenza and common cold.

Chapter 17
An Assessment of Media Contribution to Behaviour Change and HIV Prevention in Nigeria

Olugbenga Elegbe
University of Ibadan, Nigeria

ABSTRACT

This chapter explores media influence on behaviour change and its implication for HIV prevention programmes in Nigeria. Using the agenda setting, and the uses and gratifications theories, it analyzes how the media set agenda for public opinion and the gratification that audience receive from media information vis-a-vis behaviour change initiatives in reducing HIV epidemic in Nigeria. The chapter through a descriptive and literature review approach assesses effective communication framework for behaviour change and some of the factors contributing to HIV prevalence in Nigeria. The chapter sums up case studies of interventions that have documented the effectiveness of the media in HIV prevention initiatives. The chapter concludes on the positive influence of media intervention in sexual behaviour change programmes but notes that behaviour change still lies with individual decisions. Therefore it recommends other forms of communication like interpersonal communication, community mobilization and advocacy to support media strategies for effective HIV prevention initiatives in Nigeria.

INTRODUCTION AND BACKGROUND

Communication has a marked effect on behaviour but research shows that behaviour change rarely happens immediately upon exposure to a message. There are five models of communication effects (*knowledge, approval, intention, practice and advocacy)* usually employed during behaviour change intervention. However, studies shows that applying these models in changing behaviour especially sexual behaviour is often problematic. Hence, in addressing issues related to sexual behaviour in Nigeria, majority of communicators often utilize the media because of their ability to reach huge and diverse audiences with

DOI: 10.4018/978-1-5225-1859-4.ch017

the notion of developing positive behaviour among their target audiences. The mass media have been a potent strategy used continuously in various interventions to help reduce the spread of HIV in Nigeria and around the world. These mass communication campaigns have employed single or multiple media at the national, state and community levels, either as stand-alone efforts or as part of multi-component programmes. Basically, conventional strategies have often used television, radio, and/or print media while modern campaigns are increasingly incorporating "new media" such as Internet websites. Multi-component campaigns have combined media with numerous "interpersonal" channels such as peer education and outreach, community coalitions, counselling, skill-building workshops, and/or support groups. However, the mass media campaigns are often utilized because of their ability to reach huge and diverse audiences in a cost-effective manner, giving such campaigns tremendous potential as a tool in fighting the spread of HIV and AIDS (Seth, 2007). A key question that often arises about media campaigns, however, is whether or not they are effective in impacting HIV and AIDS knowledge, attitudes, and/or behaviours. Unfortunately, many HIV and AIDS mass media campaigns have been evaluated using weak research designs which can lead to unreliable or inconclusive results regarding the impact of such campaigns (Seth, 2007).

Moradi, Honari, Naghshbandi, Jabari and Kholouse (2012) discuss the five social functions of the media by McQuail's thus: *information, correlations, continuity, entertainment* and *mobilization* (McQuail, 2010). *Information* deals with dissemination of issues relating to world events, power relations, progress and innovation. *Correlations* are referred to explaining and interpreting events, and consensus building. *Continuity* describes expressing culture and national identity. *Entertainment* is devoted to relaxation and reducing social tension. Finally, *mobilization* includes social objectives, politics, war, and the like. Hence, the mass media can manage integration, coordination, control, stability, mobilization, tension, thoughts, values, discipline, and conformity (Mehdizadeh, 2010). Thus the mass media has the power of mobilization of thoughts, values and ability to control people to conform to ethics or change in the society. According to the findings of Moradi et al (2012), the effect of the media can significantly change individuals' attitudes and behaviour. Based on this description of the media, media content must be designed to appeal to the audience desire if any meaningful behaviour change must take place.

Effective Communication Framework for Behaviour Change

The need to design framework for effective communication for health programmes targeted at behaviour change cannot be overemphasized. Suggs, McIntyre, Warburton, Henderson, & Howitt (2015) recommended a framework that was documented in a report on "*Communicating Health Messages: A Framework To Increase The Effectiveness Of Health Communication Globally*". The framework has been created to guide communication design and decisions in order to improve the efficacy of health communication. The framework lays out process steps and key questions that guide the health communicator in producing coherent, understandable and effective messages. Suggs, et al (2015, p.11-12) suggest three phases or steps in the framework: Assess, Do, and Describe (ADD).

Step 1: Asses the Communication.

Assess sets the requirements and considerations for the function of the communication. This involves understanding the health issue, the aim of the communication and the role communication can play in improving the situation. This includes how the communication will integrate with wider efforts to improve health and influence health behaviour. It requires knowing the evidence base about the issue, the causes of the issue, and how communication has been used successfully and unsuccessfully before. Legal and ethical considerations, including any regulations about communication in the specific context will determine what can and cannot be communicated and what might be the unintended consequences of communication. This should be fully explored before moving to the next stage which is the Do phase (Suggs, et al, 2015, p.11-12).

Step 2: Do the Communication.

Do requires conducting a contextual analysis, designing messages, testing and refining, and implementing the communication. The contextual analysis measures the circumstances in which the communication will be delivered. It includes understanding any political, environmental, or social imperatives affecting the communication, how it may be received, acted on, and how it could be misreported or distorted by others. It incorporates a target audience analysis, including socio-demographic, psychosocial, behavioural characteristics, trusted sources and media channels used, education and health literacy, motives for health, and current knowledge and beliefs. Evidence about the target audience is collected from both secondary sources, such as literature and research reports, and primary ones, such as interviews, focus groups, and surveys, to develop relevant messages that are in line with the objectives set out in the Assess phase. Messages should be pre-tested, refined and improved, monitored and evaluated for the necessary effects on target audience (Suggs, et al, 2015, p.11-12).

Step 3: Describe the Communication.

Describe involves the process of documenting, evaluating and sharing to add to the health communication evidence base. All steps in the first two stages (Assess and Do phases) should be documented to ensure accountability, transparency, and to improve evidence. The communication and process should be evaluated for desired and unintended effects so that lessons can be learned about what worked well and what could be improved. All should be shared electronically evidence base for other communicators to learn for any implementation for future behaviour change initiatives in the programmes (Suggs, et al, 2015,p.11-12).

Suggs, et al, (2015, p.13) further suggest questions that the communicator should ask while designing messages for behaviour change. They gave five Ws of message design below:

- *Why are you communicating?* Know exactly why you are communicating. Is it to inform, invoke an action, change behaviour, change social norms?
- *Who do you need to communicate with?* Be clear on all the key groups in your communication. Who do you need to reach with the communication? Who are the gatekeepers to access your target audiences?
- *Where will you reach the target of your communication?* Understand all aspects of your communication channels. Where will you reach people? This will include the technology used and settings (e.g. school, homes, work, clinic, community).
- *What strategies will you use to communicate?* Have absolute clarity on what your message is: Make it consistent with the predetermined objectives set in the Assess phase. Make it simple. Use language, visuals, and ideas which are easy to process, quickly. Make it intuitive. People need to

be able to understand what you are asking them to do and why it matters, and be able to act accordingly. Also there is need to consider who is best placed to deliver your message: Who are the most trusted and influential sources of health information for your audience? Consider how your message framing will affect your audience response: Is your message framed in terms of the benefit gained or the loss of something? Does it play to people's drive to conform to norms? Consider the tone and look of the communication: Is it serious, inspiring and humorous? Is the presentation of the communication (colours, type of visual) consistent with the objectives and target audience characteristics?

- *When will you communicate to whom?* For each target audience: Know the right communication dose to be applied at the right time. When will the dissemination start and finish? How many messages, how often and when exactly? When will your audience be most receptive to your message? This is where the media come into play.

The media play a critical role in how people receive messages about health. However, practices are characterized by norms that sometimes hinder the delivery of comprehensive, accurate and relevant messages. The personalization and dramatization of information can distort scientific debate, and contribute to confusion among news consumers. The media can be used to reduce complex issues to simple understanding of the information. Suggs, et al, (2015, p.13) however stated that "health messengers throughout the world also have to contend with a politicized and commercialized media that will select and adapt messages in line with the aims or political views of their owners. Even the desire for 'balanced' coverage of an issue can distort messages: it may mean that the media incorrectly give equal weight to competing views, although expert opinion is heavily on one side of the debate" The media is an important channel to be considered in dissemination health messages for behaviour change.

Importance of Mass Media

The mass media performs four basic functions among others; surveillance of the environment, interpreting information about the environment, the transmission of the social heritage from one generation to the next and provision of entertainment to mass audience. Hence, mass media are important and indispensable to the development of modern society. The individual citizens, social organizations and businesses rely on the mass media to tell them what is going on in their community, the nation and the world at large. Governments rely on the media not only to inform the people of their activities but also to be informed about what various sectors of society believe about, and want from the state. Thus, the mass media are a vital link in an information system that encompasses all levels of society and its governance. However, they do not operate as neutral conveyors of information but as social institutions that function within particular political and legal constraints, employing a distinctive mixture of trained personnel, following specific information-gathering procedures, to create materials for either visual, auditory or conceptually oriented media. They have a particular view of their role in society and work under particular financial constraints. Despite these and other constraints, the mass media play a crucial role in all modern societies. The more open a society is to all voices, the more vital is the role of the media in keeping it open and accessible to the many and varied interests of individuals and groups at all times (Severin & Tankard, 1992). Supporting this view, Curran (1977) stated that the mass media contribute information and analysis within a wide context in which a variety of individuals and groups compete to interpret the real meaning of events.

In another vein, the social power of the media was further elaborated as machinery that can attract and direct attention to problems and provide solutions, confer status and confirm legitimacy, be a channel for persuasion and mobilization. Hence, it is assumed that the more attention the media give to a topic, the greater is the importance attributed to it by the news audience (McQuail, 2010). This means that if the media give much air-time or space (in a repetitive manner) to an issue or event it is seen by information consumers as very important (Asogwa, Iyere & Attah, 2012). The increase in the establishment of media of mass communication around the globe shows the social impact of mass media as opined by McQuail (2010). Hence, Graber (1988, pp.2-3) writes that "not only are the media the chief source of most people's views of the world, but they are also the fastest way known to disperse information throughout an entire society." This shows the importance of media of mass communication as a behaviour change strategy in HIV and AIDS prevention programmes. In various documented studies, the significant role of the media in HIV prevention as it relate to behaviour change shows that mass media interventions are always much more cost-effective than alternative interventions if they reach a large percentage of the population and can reach more people faster than any other type of health intervention (Kincaid, Storey, & Babalola, 2009). It is expedient to examine theoretical framework relevant to media effect on behaviour change initiatives.

Theoretical Framework for Media Effects on Behaviour Change

Communication theories used in programmes to support social change focused primarily on the individual and interpersonal levels of communication (Freimuth & Quinn, 2004). Contemporary research, however, shows that all four levels of communication are interconnected: society and its institutions influence individuals, who, in turn, influence the larger social and institutional groups of which they are members (Freimuth & Quinn, 2004). For example, prevailing social values undoubtedly influence individual behaviour, while individual behaviour and expressions of opinion help shape social values (Freimuth & Quinn, 2004). Media effects research investigates not only how the media influence the knowledge, opinions, attitudes, and behaviours of audience members, but also how audience members affect the media (Freimuth & Quinn, 2004). Because audience members are active seekers and users of health information, the content transmitted through the media reflects their needs, interests, and preferences. Two questions are central to understanding the effects of media on audience members:

1. What factors affect the likelihood that a person will be exposed to a given message?
2. How do media effects vary with the amount of exposure to that message?

Hence, in order to address these questions, the following media theories that have been used successfully in various behaviour change initiatives and their impact as it relate to HIV and AIDS intervention in Nigeria shall be examined.

Agenda Setting Theory

The agenda setting theory postulates that the mass media can illuminate and focus attention on issues, helping to generate public awareness and momentum for change. Agenda setting involves setting the media agenda (what is covered), the public agenda (what people think about), and the policy agenda (regulatory or legislative actions on issues). Research on agenda setting has shown that the amount of media coverage an issue receives correlates strongly with the public's opinion of how important that issue is (National Institute of Health, 2005). De Fleur and Ball-Rokeach (1982) citing Funkhouser (1973) on the same issue examined the relationship between public opinion and media content, and the relationship between media content and reality. Funkhouser (1973) study shows that many people believed in the media as reliable source of information because their perception and that of media practitioners on content of the media are the same. This theory can also be linked to the Cultivation theory of mass media, proposed by George Gerbner (1997) which specifies that repeated, intense exposure to deviant definitions of "reality" in the mass media leads to perception of that "reality" as normal. The result is a social legitimization of the "reality" depicted in the mass media, which can influence behaviour (Gerbner, 1997). Mass media promotion of condom use to prevent HIV has been pervasive in South Africa for well over 10 years, so it should not be surprising that some, if not most, of the increased condom use among youth is the result of media interventions, and thus to some of the measured decline in HIV incidence. (Kincaid, Storey, & Babalola, 2009).The focus of many communications research has been on how the mass media influences public opinion, especially about politics, policymaking and other issues including health issues like HIV and AIDS. Hence, agenda setting is one of the possible ways that the mass media can have an effect on the public.

Uses and Gratification Theory

The Uses and Gratification theory has five elements. The first is that the audience is active; that media exposure has a "causal origin". The second element sees the member of the audience as the initiator of the link between need gratification and media choice. The third element acknowledges the fact that "the media compete with other sources of need satisfaction". The fourth element is that the goals of media use are better supplied on the methodological level - by the audience themselves, while the fifth states that "audience orientations" should be "explored on their own terms"(McQuail, 1994:235). Further studies reveal that the focus this theory is that audience members have certain needs that are satisfied by using varied media sources (McQuail, 2010).

Audience gratifications can be derived from at least three distinct sources: media context, exposure to the media per se and the social context that typifies the situation of exposure to different media." (Katz, Blumler & Gurevitch (1974:11) Specific attributes of the media are recognized by proponents of the gratification model, as capable of serving different needs. This submission shows that the uses and gratification approach involves a shift of focus from the purposes of the communicator to the purposes of the receiver. It attempts to determine what functions mass communication is serving for audience members. The uses and gratification approach was conceptualized to know "what people do with the media (Katz, Blumler & Gurevitch, 1974). This shows that the development of information campaigns should begin with a study of the potential information users. Hence, mass media campaign for HIV and AIDS education must take into cognizance the end user of the information, who the audiences are, before such information is diffused into the community. The need for audience research becomes highly expedient for information dissemination on HIV prevention to have the required impact, which is behaviour change.

HIV and AIDS Situation in Nigeria

Globally, an estimated 35.3 (32.2-38.8) million people were living with HIV in 2012 (Joint United Nations Programme on HIV/AIDS (UNAIDS) (2013). There were 2.3 (1.9-2.7) million new HIV infections globally, showing a 33% decline in the number of new infections from 3.4 (3.1-3.7) million in 2001. At the same time the number of AIDS deaths is also declining with 1.6 (1.4-1.9) million AIDS deaths in 2012. The epidemic continues to disproportionately affect sub-Saharan Africa, home to 70% of all new HIV infections in 2012. Current surveys in several countries in sub-Saharan Africa have detected decreases in condom use and/or an increase in the number of sexual partners (Joint United Nations Programme on HIV/AIDS (UNIADS), 2013; National Agency for the Control of AIDS (NACA), (2015).

The epidemiology of HIV in Nigeria is mainly informed by two national surveys conducted every two years by Federal Ministry of Health (FMOH), the Antenatal Clinic (ANC) survey (2013) among pregnant women and the National HIV/AIDS and Reproductive Health Survey (NARHS) (2012) which is a general population based survey in line with World Health Organization (WHO) guidelines. Sentinel survey data showed that the HIV prevalence increased from 1.2% in 1991 to 5.8% in 2001. After 2003, the prevalence declined to 4.4% in 2005 before slightly increasing to 4.6% in 2008. Results from the latest round of sentinel survey show that the national prevalence was 4.1% in 2010 (Federal Ministry of Health (FMOH), 2013).

Trend analysis of HIV prevalence from sentinel surveillance in Nigeria indicates that the epidemic has halted and is showing signs of stabilizing at about 4% from 2005 till date. The ANC 2010 survey reported a national HIV prevalence of 4.1% and the States' prevalence ranged from 1% in Kebbi State to 12.7% in Benue State. Similarly, based on projected HIV estimates of 2013, about 3,229,757 people now live with HIV while it is estimated that 220,394 new HIV infections occurred in 2013. A total of 210,031 died from AIDS related cases. It is also estimated that a total of 1,476,741 required anti-retroviral drugs (ARV) in 2013 (NACA, 2015).

Although most-at-risk populations contribute to the spread of HIV, heterosexual sex, particularly of the low-risk type, still makes up about 80 percent. Mother-to-child transmission and transfusion of infected blood and blood products, on the other hand, account for the other notable modes of transmission. Another National HIV/AIDS and Reproductive Health Survey (NARHS) conducted in 2012 reported a national prevalence of 3.4%. There was a slight decline from the previous estimates of 2007 which was 3.6%.The overall national prevalence also masks several nuances and variations in Nigeria's epidemic at the sub-national (state) levels and among population groups. The 2012 NARHS HIV prevalence was highest among those aged 35 to 39 (4.4%), and lowest among the 15- 19 age group (2.9%). The prevalence for males aged 35 to 39 years was highest at 5.3%, while women aged 30 to 34 years was 4.2%.

Geographically, the HIV prevalence was highest in the South-South zone with the prevalence rate of 5.5% recorded, in the NARHS 2012, depicting an increase from 3.5% in 2007 (NARHS2012). Key drivers of the HIV epidemic in Nigeria include low personal risk perception, multiple concurrent sexual partnerships, transactional and inter-generational sex, ineffective and inefficient services for sexually transmitted infections (STIs), and inadequate access to and poor quality of healthcare services(FMOH, 2013).. Entrenched gender inequalities and inequities, chronic and debilitating poverty, and persistence of HIV/AIDS-related stigma and discrimination also significantly contribute to the spread of the infection (FMOH, 2013).

Factors Contributing to the Spread of HIV in Nigeria

There are various factors contributing to the spread of HIV in the world with Nigeria having its own peculiarities based on socio cultural and behavioural modifications (Pennington, 2005). The findings of various studies will assist us in reviewing some of these factors.

1. **Lack of Sexual Health Information and Education:** Information is an important tool for providing access to Sexual Reproductive Health services in Nigeria. A lot of people do not have access to information and if information is not shared a lot of people will be ignorance of what goes on around them which will eventually result into poor health seeking behaviour among the people. Also, the inadequacy of sexuality education and information about sex among young people and the sexually active population is a critical problem. The sacrosanct nature of disseminating sexual information in Nigeria is basically for cultural and religious reasons. The discussion of sex with teenagers, girls in particular, is seen as indecent. Studies have shown that there is limited sexual health education for young people which has led to high rate of sexually transmitted infection (STI) and HIV among them. Lack of accurate information about sexual health has meant there are many myths and misconceptions about sex and HIV which has been contributing to increasing transmission rates among young people (WHO, 2011;Yankah & Aggleton, 2008).
2. **Stigma and Discrimination:** The issue of stigma and discrimination against people living with HIV/AIDS in Nigeria is one of the ways in which the infection rates keep on increasing. The Christians and Muslims religions believed that those who contract HIV are reaping the consequences of their immoral behaviour. This affects attitudes towards people living with HIV/AIDS (PLWHA) and HIV prevention. PLWHAs often lose their jobs or are denied healthcare services because of the ignorance and fear about HIV and AIDS. There is so much ignorance about HIV/AIDS that 60% of healthcare workers think HIV positive patients should be isolated from other patients (Pennington, 2005).
3. **Poor Healthcare Services:** The healthcare care system in Nigeria has deteriorated because of political instability, corruption and a mismanaged economy. Large parts of the country lack even basic healthcare provision, making it difficult to establish HIV testing and prevention services such as those for the prevention of mother-to-child transmission (MTCT). Sexual health clinics providing contraception and testing and treatment for other STIs are also few and far between. (Sofo, Ali-Akpajiak & Pyke, 2003). Lack of adequate human resources (doctors, nurses, midwives and Community Health Workers (CHWs) is a barrier to accessing health services. Government build health centres, clinics and hospital with inadequate staff to provide services hinder access to adequate health services by the people (WHO, 2011). The management of the health services often show major weaknesses resulting in waste and inefficiency. The inputs in the delivery of reproductive health care by the different levels of governments, voluntary organisations and other agencies are generally poorly coordinated (FMOH, 2010).
4. **Gender Imbalance:** The traditional Nigerian culture shows masculinity as strength which is a notion that men must dominate women in all aspect of life (Odimegwu, 2008).This traditional model of masculinity encourages men, young and old, to dominate relationships with women (Odimegwu, 2008). In male-dominated relationships, men may be less likely to accept a woman's request to use a condom or her desire to abstain from sexual engagement entirely, thereby increasing sexual and reproductive health risks for both partners (Odimegwu, 2008). A young woman's difficulty in negotiating condom use is further exacerbated in cross-generational relationships, which are fairly common in parts of Nigeria. The age gap limits a young woman's autonomy and her ability to make decisions, including her ability to negotiate condom use or refuse to have sex with a husband who is known to be unfaithful or have and Sexually Transmitted Infections (STIs) (NPC, 2009). The influence of masculinity and perceived "machismo" has also limited the use of sexual and

reproductive health services by young men. Young men, often, believe that use of health services and other positive-seeking behaviours signifies a sign of weakness (Odimegwu, 2008).

5. **Marriage Practices:** Harmful marriage practices violate women's human rights and contribute to increasing HIV rates in women and girls. In Nigeria there is no legal minimum age for marriage and early marriage is still the norm in some areas (National Population Commission, 2009). Parents see it as a way of protecting young girls from the outside world and maintaining their chastity. Girls get married between the ages of 12 and 13 and there is usually a large age gap between husband and wife (National Population Commission, 2009). Young married girls are at risk of contracting HIV from their husbands as it is acceptable for men to have sexual partners outside marriage and some men have more than one wife (polygamy). Because of their age, lack of education and low status, young married girls are not able to negotiate condom use to protect themselves against HIV and STIs (Population Council, 2004).
6. **Female Genital Mutilation:** Female genital cutting (FGC), also known as female circumcision or female genital mutilation (FGM) is a cultural practice which involve the partial or total removal of the external female genitalia or any other injury to the female genital organ for nonmedical reasons(WHO) 1997). An estimated 19.9 million Nigerian women have undergone FGM/C meaning that approximately 16% of the 125 million FGM/C survivors worldwide are Nigerians (NPC, 2014) Overall prevalence of FGM/C among girls and women aged 15-49 years in Nigeria (27%) is lower than in many countries However, due to its large population, Nigeria has the third highest absolute number of women and girls (19.9 million) who have undergone FGM/C worldwide (after Egypt and Ethiopia) (National Demographic and Health Survey (NDHS) 2013). In many cultures including Nigeria, FGC is a cultural practice that is considered important for the socialization of women, curbing their sexual appetites, and preparing them for marriage. Despite its cultural importance, FGC has drawn considerable criticism because of the potential for both short- and long term medical complications, as well as harm to reproductive health. FGC puts women and girls at risk of contracting HIV from unsterilized instruments.
7. **Sex Work:** Prostitution is illegal in Nigeria but there are more than 1 million female sex workers. Hence, sex work is a highly stigmatized activity in Nigeria. As a result, many sex workers operate clandestinely, although an increased presence of young women engaged in sex work are noticeable in designated spots in many Nigerian cities such as Abuja, Lagos and Port Harcourt. Female sex workers (FSWs) in Nigeria broadly consist of two groups – brothel-based and non-brothel based. FSWs are usually between late teenage years and forty in age. FSWs are generally a highly mobile population, sometimes moving from one location to the other within days (SFH, 2004).

The findings of various surveys carried out in Nigeria suggest very high level of awareness of HIV and AIDS among FSWs (SFH, 2004). However the high level of awareness does not necessarily translate into high level of knowledge (SFH, 2004). HIV infection rates among sex workers have been estimated to be as high as 30% in some areas. There are low levels of condom use among sex workers because of a lack of knowledge about HIV transmission and poor acceptance by male clients (USAID, 2003). The main risk of HIV in sex works relates to lack of condom use or inconsistent condom use by FSWs despite the high number of partners they have sex with. Reason why most FSWs would not use condom is as a result of the competition in sex works. Clients' dislike for condom is another major reason that FSWs may not use condoms. Many clients have been said to dislike condom on the ground that it diminishes their sexual pleasure. Despite their risky sexual behaviour, risk perception among FSWs is low. Fifty-five percent of FSWs surveyed in 2004 believed that they were not at the risk of contracting HIV (UNAIDS, 2007, ENHANSE, 2007).

The notion of sex work which is usually commercialized are now been seen among sex workers as a social service. Average sex workers believed they are offering social services to their client; hence they regard themselves as social workers. This view of their trade constitutes high risk for the sexually active population who patronize them.

These factors, lack of sexual health information, stigma and discrimination, poor health services, gender imbalance, marriage practices, female genital mutilation and sex work, as discussed are critical issues that must be examine if any meaningful behaviour change intervention must take on HIV and AIDS prevention initiatives. This brings the role of the media to the fore front as agent of change and information dissemination. But with depth of the intervention what role could the media play in this regard?

Media Effect on Behaviour Change

One of the major concerns of communication theorists for years has been to investigate the effect of mass communication. Mass media have become a major force to reckon with in society but the questions still stand out. Are the effects of mass communication large or small? Are they malevolent or benign? Are they obvious or subtle? Are they potent for behaviour change? These are some of the questions programme implementer put into consideration especially with the emergence of HIV and AIDS epidemic in sub Saharan Africa.

The media are interconnected, large-scale organizations that gather, process, and disseminate news, information, entertainment, and advertising worldwide. Whether they are small operations, such as a neighbourhood newspaper, or large corporations employing tens of thousands of people, the media influence almost every aspect of human life: economic, political, social, and behavioural. The outcomes of media dissemination of ideas, images, themes, and stories are termed media effects (National Institute of Health, 2005 citing Babara & Karen, 2005)). Media effects research investigates not only know how the media influence the knowledge, opinions, attitudes, and behaviours of audience members, but also how audience members affect the media. Because audience members are active seekers and users of health information, the content transmitted through the media reflects their needs, interests, and preferences. Two questions are central to understanding the effects of media on audience:

1. What factors affect the likelihood that a person will be exposed to a given message?
2. How do media effects vary with the amount of exposure to that message?

Funding is a primary factor that determines whether or not audience will be exposed to a message through the mass media. Since money is needed to buy media time and space, many public health programmes do not have large budgets, so they often must rely on strategies for free distribution. Options may include public service announcements, embedding health messages in entertainment programmes (e.g., soap operas), or promoting news coverage of public health topics in print and electronic media. Community institutions can adopt and disseminate messages, and social networks can also generate excitement about some messages, depending on their content (Freimuth & Quinn, 2004). How often do people need to hear a message before it influences their beliefs or behaviours? This depends on several factors. Characteristics of target audiences (e.g., their readiness for change, the ways they process information), the complexity of the health issue, the presence of competing messages, and the nature of the health message influence the relationship between exposure to a health message and an outcome effect. Repeated exposure to a message, especially when it is delivered through multiple channels, may intensify its impact on audience (National Institute of Health, 2005).

Planners often assume that a certain percentage of the target audience will be exposed to a message and that another fraction of those who receive the message will be engaged by it (National Institute of Health, 2005). Yet there are several possible paths through which a health communications message can influence someone's beliefs and/or behaviours. These include immediate learning (people learn directly from the message), delayed learning (the impact of the message is not processed until sometime after it has been conveyed), generalised learning (in addition to the message itself, people are persuaded about concepts related to the message), social diffusion (messages stimulate discussion among social groups, thereby affecting beliefs), and institutional diffusion (messages instigate a response from public institutions that reinforce the message's impact on the target audience) (National Institute of Health, 2005). This shows that the nature of the media and its effect on the target audience must be examined for the required impact especially on issue of HIV and AIDS that is not a popular message with the people. We shall examine case studies of media use for HIV prevention interventions in Nigeria.

Case Studies of Media Use for HIV Prevention Interventions in Nigeria

The former Africa Regional Director of International Federation of Journalist (IFJ) Gabriel Ayite Baglo stated the roles that the media should play in HIV/AIDS prevention:

... the media has a crucial role to play in the battle against HIV/AIDS. As with other development issues in the world and in Africa in particular, the media is an integral and important part of the discourse. Journalists have been and continue to be in the forefront either individually or collectively because first and foremost they are members of society and HIV/AIDS affects every member of society. (FAMEDEV Bureau RégionalAfrique, 2005)

The use of the media to disseminate messages on health issues like HIV and AIDS has been documented in various researches to be a potent strategy for behaviour change (Osakue, Kayode, Marcel & Adekunle, 2010; Obono, 2011; Ogbabo, 2011),. This has been shown in their combination of messages with entertainment, which promote specific programmes. Television media with its combination of visual images and sound make its message convincing while radio is most effective when used in conjunction with other channels of communication. Several media are best used together to reinforce each other or to compensate for each other's shortcomings because different media reach different audiences. The media has contributed to the AIDS response in the following ways:

- Providing information and education on the epidemic,
- Advocating for a cause in relation to HIV by giving voices to the 'voiceless,
- Contributing to Behaviour Change through continuous information dissemination,
- Raising public awareness by mobilizing support for the epidemic,
- Promoting gender balance reporting and literacy among PLWHAs,
- Promoting hope that being HIV positive is not a death sentence,
- Setting agenda by putting issues in the front burner of public discourse,
- Investigating cure claims by medical researchers and allaying public fears,
- Supporting and monitoring government and community efforts in HIV/AIDS prevention.

The amount of time and attention that people give to mass media provides an ideal opportunity for communicating about sexual health. Hence, the media have been used effectively to promote sexual responsibility in most countries for decades. The media have a pivotal role to play in the fight against AIDS. Many media organizations are rising to the challenge by promoting awareness of HIV/AIDS and educating listeners and viewers about the facts of the epidemic and how to stop it. A study on the media and HIV and AIDS in the United States of America, United Kingdom and other countries in Africa shows that 72% of Americans identify television, radio, and newspaper as primary sources of information about HIV and AIDS while similar statistics was reported in UK and India. Clearly, media organizations have an enormous influence in educating and empowering individuals to avoid contracting HIV. Doing so with maximum efficiency, however, requires a clear understanding of the challenges and the obstacles to widespread and effective HIV-prevention education (UNAIDS, 2004). Other studies have revealed the contribution of the media to behaviour change and HIV prevention in Nigeria. We shall review three case studies in this regard.

CASE STUDY 1

Enhancing Nigeria's Response to HIV/AIDS 2009-2014: BBC Media Action Programme Awards

The BBC Media Action broadcast radio programmes and supported local radio stations across Nigeria to produce content and public information that aimed to control the spread of HIV and mitigate the impact of HIV and AIDS. The project included:

- Youth-focused radio magazine programmes *Flava* and *Ya Take Ne* (What's happening?) broadcast in English and Hausa on a network of partner radio stations across Nigeria.
- Short films and TV and radio public service announcements (PSAs) about HIV and AIDS.
- Training for journalists, TV and radio producers and station managers, as well as film and music producers.
- Annual media awards for radio and television stations.

The project BBC Media Action helped to tackle misconceptions about HIV and AIDS, encouraged people to seek testing and treatment, and challenged stigma surrounding people living with HIV and AIDS. In addition to making radio magazine programmes and radio and TV PSAs, BBC Media Action worked with Nigeria's booming Nollywood film industry, musicians and other creative producers. This included staging workshops for more than 50 producers, directors, singers and scriptwriters during which they learned more about the science behind HIV and AIDS as well as meeting people living with the virus. Four short films (Wetin Dey, Let's Talk HIV, Shuga and I need to Know) tackling issues relating to HIV and AIDS were by created by Nollywood producers and directors as a result of the collaboration. To encourage media managers to commission radio programmes about HIV and AIDS, BBC Media Action also ran "master classes" with senior managers from partner radio stations. This training covered editorial guidelines, broadcast management and revenue generation (BBC Media Action, 2015)

CASE STUDY 2

Journalist Against AIDS (JAAIDS) 1988-2015

Journalists Against AIDS in Nigeria (JAAIDS) an NGO on advocacy and communications was formed by Omololu Falobi, a features editor with the Sunday Punch. The most visible aspect of JAAIDS HIV prevention activity was the daily Nigeria AIDS e-forum which was used as a communications tool on HIV/AIDS in Nigeria and across Africa. Health and development professionals around the world use the daily list-serve to share information, news, contacts, and debates on HIV/AIDS in Nigeria. Omolulu's leadership, his long-time commitment to pushing for Africa's own community leadership over its epidemic, his example of how journalists can be stronger social activists, his vision and championing of the Nigeria e-forum, and his ability to draw other activists together for positive change have inspired and motivated people from around the globe into action. Hence, JAAIDS became a leading global organization in advocate for HIV/AIDS treatment, care, and prevention (Journalist Against AIDS (JAAIDS), 2007)

CASE STUDY 3

HIV/AIDS and the Broadcast Media in Urban Communities in Edo State, Nigeria

The study examines the role of the broadcast media particularly the television medium in educating the public on Human Immunodeficiency Virus/Acquired Immune Deficiency Syndrome (HIV/AIDS) as well as caring for the HIV/AIDS infected persons in Edo State. The result showed that television, as a medium of mass communication has been able to raise awareness on HIV/AIDS through its various programmes, which are geared towards enlightening people about HIV/AIDS or helping the infected persons to live positively. It was also found that a partial disconnect exist between the people's sense of awareness and sexual behaviours as could be gleaned from their attitude in terms of risky sexual behaviours indicated by the increase in HIV/AIDS infection in Edo state (Osakue, Kayode, Marcel & Adekunle, 2009).

DIRECTIONS FOR FUTURE RESEARCH

Interventions designed to modify people's behaviour need to be based on sound principles of behaviour change. Several media strategies have been mention in the chapter to understanding dissemination of information for HIV prevention. However, no single media strategy is sufficiently potent to provide all the essential messages or information to change behaviour. Based on this understanding there is need for a complementary combination of media strategies that can be incorporated into behaviour change program design. Hence future studies can be explores into developing media strategies for targeting specific high risk groups, for reinforce individual behaviour change efforts and maintenance of behaviour change over time.

CONCLUSION AND RECOMMENDATION

Improving the Role of the Media in HIV Prevention in Nigeria

The role of the media is undermined by a variety of factors including creating more awareness that are not translating to *"required behaviour change"*, reaching large audience with no potent feedback mechanism, reporting about HIV which seems monotonous, competing of other health problems with HIV, inadequate training of journalist about HIV reporting, ideological constraints as determined by information gatekeepers and the commercial interests of media formations(mailto:jaaidsng@nigeria-aids.org 2007,CADRE, 2001,p. 11). Hence, there should be guiding principles for project communicator using the media for HIV prevention in Nigeria as is the case in South Africa (CADRE, 2001). Media use should include: accuracy and consistency in reporting, the use of non-alarmist, non-moralistic and non-discriminatory language implicitly or explicitly blaming those with HIV or involved in high-risk activities, the promotion of basic human rights and gender equity, the use of acceptable local languages, and the promotion of HIV related services and resources (CADRE 2001p.11). In addition, "greater attention should be paid to working within existing media institutional frameworks to train journalists and promote editorial policy conducive to the production of good quality reporting in line with health promotion aims" (CADRE 2001p.12).Hence, the media should be involved in HIV project implementation from the inception rather than a situation of just taking HIV news to media houses. On the part of media institutions, they urgently need to develop codes of good practice for reporters to safeguard against reporting which simply counts the numbers of the infected and the dead and which explicitly or implicitly presents AIDS as an issue of public morality rather than of public health.

Communication efforts that use multiple media, reach most people and best reinforce the message. When closely tied to behaviour change, such multimedia efforts can promote and sustain positive behaviour among the target population especially in Nigeria. The effectiveness of communication channels like radio and television depends on the accessibility of such media by intended audiences (De Fossard, Baptiste, Corrales &Bosch, 1993). Hence, there is great potential for the media to contribute more constructively to the conceptualization of the HIV/AIDS epidemic for behaviour change and to be proactive in shaping societal and government response to the epidemic (CADRE, 2001), However, various studies and experiences have shown that irrespective of the contribution of the media and various intervening strategies, behaviour change still lies with the individual decisions.

In a study by the Global HIV Prevention working group (2008), their findings reveal that there are limitations to the potency of applying the existing theories of behaviour change interventions. Hence, they stated thus

> *... existing models of behavioral interventions are based on various cognitive behavioral theories that assume that individuals will take steps to avoid risks if they are fully informed and sufficiently motivated. While such approaches may work well for many people, they are unlikely to address the needs of the myriad populations at risk of infection. Because human sexuality and drug dependence are phenomena that are not always subject to cognitive control or mediation, cognitive approaches alone will not produce behaviour change in many people. (Global HIV Prevention working Group, 2008, p.13)*

They concluded that "many individuals confront exceptionally elevated risk of infection not primarily as a result of their own risky behaviour, but rather because of the behavioural characteristics of their partners or the particular structure and functioning of the social networks to which they belong. Hence, influencing individual behaviour in such cases will have only a limited impact on infection rates (Global HIV Prevention working Group, 2008, p.15). This submission revealed that, there is need to evolve other form of communication like interpersonal communication, community mobilization and advocacy to support media strategies for effective HIV prevention initiatives in Nigeria for the required impact on the target population who are diverse in socio cultural and behavioural characteristics.

REFERENCES

Adeokun, L. A., Ladipo, O. A., Kanki, P., Dealno, G. E., Carrington, A., & Odimegwu, C. O. (2005). The Role of HIV/AIDS Prevention Campaigns on HIV-Related Behavioural Change in Ibadan, Nigeria. *Journal of Population Studies*, *20*(2), 1–21.

AIDSCAP/Family Health International. (1996). *Behavior Change through Mass Communication: Using Mass Media for AIDS Prevention*. Arlington, VA: Author.

AIDSCAP/Family Health International. (1997). Behavior Change Communication for the Prevention and Treatment of STDs. Arlington, VA: Author.

Asogwa, C. E., Iyere, J. I., & Attah, C. O. (2012). The Mass Media Reportage of Crimes and Terrorists Activities: The Nigerian Experience. *Journal of Culture and History*, *4*(2), 175–181.

Bandura, A. (1986). *Social Foundations of Thought and Action: A Social Cognitive Theory*. Englewood Cliffs, NJ: Prentice-Hall.

BBC Media Action. (2015). *Enhancing Nigeria's Response to HIV/AIDS 2009-2014: BBC Media Action Programme Awards*. Retrieved from http://www.bbc.co.uk/mediaaction/contact

Burgoon, M., & Ruffner, M. (1978). *Human Communication, an introduction*. Holt, Rinehart and Winston.

Center for Communication Programs Population Report. (1995). *Helping the News Media cover Family Planning*. Johns Hopkins Bloomberg School of Public Health.

Center for Communication Programs Population Report. (2001). *HIV/AIDS communication strategies: Communication is a key factor in combating the HIV/AIDS epidemic*. Johns Hopkins Bloomberg School of Public Health.

Centre for AIDS Development, Research and Evaluation. (2001). *HIV/AIDS and the Media: A literature review*. Johannesburg: CADRE.

Chew, F., & Palmer, S. (2005). Television Health Promotion in Four Countries. *Nutrition Journal*, *21*(5), 634–638. doi:10.1016/j.nut.2004.09.022 PMID:15850972

Curran, J. (1977). *Mass Communication and Society*. London: Edward Arnold Publisher Ltd and Open University Ltd and Open University Press.

De Fossard, E., Baptiste, J., Corrales, C., & Bosch, A. (1993). *Interactive Radio Instruction*. Washington, DC: USAID.

Enabling HIV/AIDS+TB and Social Sector Environment project (ENHANSE). (2007). *Most at Risk Population foe HIV and AIDS in Nigeria*. Abuja: ENHANSE Project.

Family Health International. (2006). Strategic Behavior Communication for HIV and AIDS: A Framework. Arlington, VA: Author.

Federal Ministry of Health. (2005). *Behavioural Surveillance Survey. Draft Report*. Abuja: Federal Ministry of Health.

Federal Ministry of Health. (2010). *NationalHIV Sero-prevalence Sentinel Survey*. Abuja: NationalAIDS/STI Control Programme.

Federal Ministry of Health. (2012). *National HIV/AIDS and Reproductive Health Survey (NARHS)*. Abuja: FMOH.

Federal Ministry of Health. (2013). *NationalHIV Sero-prevalence Sentinel Survey*. Abuja: NationalAIDS/STI Control Programme.

Federation of African Media and Development (FAMEDEV). (2005). *HIV and AIDS Research, Reports and Media Guide: Inter-African network for women, media, gender equity and development*. Senegal: Federation of International Journalist, Bureau RégionalAfrique.

Freimuth, V., & Quinn, S. C. (2004). The Contributions of Health Communication to Eliminating Health Disparities. *American Journal of Public Health*, *94*(12), 2053–2054. doi:10.2105/AJPH.94.12.2053 PMID:15569949

Freimuth, V. S. (1992). Theoretical Foundations of AIDS Media Campaigns. In AIDS: A Communication perspective. Hillsdale, NJ: Eribaum.

Freire, P. (1971). *Education for Critical Consciousness*. New York: The Continuum Publishing Company.

Gerbner, D. (1997). *Mass Media Policies in Changing Cultures*. New York: John Wiley & Sons.

Global HIV Prevention working Group. (2008). *Behavior Change and HIV Prevention: (Re)Considerations for the 21st Century*. Bill & Melinda Gates Foundation and the Henry J. Kaiser Family Foundation. Retrieved from www.GlobalHIVPrevention.org

Graber, D. A. (1988). *Processing the News: How people Tame the Information Tide*. New York: Longman.

Joint United Nations Programme on HIV/AIDS (UNAIDS). (2013). *Global Report: UNIAIDS Report on global AIDS epidemic 2013*. Author.

Joint United Nations Programme on HIV/AIDS (UNAIDS). (2004). *The media and HIV/AIDS: Making a Difference*. Geneva: UNAIDS.

Joint United Nations Programme on HIV/AIDS (UNAIDS). (2007). A Framework for Monitoring and Evaluating HIV Prevention Programmes for Most-At-Risk Populations. Geneva: UNAIDS/07.15E / JC1338E.

Joint United Nations Programme on HIV/AIDS (UNAIDS) and World Health Organisation (WHO). (2006). AIDS Epidemic Update. Geneva: UNAIDS/06.29E.

Katz, D. (1961). The Functional Approach to the Study of Attitudes. *Public Opinion Quarterly*.

Katz, E., Blumler, J., & Gurevitch, M. (1974). Uses of Mass Communication by the Individual. In W. P. Davidson & Yu. Fredrick (Eds.), *Mass Communication Research: Major Issues and Future Directions* (pp. 11–35). New York: Praeger.

Keating, J., Meekers, D., & Adewuyi, A. (2006). Assessing effects of a media campaign on HIV/AIDS awareness and prevention in Nigeria: Results from the VISION Project. *BMC Public Health*, *6*(1), 123. doi:10.1186/1471-2458-6-123 PMID:16672067

Kincaid, D. L., Storey, D., & Babalola, S. (2009). *Effect of Communication on HIV Prevention and Living with HIV/AIDS: Key points for discussion*. Baltimore, MD: Johns Hopkins Center for Communication Programs.

Klapper, J. T. (1960). *The Effect of Mass Communication*. New York: The Free Press.

Lorimer, R. (1994). *Mass Communication; A Comparative introduction*. Manchester, UK: Manchester University Press.

McQuail, D. (1980). *Towards a sociology of Mass Communication*. London: Collier Macmillan Ltd.

McQuail, D. (1982). *Mass Communication Theory: An Introduction*. London: Constable.

McQuail, D. (2010). *Mass Communication Theory* (6th ed.). London: Sage.

McQuail, D., Blumler, J. G., & Brown, J. R. (1972). The Television Audience: A Revised Perspective. In *Sociology of Mass Communications*. Harmondsworth, UK: Penguin.

McQuail, D., & Gurevitch, M. (1974). *Explaining Audience Behavior: Three Approaches Considered. In The uses of Mass Communication: Current Perspectives on Gratifications Research*. London: Sage Publication.

McQuail, D., & Windahl, S. (1981). *Communication Model for the Study of Mass Communication*. London: Longman.

Mehdizadeh, M. (2010). *The theories of media: Current thoughts and critical attitudes*. Tehran: Hamshahri Press.

Moradi, M., Honari, H., Naghshbandi, S., Jabari, N., & Kholouse, P. (2012). Social Functions of Mass Media in Developing Citizen Sports. *International Journal of Academic Research in Business and Social Sciences*, *2*(6), 2222–6990.

National Agency for the Control of AIDS (NACA). (2015a). *National HIV/AIDS Prevention Plan 2014-2015*. Abuja: NACA.

National Agency for the Control of AIDS (NACA). (2015b). *Federal Republic of Nigeria Global AIDS Response: Country Progress Report 2014*. Abuja: NACA.

National Demographic and Health Survey (NDHS). (2013). *National Population Commission*. Abuja: NPC.

National Institute of Health. (2005). Theory at a Glance: A Guide for Health Promotion Practice. Washington, DC: National Institute of Health Publication. No. 05-3896.

National Population Commission. (2009). *Nigeria Demographic and Health Survey 2008*. Abuja: NPC.

National Population Commission. (2013). *Nigeria Demographic and Health Survey 2013*. Abuja: NPC.

National Population Commission. (2014). *Nigeria Demographic and Health Survey 2013*. Abuja: NPC.

Nonoyama, M., Tsurugi, Y., Shirai, C., Ishikawa, Y., & Horiguchi, M. (2005). Influences of Sex-Related Information for STD Prevention. *The Journal of Adolescent Health*, *36*(5), 442–445. doi:10.1016/j.jadohealth.2004.01.017 PMID:15837349

Odimegwu, C., & Okemgbo, C. N. (2008). Mens Perceptions of Masculinities and Sexual Health Risks in Igboland, Nigeria. *International Journal of Mens Health*, *7*(1), 21–40. doi:10.3149/jmh.0701.21

Okonta, P. I., & Oseji, M. I. (2006). Relationship between knowledge of HIV/AIDS and sexualbehaviour among in school adolescents in Delta State. *Nigerian Journal of Clinical Practice*, *9*(1), 37–39. PMID:16986287

Osakue, S. O., Kayode, R. A., Marcel, A. O., & Adekunle, A. A. (2009). 'HIV/AIDS and the Broadcast Media in Urban Communities in Edo State. *The Nigeria International Journal of Research and Review*, *1*(3), 111–116.

Owens-Ibie, N. (1989). *A Comparative Study of Mass & Specialized Media Use by Researchers in Nigeria* (Unpublished PhD Thesis). University of Ibadan.

Pennington, J. (2005). Retrieved from www.avert.org/aids-countries

Population Council. (2004). *Child Marriage Briefing: Nigeria*. New York, NY: Author.

Report, P. (1989). *Light! Camera! Action! Promoting Family Planning with TV, Video and Film*. Johns Hopkins Bloomberg School of Public Health.

Rogers, E. M., & Shoemaker, F. F. (1971). *Communication of Innovations: A cross cultural approach*. New York: Free Press.

Seth, M. N. (2007). HIV/*AIDS Mass Media Campaigns*. Indiana University.

Severin, W. J., & Tankard, J. W. (1992). *Communication Theory: Origin and Uses in the Mass Media*. New York: Longman.

Society for Family Health. (2004). *National Behavioural Survey II: Brothel-based sexwork in Nigeria*. Abuja: Society for Family Health.

Sofo, C. A., Ali-Akpajiak, & Pyke. (2003). *Social development and poverty in Nigeria: Measuring poverty in Nigeria*. Oxfam Working Paper.

Soola, E. O. (1988). *An Evaluation Study of The Dissemination of Family Planning Information in Ogun State* (Unpublished PhD Thesis). University of Ibadan.

Suggs, L. S., McIntyre, C., Warburton, W., Henderson, S., & Howitt, P. (2015). *Communicating Health Messages: A Framework To Increase The Effectiveness Of Health Communication Globally*. Report of the WISH Communicating Complex, Institute of Global Health Innovation, Imperial College London.

World Health Organisation (WHO). (2011). *The sexual and reproductive health of younger adolescents: Research issues in developing countries*. Geneva, Switzerland: WHO Press.

Yankah, E., & Aggleton, P. (2008). Effects and effectiveness of life skills education for HIV prevention in young people. *AIDS Education and Prevention*, *20*(6), 465–485. doi:10.1521/aeap.2008.20.6.465 PMID:19072524

Yazdi, C. A., Aschbacher, K., Arvantaj, A., Naser, H. M., Abdollahi, E., Asadi, A., … Moghadam, A. K. (2006). Knowledge, attitudes and sources of information regarding HIV/AIDS in Iranian adolescents. *AIDS Care, 18*(8), 1004-10.

KEY TERMS AND DEFINITIONS

Behaviour Change: Behavior change can refer to any transformation or modification of human behavior.

Behaviour Change Communication (BCC): An interactive process of any intervention with individuals, communities and/or societies (as integrated with an overall program) to develop communication strategies to promote positive behaviors which are appropriate to their settings.

Health Services: Include all services dealing with the diagnosis and treatment of disease, or the promotion, maintenance and restoration of health. They include personal and non-personal health services. Health services are the most visible functions of any health system, both to users and the general public. Service provision refers to the way inputs such as money, staff, equipment and drugs are combined to allow the delivery of health interventions.

HIV and AIDS: HIV (human immunodeficiency virus) is a virus that attacks the immune system, the body's natural defense system. Without a strong immune system, the body has trouble fighting off disease. Both the virus and the infection it causes are called HIV.The last stage of HIV infection is AIDS (acquired immunodeficiency syndrome). People with AIDS have a low number of CD4+ cells and get infections or cancers that rarely occur in healthy people. These can be deadly.

Mass Media: The use of technology that is intended to reach a mass audience. It is the primary means of communication used to reach the vast majority of the general public. The most common platforms for mass media are newspapers, magazines, radio, television, and the Internet. The general public typically relies on the mass media to provide information regarding political issues, social issues, entertainment, and news in pop culture.

Sexual Behaviour: Sexual behaviour deals with all things relating to sex, conception and satisfaction. Also this deals with any activity—solitary, between two persons, or in a group—that induces sexual arousal. There are two major determinants of human sexual behaviour: the inherited sexual response patterns that have evolved as a means of ensuring reproduction and that are a part of each individual's genetic inheritance, and the degree of restraint or other types of influence exerted on the individual by society in the expression of his sexuality.

Sexual Health Information: Information about a person state of physical, mental and social well-being in relation to sexuality. It contain information about a positive and respectful approach to sexuality and sexual relationships, as well as the possibility of having pleasurable and safe sexual experiences, free of coercion, discrimination and violence.

Sexual Practices: The manner in which humans experience and express their sexuality. People engage in a variety of sexual acts, ranging from activities done aloneto acts with another person in varying patterns of frequency, for a wide variety of reasons.

Compilation of References

Abara, C. J. (2012). *Inequality and discrimination in Nigeria tradition and religion as negative factors affecting gender.* Paper presented at the Federation of International Human Rights Museums.

Abone, G. O., & Kur, J. T. (2014). Perceptual influence of freedom of information Act on Journalism Practice in Nigeria. *Arabian Journal of Business and Management Review*, *3*(7), 23–35.

Accenture, V. (2011). *Connected agriculture: The role of mobile in driving efficiency and sustainability in the food and agriculture value chain*. Berkshire: Vodafone.

Acs, Z., Arenius, P., Hay, M., & Minniti, M. (2005). *Global Entrepreneurship Monitor, 2004 Executive Report, Babson College and London BusinessSchool*. Retrieved March 17, 2005, from www.gemconsortium.org

Acs, Z. J., & Audretsch, D. B. (2003). *Innovation and technological change*. Springer, US.

ActionAid. (2015). *Democracy watchdogs: Youth-driven participatory monitoring and accountability in the Post-2015 development agenda*. Author.

Adams, J., & Jaffe, A. (1996). Bounding the effects of R&D. *The Rand Journal of Economics*, *27*(4), 700–721. doi:10.2307/2555878

Adedeji, A., & Ajibogun, G. (2012). Gender issues and democracy: The Nigeria Experience. *International Journal of Academic Research in Progressive Education and Development.*, *2*(1), 117–127.

Adedimeji, M. A., & Abdulbaqi, S. S. (2007). *A communicative approach to the prevention and control of HIV/AIIDS pandemic in Nigeria*. Paper presented at the Faculty of Business and Social Sciences, University of Ilorin Second National Conference themed "Nigeria Beyond 2007: Issues, Perspectives and Challenges" held at the FBSS Lecture Theatre, University of Ilorin, Ilorin, Nigeria.

Adegbija, M. V., Fakomogbon, M. A., & Adebayo, M. S. (2013). Roles of Broadcast Media for Instructional Delivery in Open and Distance Learning: Nigeria as a Case Study. *European Scientific Journal*, *9*(23), 279–290.

Adegoju, A., & Oyebode, O. (2015). Humour as discursive practice in Nigerias 2015 presidential election online campaign discourse. *Discourse Studies*, *17*(6), 643–662. doi:10.1177/1461445615602378

Adenle, G., & Oso, L. (2014). *Muted gender: How Nigeria press portrays female politicians. In Women's political visibility and media access: The case of Nigeria*. Cambridge Scholar.

Adeokun, L. A., Ladipo, O. A., Kanki, P., Dealno, G. E., Carrington, A., & Odimegwu, C. O. (2005). The Role of HIV/AIDS Prevention Campaigns on HIV-Related Behavioural Change in Ibadan, Nigeria. *Journal of Population Studies*, *20*(2), 1–21.

Aderemi, H. O., Ilori, M. O., Siyanbola, W. O., Adegbite, S. A., & Aberiojo, I. O. (2008). A an assessment of the choice and performance of women entrepreneurs in technological and non-technological enterprises in Southwest Nigeria. *African Journal of Business Management.*, *2*(10), 165–176.

Adesoji, A. O., & Hahn, H. P. (2011). When (not) to be a proprietor: Nigerian newspaper ownership in a changing polity. *African Study Monographs*, *32*(4), 177–203.

Adewunmi, A. (n.d.). *Nigerian women Agro-entrepreneurship development: Issues and challenges.* Retrieved on 2 August, 2015 from http://peda.cc/nigerian-women-agro-entrepreneurship-development-issues-and-challenges/

Adeyemi, A. (2013). Nigerian media and corrupt practices: The need for paradigm shift. *European Scientific Journal*, *9*(1), 125–126.

Africa Partnership Forum. (2007). *Gender and economic empowerment of women, APF Support Unit and NEPAD, Briefing Paper 3.* Retrieved from http://www.oecd.org/dataoecd/16/42/39921766.pdf

Africapractice. (2014, April). Retrieved from http://www.africapractice.com/wp-content/uploads/2014/04/Africa-Practice-Social-Media-Landscape-Vol-1.pdf

Agardh, A., Cantor-Graae, E., & Ostergren, P. (2012). Youth, sexual risk-takingbehaviour, and mental health: A study of University students in Uganda. *International Journal of Behavioral Medicine*, *19*(2), 208–216. doi:10.1007/s12529-011-9159-4 PMID:21590465

Agbalajobi, D., (2010) Women's participation and the political process in Nigeria: Problems and prospects. *African Journal of Political Science and International Relations, 4*(2), 75-82.

Agbanu, V. N. (2014). *Propaganda and public opinion: A discourse on political communication and mind management.* Enugu: Rhyce Kerex.

Agbon, I. (2012, January 8). The Real Cost of Nigeria Petrol. *The Guardian*, p. 7.

Ahmed, S., Hassali, M., & Abdul-Aziz, N. (2009). An assessment of the knowledge, attitudes and risk Perceptions of Pharmacy Students regarding HIV/AIDS. *American Journal of Pharmaceutical Education*, *73*(15), 1–7. PMID:19513153

AIDSCAP/Family Health International. (1996). *Behavior Change through Mass Communication: Using Mass Media for AIDS Prevention.* Arlington, VA: Author.

AIDSCAP/Family Health International. (1997). Behavior Change Communication for the Prevention and Treatment of STDs. Arlington, VA: Author.

Ajala, V. (2014). *Exploring the indifference of women and minorities in Nigeria's politics. In Women's political visibility and media access: The case of Nigeria.* Cambridge Scholar.

Akbulut, A. (2014). Conflict in the global age: The case of danish cartoon crisis. *European Scientific Journal*, *10*(14), 13–23.

Ake, C. (1985). *Political economy of Nigeria.* Lagos: Longman.

Akeem, R. A., Oyeyinka, O. T., Qasim, A. O., Lateef, O., Omolayo, J., & Onyinyechi, N. (2013). Deregulation of broadcast media in Nigeria: An appraisal. *Kuwait Chapter of Arabian Journal of Business and Management Review*, *3*(3), 6–13. doi:10.12816/0017469

Aker, J. C., & Mbiti, I. M. (2010). Mobile phones and economic development in Africa. *The Journal of Economic Perspectives*, *24*(3), 207–232. doi:10.1257/jep.24.3.207

Akin, J. (2005). *Mass media*. Retrieved from http://www.beyondintractability.org/essay/mass-communication

Akinfeleye, R., & Amobi, I. (2011). Nollywood video films as a medium for reconstructing the African cultural identity. *Communication Review*, *5*(1), 1–23.

Akintola, O., Ngubane, L., & Makhaba, L. (2012). I did it for him, not for me An exploratory study of factors influencing sexual debut among female university student in Durban, South Africa. *Journal of Health Psychology*, *17*(1), 143–153. doi:10.1177/1359105311410512 PMID:21672963

Akinwale, A. A. (2010). Repression of press freedom in Nigerian democratic dispensations. *Africa Development, 35*(3), 47–70.

Akpabio, E. (2003). Themes and content of Nigerian home video films. *Unilag Sociological Review*, *4*(October), 138–139.

Akpabio, E. (2007). Attitude of audience members to Nollywood films. *Nordic Journal of African Studies*, *16*(1), 90–100.

Alabi, S. (2003). Press freedom: Background, legal bases and constraints. *UNILAG Communication Review*, *4*(1), 52–65.

Alade, A. (2015, December 6). Social media censorship bill: We'll take your input, Senate assures Nigerians. *The Vanguard.* Retrieved from http://www.vanguardngr.com/2015/12/social-media-censorship-bill-well-take-your-input-senate-assures-nigerians/

Aladese, T. (2013). *Gender stereotyping in Nigerian home videos* (Unpublished undergraduate thesis). Department of Mass Communication, College of Development Studies, Covenant University, Ota, Ogun State, Nigeria.

Alawode, S. O., & Adesanya, O. O. (2016). Content Analysis of 2015 election political advertisements in selected national dailies of Nigeria. *European Scientific Journal*, *12*(5), 234–259.

Alawode, S. O., Atofojomo, O., & Fatonji, S. (2015). Home video depictions of health and moral decadences as correlates of entertainment-education in Nigeria. *Mediterranean Journal of Social Sciences.*, *6*(2), 293–307.

Alberts, B., Alexander, J., Lewis, J., Raff, M., Roberts, K., & Walter, P. (2002). *Molecular biology of the cell* (4th ed.). New York: Garland Science.

Aldrich, H., & Zimmer, C. (1986). Entrepreneurship through Social Networks. In D. L. Sexton & R. W. Smilor (Eds.), *The Art and Science of entrepreneurship*. Cambridge, MA: Ballinger.

Ali, G., & Khan, L. (2012). A language and construction of gender: A feminist critique of Sms Discourse. *British Journal of Arts and Social Sciences*, *4*(2), 342–360.

Alika, I. H., & Egbochuko, E. O. (2009). Drop Out from School among Girls in Edo State: Implications for Counselling. *Edo Journal of Counselling*, *2*(2), 135–141.

Al-Kandari, A., & Hasan, M. (2012). The impact of the internet on political attitudes in Kuwait and Egypt. *Telematics and Informatics, 29*, 245-253. 10.1016/j.tele.2012.02.005

Allen, B., & Reser, D. (1990). Content Analysis in Library and Information Science Research. *Library & Information Science Research*, *12*(3), 251–260.

Allen, K. (2003). *Launching new ventures: An entrepreneurial approach* (3rd ed.). Boston: Houghton Mifflin.

Allmer, T. (2014). (Dis)like Facebook? Dialectical and critical perspectives on social media. *Javnost-The Public*, *21*(2), 39–56. doi:10.1080/13183222.2014.11009144

Althusser, L. (2001). Ideology and ideological state apparatuses. In G. D. Meenakshi & D. M. Kellner (Eds.), *Media and Cultural Studies Reader*. Malden, MA: Blackwell Publishing.

Amadi, F. A. (2014). Reportorial activism and Nigerian journalism after the enactment of Freedom of Information Act. *New Media and Mass Communication*, *21*, 10–15.

Amienyi. (2004). Broadcasting and national development in Africa. In C. C. Okigbo & F. Eribo (Eds.), *Development communication in Africa* (pp. 106-114). Lanham, MD: Rowman & Littlefield.

Amobi, I. (2010). *Audience interpretation of the representation of women in Nigerian Nollywood films: a study of women from different social contexts in Nigeria*. Retrieved from http://profteri.wordpress.com/2010/11/12/online-journalism-syllabus/

Amobi, I. (2013). *Portrayal and participation of women in Nigerian media*. Retrieved from https://profreri.wordpress.com/2013/07/27

Amoroso, E. G. (2011). *Cyber Attack: Protecting National Infrastructure*. Butterworth-Heinemann.

Anaeme, F. (2012). Reducing gender discrimination and violence against women through library and information service. *Library philosophy and practice*. Retrieved on 12 January, 2015 from https://www.researchgate.net/profile/Francis_Anaeme/publication/267264193

Anaeto, S. G., Onabanjo, S. O., & Osifeso, B. J. (2008). *Models and theories of communication*. African Renaissance Books Incorporated.

Anyanwu, C. (2015). Communication and society an introductory overview. In J. E. Aliede (Ed.), *Studies in communication, mass media and society: Discourses and debates* (pp. 28–37). Lagos: PR-MIX.

Appadurai, A. (2002). *Globalization*. Available from: http://www.polity.org.za/pol/home/

Arimoro, A. E. (2015). Applying the law to tackle the menace of revenge porn. Creating Global. *Journal of Humanities and Social Science*, 75-80.

Aririguzoh, S. (2005). The function and structure of communication in entrepreneurship. *Nssuka Journal of the Humanities*, (15), 160-178.

Asante, K. (2013). HIV/AIDS knowledge and uptake of HIV counseling and testing among undergraduate private University Students in Accra, Ghana. *Reproductive Health Journal*, *10*(17), 1–8.

Ascher, W. (1999). *Why governments waste natural resources: Policy failures in developing countries*. Baltimore, MD: Johns Hopkins University Press.

Asemah, E. S. & Edegoh, L. O. N. (2014). Viewers' perception of ritualism in Nigerian home videos. *ANSU Journal of Arts and Social Sciences, 3*(1), 18-30.

Ashford, W. (2015). *Cyber-crime is a threat to global economy*. Retrieved from: http://www.computerweekly.com/news/2240239300/Cyber-crime-is-a-threat-to-global-economy-says-researcher

Ashikpe, J. A. (2015). Perceived influence of social media use patterns on the social development of Nigerian adolescents and youths. In J. E. Aliede (Ed.), *Studies in communication, mass media and society: Discourses and debates* (pp. 204–222). Lagos: PR-MIX.

Asika, N. (1991). *Research Methodology in the Behavioural Science*. Ikeja: Longman.

Aslama, M. (2009). *Public Spheres, networked publics, networked public Spheres*. New York: Donald McGannon.

Asogwa, C. E., Iyere, J. I., & Attah, C. O. (2012). The Mass Media Reportage of Crimes and Terrorists Activities: The Nigerian Experience. *Journal of Culture and History*, *4*(2), 175–181.

Assay, B. E., & Adigu, R. O. (2008). The impact of mass media sensitisation on the control of tobacoo abuse in selected local government of Delta State. In E. M. Mojaye, O. O. Oyewo, R. M'Bayo, & I. A. Sobowale (Eds.), *Health communication, gender violence and ICTs in Nigeria* (pp. 65–76). Ibadan: Ibadan University Press.

Association of Media Women in Kenya. (2015). Media monitoring analysis of the coverage of the 2/3rds gender principle. Nairobi, Kenya: Author.

Aterido, R., & Hallward-Driemeier, M. (2011). Whose business is it anyway? *Small Business Economics, Springer, 37*(4), 443–464. doi:10.1007/s11187-011-9375-y

Attoh, N. R. (2016). The Frivolous Petitions Bill and permissible restrictions on freedom. *Social Science and Research Network*. Retrieved March 28, 2016 from http://papers.ssrn.com/sol3/papers.cfm?abstract_id=2713169

Autio, E. (1994). New, technology-based firms as agents of R&D and innovation: An empirical study. *Technovation, 14*(4), 259–273. doi:10.1016/0166-4972(94)90010-8

Autio, E. (2005). Creative Tension: The Significance of Ben Oviatts and Patricia McDougalls Article Toward a Theory of International New Ventures. *Journal of International Business Studies, 36*(1), 9–19. doi:10.1057/palgrave.jibs.8400117

Avis, W. (2015). *Mobile phone and social media interventions for youth development outcomes*. Birmingham, UK: GSDRC, University of Birmingham. Retrieved from http://reliefweb.int/sites/reliefweb.int/files/resources/HDQ1317.pdf

Awobamise, A., & Aderibigbe, A. (2015). An analysis of Western and Nigerian media depictions of President Jonathan's presidential bid. Global Media Journal (African Edition), 9(1), 63-79.

Baden and Reeves. (2000). *Gender and development: concepts and definitions*. Bridge (the development-gender) Report No 55. Retrieved on 27 May, 2015 from www.bridge.ids.ac.uk/report

Badenhorst, G., Van Staden, A., & Coetsee, E. (2008). HIV/AIDS risk factors among residence students at the University of the Free State. *Curationis, 31*(3), 27–35. doi:10.4102/curationis.v31i3.1028 PMID:19177968

Baker, K., & Raney, A. A. (2007). Equally super?: Gender-role stereotyping characters in childrens animated programs. *Mass Communication & Society, 10*(1), 25–41. doi:10.1080/15205430709337003

Bakesha, S., Nakafeero, A., & Okello, D. (2009). ICTs as agents of change: A case of grassroots women entrepreneurs in Uganda. In I. Buskens & A. Webb (Eds.), *African Women & ICTs: Investigating Technology, Gender and Empowerment*. Pretoria: Unisa Press.

Balatchandirane, G. (2007). Gender discrimination in education and economic Development: A study of Asia. *International Studies, 40*(4), 349–378. doi:10.1177/002088170304000403

Ballack, X. (2009). *Cultural norms theory*. Retrieved from http://ballack.xomba.com/cultural_norm_theory_mass_communication

Balogun, M. (2012, January 1). Bridging the Wealth gap. *The Punch*, p. 16.

Bandura, A. (1986). *Social foundations of thought and action*. Englewood Cliffs, NJ: Prentice-Hall.

Bandura, A. (1986). *Social Foundations of Thought and Action: A Social Cognitive Theory*. Englewood Cliffs, NJ: Prentice-Hall.

Baran, J. S., & Davis, K. D. (2003). *Mass communication theory: foundations, ferment, and future*. Belmont, CA: Thomson Wadsworth.

Baran, S. (2008). *Introduction to Mass communication. In Media literacy and culture* (5th ed.). London: McGrill Hill.

Baran, S. J. (2009). *Introduction to mass communication: Media literacy and culture* (5th ed.). New York: McGraw Hill.

Barbera, P., & Zeitzoff, T. (2014). *The Empirical Determinants of Social Media Adoption by World Leaders and its Political Consequences*. A Working Paper.

Baren, S. J., & Davis, D. K. (2003). *Communication theory. In Foundation, ferment, and future* (3rd ed.). Wadsworth.

Barrientos, S., Gereffi, G., & Rossi, A. (2010). *Economic and social upgrading in global production networks: Developing a framework for analysis*. Capturing the Gains Working Paper 2010/03.

Barringer, B., & Ireland, D. (2005). *Entrepreneurship: Successfully launching new ventures*. Pearson Education.

Barton, F. (1979). *The press in Africa: Persecution and perseverance*. London: Macmillan. doi:10.1007/978-1-349-04075-9

Barz, G., & Cohen, J. (2011). *The culture of AIDS in Africa: Hope and healing through music and arts*. Oxford, UK: Oxford University. doi:10.1093/acprof:osobl/9780199744473.001.0001

Baumol, W. J. (2002). *The Free-Market Innovation Machine: Analysing the Growth Miracle of Capitalism*. Princeton, NJ: Princeton University Press. doi:10.1515/9781400851638

BBC Media Action. (2015). *Enhancing Nigeria's Response to HIV/AIDS 2009-2014: BBC Media Action Programme Awards*. Retrieved from http://www.bbc.co.uk/mediaaction/contact

Beck, U. (1992). *Risk Society: Towards A New Modernity*. London: Sage.

Bem, S. L. (1974). The measurement of psychological androgyny. *Journal of Consulting and Clinical Psychology*, *42*(2), 155–162. doi:10.1037/h0036215 PMID:4823550

Benford, R. D., & Snow, D. A. (2000). Framing processes and social movements: An overview and assessment. *Annual Review of Sociology*, *26*(1), 611–639. doi:10.1146/annurev.soc.26.1.611

Benson-Eluwa, V. (2013). *Media and society*. Enugu: Virgin Creation.

Berelson, B. (1952). *Content Analysis in communication Research*. Glencoe, IL: The Free Press.

Berge, S. (1951). Subsidies and Competition as Factors in Air Transport Policy. *The American Economic Review*, *41*(2), 519–529. Retrieved from http://www.jstor.org/stable/1910826

Birkinshaw, J. (1997). Entrepreneurship in multinational corporations: The characteristics of subsidiary initiatives. *Strategic Management Journal*, *18*(3), 207–229. doi:10.1002/(SICI)1097-0266(199703)18:3<207::AID-SMJ864>3.0.CO;2-Q

Bittner, J. R. (1989). *Mass communication an introduction* (5th ed.). Prentice-Hall.

Black, S., & Juhn, C. (2000). The rise of female professionals: Are women responding to skill demand? American Economic Association.

Bosch, T. (2013). Youth, Facebook and politics in South Africa. *Journal of African Media Studies*, *5*(2), 119–130. doi:10.1386/jams.5.2.119_1

Bosshart, L., & Hellmuller, L. (2009). Pervasive entertainment ubiquitous entertainment. *Communication Research Trend*, *28*(2), 1–18.

Botlhomilwe, M. Z., Sebudubudu, D., & Maripe, B. (2011). Limited freedom and intolerance in Botswana. *Journal of Contemporary African Studies*, *29*(3), 331–348. doi:10.1080/02589001.2011.581501

Bowen, H. (2010). *Information at the grassroots: Analyzing the media use and communication habits of Kenyans*. Washington, DC: InterMedia Survey Institute. Retrieved from http://audiencescapes.org/sites/default/files/AudienceScapes%20/Kenya%20Survey%20Research%20Report%202010.pdf

Boyd, D. (2014). *Its Complicated*. Yale University Press.

Bradely, J. L. (2016, March 29). *Digital Vortex. How Digital Disruption is redefining industries.* Retrieved from Cisco.com: http://www.cisco.com/c/dam/en/us/solutions/collateral/industry-solutions/digital-vortex-report.pdf

Bratton, M. (2013). Briefing: Citizens and ccll phones in Africa. *African Affairs*, *112*(447), 304–319. doi:10.1093/afraf/adt004

Brazeal, D. V., & Herbert, T. T. (1999). The genesis of entrepreneurship. Entrepreneurship Theory and Practice, 23(3), 29-45.

Brinded, L. (2014). Cybercrime Costs $445bn Worldwide and Hits 150,000 EU Jobs. *International Business Times*. Retrieved from: http://www.ibtimes.co.uk/cybercrime-costs-445bn-worldwide-hits-150000-eu-jobs-1451895

Brinthaupt, T. M., & Lipka, R. P. (2002). *Understanding Early Adolescent Self and Identity: Applications and Interventions*. Albany, NY: State University of New York Press.

Broadcasting Act 1998. 72:04 (Botswana).

Broadcasting Regulations 2004. 72:04 (Botswana).

Brown, J. D., Steele, J. R., & Walsh-Childers, K. (2002). Sexual teens, sexual media: Investigating media's influence on adolescent sexuality. London: Lawrence Erlbaum.

Brown, W. (1992). The use of entertainment television programs for promoting prosocial messages. *The Howard Journal of Communications*, *3*(3&4), 253–266. doi:10.1080/10646179209359754

Brown, W. J., & Singhal, A. (1999). Entertainment-education media strategies for social change. In D. P. Demers & K. Viswanath (Eds.), *Mass media, social control and social change* (pp. 263–280). Ames, IA: Iowa State University Press.

Brummet, Q. (2008). The effect of Gender Inequality on growth: A cross-country Empirical Study. *The Park Place Economist, 16*(1), 13-23.

Bryant, J., & Thompson, S. (2002). *Fundamentals of media effects*. New York: McGraw Hill.

Buchanan, E. (2016). *Kenya: Anti-election protesters hit Nairobi streets despite police threats of lethal.* Retrieved from International Business Times: http://www.ibtimes.co.uk/kenya-anti-election-protesters-hit-nairobi-streets-despite-police-threats-lethal-force-1563821

Buckingham, D. (Ed.). (2008). *Youths, Identity, and Digital Media*. MIT Press.

Buckley, P. J., & Casson, M. C. (1976). *The Future of Multi-national Enterprise.* Basingstoke, UK: Burgelman. doi:10.1007/978-1-349-02899-3

Burgoon, M., & Ruffner, M. (1978). *Human Communication, an introduction*. Holt, Rinehart and Winston.

Burton, G. (2010). *Media and society: Critical perspectives* (2nd ed.). McGraw-Hill.

Buskens, I., & Webb, A. (2009). *African women and ICTs: Investigating technology, gender and empowerment*. Zed Books.

BuzzGhana. (2015, October 14). *Ghana's Kente: Fascinating Stories and Meanings behind the Cloth's Designs and Colours.* Retrieved from http://buzzghana.com/ghanas-kente-fascinating-stories-and-meanings-behind-the-cloths-designs-and-colours/

Cairncross, I. (1997). *The death of distance: how the communications revolution will change our lives.* Cambridge, MA: Harvard Business School Press.

Caldas-Coulthard, C., & Leeuwen, T. (2003). Critical social semiotics. *Social Semiotics, 13*(1), 3–4. doi:10.1080/1035033032000133481

Calsyn, D., Peavy, M., Wells, E., Campbell, A. et al, 2013. Differences between men and women in Condom use, attitudes and skills in substance abuse treatment seekers. *American Journal of Addicts, 22*(2), 150-157. Doi:101111/j.1521-0391.000312

Calvert, S. L. (2002). Identity Construction on the Internet. In S. L. Calvert, A. B. Jordan, & R. R. Cocking (Eds.), *Children in the Digital Age: Influence of Electronic Media on Development* (pp. 57–70). Westport, CT: Praeger.

Cambridge, V. C. (2007). Milestones in communication and national development. In Y. R. Kamalipour (Ed.), *Global communication* (pp. 181–200). Belmont, CA: Thomson Wadsworth.

Campbell, S. C., & Kwak, N. (2012). Political involvement in mobilized society: The Interactive relationships among mobile communication, network characteristics, and political participation. *Journal of Communication, 61*(6), 1005–1024. doi:10.1111/j.1460-2466.2011.01601.x

Carree, M. A., & Thurik, A. R. (2003). The Impact of Entrepreneurship on Economic Growth. In Z. J. Acs & D. B. Audretsch (Eds.), *Handbook of Entrepreneurship Research* (pp. 437–471). Boston, MA: Kluwer Academic Publishers.

Carrier, C. (1994). Intrapreneurship in large firms and SMEs: A comparative study. *International Small Business Journal, 12*(3), 54–61. doi:10.1177/0266242694123005

Carroll, W. K., & Hackett, R. A. (2006). Democratic media activism through the lens of social movement theory. *Media Culture & Society, 28*(1), 83–104. doi:10.1177/0163443706059289

Castells, M. (2002). The Internet and the Network Society. In The Internet in Everyday Life. Blackwell Publishing Company.

Castells, M. (2001). *The Internet Galaxy: Reflections on the Internet, Business, and Society.* Oxford, UK: Oxford University Press. doi:10.1007/978-3-322-89613-1

Castells, M., Fernandez-Ardevol, M., Linchuan Qiu, J., & Sey, A. (2007). *Mobile communication and society: A global perspective.* Cambridge, MA: MIT Press.

Caulderwood, K. (2015, March 29). *Burundi 2015: How Twitter, WhatsApp and Facebook Influenced.* Retrieved from ibtimes: http://www.ibtimes.com/burundi-2015-how-twitter-whatsapp-facebook-influenced-protests-possible-coup-1921134

CCK (Communications Commission Kenya). (2012). *Quarterly sector statistics report: Third quarter of the financial year 2011/12 (January-March 2012).* Available at http://www.cck.go.ke/resc/downloads/SECTOR_STATISTICS_REPORT_Q3_11-12.pdf

Cederbaum, J., Coleman, L., Goller, G., & Jemmott, S. (2006). Understanding the HIV risk reduction needs of heterosexual African American substance-abusing men. *The Journal of the Association of Nurses in AIDS Care, 17*(6), 28–37. doi:10.1016/j.jana.2006.09.001 PMID:17113481

Center for Communication Programs Population Report. (1995). *Helping the News Media cover Family Planning.* Johns Hopkins Bloomberg School of Public Health.

Center for Communication Programs Population Report. (2001). *HIV/AIDS communication strategies: Communication is a key factor in combating the HIV/AIDS epidemic*. Johns Hopkins Bloomberg School of Public Health.

Center for Strategic and International Studies (CSIS). (2014). *Net Loses: Estimating the Global Cost of Cyber-crime*. Retrieved from http:///csis.org/

Centers for Disease Control and Prevention. (2015). *Sexually transmitted diseases (STDs)*. Retrieved from http://cdc.gov/std/

Centre for AIDS Development, Research and Evaluation. (2001). *HIV/AIDS and the Media: A literature review*. Johannesburg: CADRE.

CERT. (2013). *Australia Cyber-crime and Security Survey Report*. Retrieved from https://www.cert.gov.au/

Ceulemans, M., & Fauconnier, G. (1979). Mass media: The Image, role and Social conditions of women. A collection and analysis of research materials. *Global Media Journal*.

Chatora, A. (2012). Encouraging political participation in Africa: The potential of social media platforms. Pretoria: Institute for Security Studies. Retrieved from http://www.issafrica.org/publications/situation-reports/encouraging-political-participation-in-africa-the-potential-of-social-media-platforms

Chen, A. (2016). *Independent Work Report Fall, 2015. Understanding Pricing Effects on Mobile Data*. Retrieved from cs.princeton.edu: https://www.cs.princeton.edu/sites/default/files/uploads/ava_chen.pdf

Chen, B., & Liao, D. (2014). *Social Media, Social Movements and the Challenge of Democratic Governability*. Paper presented at Taiwan Democracy Project, Stanford University.

Chen, D. H. C. (2004). *Gender Equity and Economic Development. The Role for ICTS*. Washington, DC: World Bank Working Paper 3285.

Chew, F., & Palmer, S. (2005). Television Health Promotion in Four Countries. *Nutrition Journal*, *21*(5), 634–638. doi:10.1016/j.nut.2004.09.022 PMID:15850972

Chiluwa, I., & Adegoke, A. (2013). Twittering the Boko Haram uprising in Nigeria: Investigating pragmatic acts in the social media. *Africa Today*, *59*(3), 82–102. doi:10.2979/africatoday.59.3.83

Chiluwa, I., & Ifukor, P. (2015). War against our children: Stance and evaluation in #BringBackOurGirls campaign discourse on Twitter and Facebook. *Discourse & Society*, *26*(3), 267–296. doi:10.1177/0957926514564735

Chinedum, N., Dephne, H., Paul, T., & Sylvia, C. (2009). African Journal of Business and Economic Research. *Business and Social Profiles of Female Immigrant Entrepreneurs in Nigeria*, *4*(2-3), 85–92.

Chng, L., Eke-Huber, E., Eaddy, S., & Collins, J. (2005). Nigerian college Students: HIV Knowledge, perceived susceptibility for HIV and sexual behaviours. *College Student Journal*, *39*, 60–71.

Chong, D., & Druckman, J. N. (2007). Framing theory. *Annual Review of Political Science*, *10*(1), 103–126. doi:10.1146/annurev.polisci.10.072805.103054

cigionline. (2015). Retrieved from https://www.cigionline.org/publications/global-cyberspace-safer-you-think-real-trends-cybercrime?gclid=COrAyIbP7sYCFS7KtAodmTENRQ

Clarke, J. N. (2010). The domestication of health care: Health advice to Canadian mothers 19932008 in todays parent. *Family Relations*, *59*(2), 170–179. doi:10.1111/j.1741-3729.2010.00593.x

Clements, M. B. J., Coady, D., Fabrizio, M. S., Gupta, M. S., Alleyne, M. T. S. C., & Sdralevich, M. C. A. (2013). *Energy subsidy reform: lessons and implications*. International Monetary Fund.

Clough, J. (2010). *Principles of Cybercrime*. Cambridge, UK: Cambridge University Press. doi:10.1017/CBO9780511845123

Cobb, S. (2015). *Cyber-crime deterrence: 6 important steps*. Retrieved from: http://www.welivesecurity.com/2015/01/20/cybercrime-deterrence-6-important-steps/

Communications and Regulatory Authority Act 2012. 72:03 (Botswana).

Conard, P. (2007). *The medicalization of society: On the transformation of human conditions into treatable disorders*. Maryland University Press.

Constitution of the Federal Republic of Nigeria. (1999). Lagos: Federal Government Press.

Constitution of the Federal Republic of Uganda 1995

Corcoran, N. (2007). Communicating health: Strategies for health promotion Thousand Oaks, CA: Sage.

Corin, A. (2013). *Tracking the Cost of Cyber-crime*. Retrieved from http://fcw.com/articles/2013/07/23/cyber-crime-cost.aspx

Correa, C. M. (2000). *Intellectual property rights, the WTO and developing countries: The TRIPS Agreement and Policy Options*. London: Zed Books.

Cronkhite, G. (1976). *Communication and Awareness*. Cummings.

Croteau, D., & Hornes, W. (2000). *Media society. Thousand Daks*. Pine Forge Press.

CSIS and McAfee. (2013). *Cyber-crime costs global economy $445 billion a year: report*. Retrieved from http://www.reuters.com/article/2014/06/09/us-cybersecurity-mcafee-csis-dUSKBN0EK0SV20140609

Curran, J. (1977). *Mass Communication and Society*. London: Edward Arnold Publisher Ltd and Open University Ltd and Open University Press.

Daly, J. A. (2003). *ICT, gender equality, and empowering women*. Retrieved from www.developmentgateway.org/node/133831/sdm/blob?pid=5233

Damanpour, F. (1991). Organizational innovation: A meta-analysis of effects of determinants and moderators. *Academy of Management Journal*, *34*(3), 555–590. doi:10.2307/256406

Dana, L. P., Etemad, H., & Wright, R. W. (2009). The impact of globalization on SMEs. *Global Focus*, *11*(4), 93–105.

Danladi, K. (2015). Assessment of the use of social media by secondary school students in Zaria. In D. Wilson (Ed.), *Communication, society and the Nigerian child: Issues and emerging trends in the 21st century* (pp. 227–236). Nigeria: ACCE.

Daramola, Y. (2008). Mass media and society in Nigeria: Selected functional perspectives. In R. Akinfeleye (Ed.), *Mass Media and Society: A Multi-perspective approach* (pp. 31–45). Lagos: Department of Mass Communication, University of Lagos.

Das, J. K. (2003). *Education broadcast through radio*. Retrieved on 21 July, 2015 from http://www.aiaer.net/ejournal/vol19107/8.htm

De Fossard, E., Baptiste, J., Corrales, C., & Bosch, A. (1993). *Interactive Radio Instruction*. Washington, DC: USAID.

Deacon, D., Pickering, M., Golding, P., & Murdock, G. (2007). *Researching communications*. New York: Hodder Arnold.

Dean, M. (1999). Risk, Calculable and Incalculable. In Risk and Socio- cultural Theories: New Directions and Perspectives. Cambridge, UK: Cambridge University Press.

Delios, A., & Henisz, W. J. (2000). Japanese firms investment strategies in emerging economies. *Academy of Management Journal*, *43*(3), 305–323. doi:10.2307/1556397

Department of Broadcasting Services. (2006). *Department of Broadcasting Services-Quality radio and television: PMS strategic plan 2006-2016*. Department of Information and Broadcasting. Retrieved from http://www.dib.gov.bw

Department of Broadcasting Services. (2012). *Department of Broadcasting Services Strategy Document 2012-2016*. Gaborone, Botswana: Department of Broadcasting Services. Retrieved from http://www.dib.gov.bw

Devezin, T. F. (2016, March 29). *Enhanced Engagement and Youth Participation in Sexual Health Programming. Effects of Smartphone Communication (WhatsApp)*. Retrieved from www.africasexuality.org: http://www.africasexuality.org/wp-content/uploads/2016/02/43.pdf

deVreese, C. H. (2005). News framing: Theory and typology. Information Design Journal +. *Document Design (Amsterdam)*, *13*(1), 51–62.

Devries, N. (2015). *Sierra Leone faces setback in Ebola fight*. Retrieved from http://aljazeera.com/news/2015/09/sierra-leone-faces-setback-ebola-fight-150920094802322.html

Dheer, E., & Verdegem, P. (2014). Conversations about the elections on Twitter: Towards a structural understanding of Twitters relation with the political and the media field. *European Journal of Communication*, *29*(6), 720–734. doi:10.1177/0267323114544866

Dike, K., & Anikwudike, C. N. (2015). Evaluation of the influence of social media use on the academic performances of federal polytechnic, Bida students. In D. Wilson (Ed.), *Communication, society and the Nigerian child: Issues and emerging trends in the 21st century* (pp. 91–102). ACCE.

Director Broadcasting Services. (n.d.). *From the Director*. Botswana Television. Retrieved from http://www.btv.gov.bw/from_the_director.html

Dlodlo, N. (2014). Measuring Selected M-Texting Addiction Indicators with Gender and Self. *Mediterranean Journal of Social Sciences*, 489-499.

Dodd, N., & Hutter, B. M. (2000). Geopolitics and the Regulations of Public Life. *Law & Policy*, *22*(2), 1–24. doi:10.1111/1467-9930.00083

Dominick, J. A. (2005). *Dynamics of Mass Communication. Media in the Digital Age*. Boston: Mc Graw Hill.

Drago, E. (2015). The Effect of Technology on Face-to-Face Communication. *The Elon Journal of Undergraduate Research in Communications*, 13-19.

Drakapoulos, S. A., Theodossiou, I., & Dodd, S. (2008). Gender Discrimination and Institutional Frameworks: Evidence from Four European Union Countries. *International Journal of Economic Sciences and Applied Research*, *1*(1), 6–36.

Drucker, P. (1985). *Innovation and entrepreneurship: Practice and principles*. Amsterdam: Elsevier Butterworth-Heinemann.

Du toit, L. (2005). A Phenomenology of Rape: Forging a new vocabulary for action. In *(Un) thinking Citizenship*. Burlington, VT: Ashgate Publishing.

Dufour, A. (1995). *Internet*. Paris: Presses Universitaires de France.

Durkin, K., & Nugent, B. (1998). Kindergarten childrens gender-role expectations for television actors. *Sex Roles*, *38*(5), 387–402. doi:10.1023/A:1018705805012

Durojaiye, O. C. (2011). Knowledge, attitude and practice of HIV/AIDS: Behaviour change among tertiary education students in Lagos, Nigeria. *Annals of Tropical Medicine and Public Health*, *4*(1), 18–24. doi:10.4103/1755-6783.80516

Dyagilev, K., & Yom-Tov, E. (2014). Linguistic factors associated with propagation of political opinions in Twitter. *Social Science Computer Review*, *32*(2), 195–204. doi:10.1177/0894439313506850

Ebeniro, C. (2010). Knowledge and beliefs about HIV/AIDS amongst male and female students of Nigerian Universities. *Journal of Comparative Research in Anthropology & Sociology*, *1*(1), 121–131.

Ebewo, P. (2007). The emerging video film industry in Nigeria: Challenges and prospects. *Journal of Film and Video, 59*(3), 46-57.

Ebo, S. J. (1999). *African mass media systems*. Enugu: Computer Edge.

Ehrenberg, A. (2010). *Weariness of the self: Diagnosing the history of depression in the contemporary age.* McGill Queen's University Press.

Eke, M. (2012). *Nollywood, women, and cultural identity*. Paper presented at the International Conference on Nollywood, Women, and Cultural Identity. Retrieved from http://www.cfplist.com/default.aspx

Ekeanyanwu, N. T., & Peter, A. S. (2015). Children, youths and mediated violence: A reflective evaluation of some selected theoretical models. *Journal of the African Council for Communication Education*, *12*(1), 50–71.

Eko, E. (2002). *Effective Writing*. Heinemann Educational Books.

Ekpo, C. E., Chime, J., & Enor, F. N. (2016). The Irony of Nigeria's Fight Against Corruption: An Appraisal of President Muhammadu Buhari's First Eight Months in Office. *International Journal of History and Philosophical Research*, *4*(1). Retrieved from http://www.eajournals.org/journals/international-journal-of-history-and-philosophical-research-ijhphr/vol-4-issue-1-march-2016/the-irony-of-nigerias-fight-against-corruption-an-appraisal-of-president-muhammadu-buharis-first-eight-months-in-office/

Ekundayo, A., & Ama, B. (2014). Nigerian women and political participation: The way forward. *International Journal of Educational Foundations and Management*, *2*(1), 71–81.

Elebute, A. (2015). Issues on the gagging of Nigerian press with obnoxious laws. *International Journal of Arts and Humanities, 4*(1), 207-223. DOI: 10.4314/ijah.v4i1.15

Elo, S., & Kyngas, H. (2008). The qualitative content analysis process. *Journal of Advanced Nursing*, *62*(1), 107–115. doi:10.1111/j.1365-2648.2007.04569.x PMID:18352969

Emechebe, G. O., Emodi, I. J., Ikefuna, A. N., Ilechukwu, A. C., Igwe, W.C., Ejiofor, O. S., & Ilechukwu, C. A. (2009). Hepatitis B virus infection in Nigeria – a review. *African Journal online, 50*(1), 8-22.

Emechebe, G. O., Emodi, I. J., Ikefuna, A. N., Ilechukwu, G. C., Igwe, W. C., Ejiofor, O. S., & Ilechukwu, C. A. (2010). Hepatitis B virus infection in Nigeria– a review. *Nigerian Medical Journal*, *50*(1), 18–22.

Enabling HIV/AIDS+TB and Social Sector Environment project (ENHANSE). (2007). *Most at Risk Population foe HIV and AIDS in Nigeria*. Abuja: ENHANSE Project.

Ende, T. S. (2011). Investing in Young People through Media Literacy. *The Journal of the African Council for Communication Education, 9*(1) 54-81.

Esidene, E. C., & Abdul, S. U. (2013). The role of women politics: Interrogating the gender question for an enhanced political representation in the fourth republic. *Afro Asian Journal of Social Sciences.*, *4*(4), 1–17.

Esteve-Volart, B. (2004). *Gender discrimination and growth: Theory and evidence from India.* Retrieved on 10th March, 2015 from http://eprints.lse.ac.uk/6641/1/Gender_Discrimination_and_Growth_Theory_and_Evidence_from_India.pdf

European Business News. (2014). *PwC survey finds economic crime rising globally, all business sectors, regions suffer from impact.* Retrieved from www.european-business.com/news/business_economy_news/economic_crime_rising_globally/

European Food Information Council. (2015). Retrieved from http://www.eufic.org/page/en/show/latest-science-news/fftid/chronic-disease/

European Union Trade Policy. (n.d.) Retrieved September 25, 2015 from http://ec.europa.eu/trade/policy/accessing-markets/trade-defence/actions-against-imports-into-the-eu/anti-subsidy/

Ewelukwa, N. (2001). *Introduction to press law.* Maranatha Press ltd.

Ezeocha, J., & Mojaye, E. (2014). Mass Media influence on discrimination and violence against women in Delta State, Nigeria. *Journal of Communication and Media Research*, *3*(1), 75–92.

Falola, T., & Heaton, M. M. (2008). *A History of Nigeria.* Cambridge University Press. doi:10.1017/CBO9780511819711

Family Health International. (2006). Strategic Behavior Communication for HIV and AIDS: A Framework. Arlington, VA: Author.

Farhana, S., & Ahmed, J. U. (2008). Entertainment-education to promote safe motherhood: In the context of Bangladesh. *Canadian Social Science*, *4*(4), 16–26.

Federal Ministry of Health. (2005). *Behavioural Surveillance Survey. Draft Report.* Abuja: Federal Ministry of Health.

Federal Ministry of Health. (2010). *NationalHIV Sero-prevalence Sentinel Survey.* Abuja: NationalAIDS/STI Control Programme.

Federal Ministry of Health. (2012). *National HIV/AIDS and Reproductive Health Survey (NARHS).* Abuja: FMOH.

Federation of African Media and Development (FAMEDEV). (2005). *HIV and AIDS Research, Reports and Media Guide: Inter-African network for women, media, gender equity and development.* Senegal: Federation of International Journalist, Bureau RégionalAfrique.

Feenstra, R., & Hanson, G. (2005). Ownership and Control in Outsourcing to China: Estimating the Property-Rights Theory of the Firm. *Quarterly Journal of Economics*, *120*(2), 729-761.

Ferreira, S., & Hebrrle, V. (2013). Text linguistics and critical discourse analysis: A multimodal analysis of amagazine advertisement. *Ilha do Desterro*, *64*, 263–267.

Fiaveh, D. Y. (2011). *Male Perspective on condom use: implications for STI/HIV.* Saarbrucken, Germany: Lambert Academic Publishing.

Fiaveh, D. Y. (2012). Condom Myths and Misconceptions: The Male perspective. *Global Journal of Medical Research*, *12*(5), 43–50.

Fin24. (2016, February 10). *Watch out for these 5 WhatsApp scams.* Retrieved from fin24.com: http://www.fin24.com/Tech/Opinion/watch-out-for-these-5-whatsapp-scams-20160210

Fischer, S. (2010). Powerful or pretty: A content analysis of gender images in children's animated films. Graduate faculty. Auburn University. Retrieved from http://etd.auburn.edu/etd/bitstream/handle/10415/2065/ThesisSabrinaFischer.pdf?sequence=2

Folarin, B. (2005). *Theories of mass communication.* Ibadan: Bakinfol Publications.

Food and Agricultural Organisation of UN. (2014). The State of food and agriculture. *Innovation in Family Farming*, 95-139.

Fourie, P.J. (2007). *Media history, Media and society.* Wetton: Juta.

Fox, N. J. (1999). Postmodern reflections on 'risk' 'hazard' and life choices. In Risk and socio-cultural theories: New directions and perspectives. Cambridge, UK: Cambridge University Press.

Franklin, B. (Ed.). (2001). *British television policy: A reader.* London: Routledge.

Freedman, D. (2008). *The politics of media policy.* Cambridge, UK: Polity Press.

Freedom-House. (2016, June 24). *Rwanda.* Retrieved from Freedom House: https://freedomhouse.org/sites/default/files/resources/FOTN%202015_Rwanda.pdf

Freimuth, V. S. (1992). Theoretical Foundations of AIDS Media Campaigns. In AIDS: A Communication perspective. Hillsdale, NJ: Eribaum.

Freimuth, V., & Quinn, S. C. (2004). The Contributions of Health Communication to Eliminating Health Disparities. *American Journal of Public Health*, *94*(12), 2053–2054. doi:10.2105/AJPH.94.12.2053 PMID:15569949

Freire, P. (1971). *Education for Critical Consciousness.* New York: The Continuum Publishing Company.

Fuchs, C. (2011, 8 June). *Critical Media and Communication Studies Today. A Conversation.* Retrieved from wordpress.com: https://dwmw.wordpress.com/2011/08/06/critical-media-and-communication-studies-today-a-conversation-between-dwayne-winseck-and-christian-fuchs-part-1/

Fuchs, C. (2013). *Social Media: a critical introduction.* London: Sage.

Galvin, K. M., & Terrell, J. (2009). *Communication works. Communication applications in the workplace.* London: National Text Book Company.

Gamble, T. K., & Gamble, M. W. (2003). *The gender communication connection.* Boston: Houghton Mifflin Company.

Gambo, D. (2006). Legal and ethical challenges for Nigerian journalists in a Democracy. *An Interdisciplinary Journal of Communication Studies*, *20*(4), 86–95.

Gandy, O. H. Jr. (2002). The real divide: citizens versus consumers. In L. Lievrouw & S. Livingstone (Eds.), *The handbook of new media* (pp. 448–460). London: Sage.

Garland, D. (2001). *Culture of control.* Oxford, UK: OUP.

Garnham, N. (2004). The media and the public sphere. In F. Webster (Ed.), *The information society reader* (pp. 357–366). London: Routledge.

Gauntlett, D. (2011). *Making is creating: the social meaning of creativity, from DIY and knitting to Youtube and Web 2.0.* London: Polity Press.

Gberevbie, D., Osibanjo, A., Adeniji, A., & Oludayo, O. (2014). An Empirical performance among Academic staff of government University in Lagos State, Nigeria. *International Journal of Social, Education, Economics and Management Engineering, 8*(1). Retrieved on 10th March, 2015 from http://www.google.com.ng/url?sa=

Gehl, R., W. (2015). The Case for Alternative Social Media. *Social Media + Society,* 1–12. DOI: 10.1177/2056305115604338

George, O., Shadare, S., & Owoyemi, O. (2012). Military interventions in the Nigerian politics: 'A timed bomb' waiting to explode? The avowal of a new management elites. *International Journal of Business. Human Technology, 2*(5), 191–198.

Gerbner, D. (1997). *Mass Media Policies in Changing Cultures.* New York: John Wiley & Sons.

Geringer, J. M., Tallman, S., & Olsen, D. M. (2000). Product and international diversification among Japanese multinational firms. *Strategic Management Journal, 21*(1), 51–80. doi:10.1002/(SICI)1097-0266(200001)21:1<51::AID-SMJ77>3.0.CO;2-K

Gever, C. V. (2014, November 25-27). *Assessing the psychological effects of horrible pictures on newspaper readers.* Paper presented at the Nigerian Psychological Association Conference, Nsukka, Nigeria. Retrieved from http://www.academia.edu/.../assessing_the_psychological_effect

Gillwald, A. (2005). Good intentions, poor outcomes: Telecommunications reform in South Africa. *Telecommunications Policy, 29*(7), 469–491. doi:10.1016/j.telpol.2005.05.005

Githui, D. M. (2011). Mobile money transfer in Kenya: An ethical perspective in research. *Journal of Finance and Accounting, 2*(2).

Global Economic Crime Survey (GECS). (2014). *The benefits—and –risks –of a networked world are here to Stay.* Retrieved from: http://www.pwc.com/gx/en/economic-crime-survey/cybercrime.jhtml

Global HIV Prevention working Group. (2008). *Behavior Change and HIV Prevention: (Re)Considerations for the 21st Century.* Bill & Melinda Gates Foundation and the Henry J. Kaiser Family Foundation. Retrieved from www.Global-HIVPrevention.org

Global Information Society Watch. (2013). *Report on Women's Rights, Gender and ICTs.* Association for Progressive Communication (APC) and Humanist Institute for Cooperation with Developing Countries (Hivos). Retrieved from http://www.giswatch.org/2013-womens-rights-gender-and-icts

Global Media Monitoring Project. (2010). *Who makes the news, 2010 report highlights.* Toronto: The World Association for Christian Communication. Retrieved from http://www.whomakesthenews.org/images/reports_2010/highlights/highlights_en.pdf

Goldstuck, A. (2014, July 29). *Massive increase in SA smartphone purchases.* Retrieved from Mail & Guardian: http://mg.co.za/article/2014-07-29-smartphone-uptake-in-south-africa-rising-dramatically

Good, K. (2008). *Diamonds, dispossession and democracy in Botswana.* Retrieved from http://UQL.eblib.com.au/patron/FullRecord.aspx?p=1069026

Good, K. (1993). At the end of the ladder: Radical inequalities in Botswana. *The Journal of Modern African Studies, 31*(2), 203–230. doi:10.1017/S0022278X00011903

Good, K. (1996). Authoritarian liberalism: A defining characteristic of Botswana. *Journal of Contemporary African Studies, 14*(1), 29–51. doi:10.1080/02589009608729580

Gordon, T. (2002). *Marshall Who?* Available from http://www.marshallmcluhan.com/biography/

Gotlop, K. (2000, July 14). Btv on July 31. *Daily News.* Retrieved from http://www.gov.bw

Governance and Social Development Resources Centre. (2011). *Helpdesk research report: Communication initiatives to change attitude and behaviours*. Retrieved from http://gsdrc.org/doc/open/pdf

Graber, D. A. (1988). *Processing the News: How people Tame the Information Tide*. New York: Longman.

Graham, G. (1993). *Philosophy of mind: An introduction*. Oxford, UK: Blackwell Publishers Ltd.

Grant, R. M. (1987). Multinationality and performance among British manufacturing companies. *Journal of International Business Studies*, *18*(1), 79–89. doi:10.1057/palgrave.jibs.8490413

Greenaway, D., Gullstrand, J., & Kneller, R. (2008). Surviving Globalisation. *Journal of International Economics*, *74*(2), 264–277. doi:10.1016/j.jinteco.2007.08.005

Green, L. (2010). *The Interne: An introduction to new media*. New York: Berg.

Green, L. (2010). *The Internet. An introduction to new media*. New York: Berg.

Griffin, E. M. (2012). *A first Look at Communication Theory*. McGraw Hill Higher Education.

Grootes, S. (2016, January 14). *MTN calls for regulation of messaging services*. Retrieved from Eyewitness News: http://ewn.co.za/2016/01/14/MTN-calls-for-regulation-of-messaging-services

Groshek, J., & Al-Rawi, A. (2013). Public sentiment and critical framing in social media content during the 2012 U.S. presidential campaign. *Social Science Computer Review*, *31*(5), 563–576. doi:10.1177/0894439313490401

GSMA. (2014). *The Role of Mobile Economy in Sub Saharan Africa*. GSMA.

GSMA. (2014). *Women and Mobile: A Global Opportunity: A Study on the Mobile Phone Gender Gap in Low and Middle Income Countries*. Retrieved from http://www.gsma.com/mobilefordevelopment/wpcontent/ uploads/2014/01/GSMA_Women_and_Mobile-A_Global_Opportunity.pdf

GSMA. (2015). *The Mobile Economy. Empowering People and Society*. GSMA Intelligence.

Guback, T. H. (1968). Review of mass entertainment by Harold Mendelsohn. *Journal of Aesthetic Education, 2*(3), 147-149.

Gupta, A., & Sinha, K. A. (2010). Health coverage in mass media: A content analysis. *Journal of Communication*, *1*(1), 9–25. Retrieved from www. krepublishers.com/02-Journals/JC/JC-

Habermas, J. (2004). The public sphere. In F. Webster (Ed.), *The information society reader* (pp. 350–356). London: Routledge.

Haider, H. (2012). *Helpdesk research report: Changing attitudes and behaviours in relation to gender equality*. Governance and Social Development Resource Centre (GSDRC), Birmingham. Retrieved from http://www.gsdrc.org/docs/open/HDQ817.pdf

Hall, S. (1973/1980). Encoding/decoding. In Culture, Media, Language: Working Papers in Cultural Studies (pp. 128-338). London: Hutchinson.

Hamelink, C. J. (1994). *The politics of world communication*. London: Sage.

Haralambos, M., & Holborn, M. (2008). *Sociology: Themes & Perspectives*. London: Collins Educational.

Haralambos, M., Holborn, M., & Heald, R. (2004). *Sociology: themes and perspectives* (6th ed.). London: HarperCollins Publishers Limited.

Harare-Bureau. (2016, March). *20 cops detained over WhatsApp Chat.* Retrieved from The Chronicle: http://www.chronicle.co.zw/20-cops-detained-over-whatsapp-chat/

Harry, I. (2015). Towards a global democratization of free speech: A critical appraisal of constitutional landmines on the trajectory of press freedom in Nigeria. *International Journal of International Relations. Media and Mass Communication Studies, 1*(2), 25–36.

Harter, S. (1999). *The construction of the self: A developmental perspective.* New York: Guilford Press.

Hassan, R. (2004). *Media, politics and the network society.* Berkshire: Open University Press.

Hawthorne, J., Houston, J. B., & McKinney, M. S. (2013). Live-tweeting a presidential primary debate: Exploring new political conversations. *Social Science Computer Review, 31*(5), 552–562. doi:10.1177/0894439313490643

HEAIDS. (2010). *HIV prevalence and Related factors – Higher Education Sector Study, South Africa, 2008–2009.* Pretoria: Higher Education South Africa.

Henry, J. V. (n.d.). Enhancing the Supervision of Community Health Workers with WhatsApp Messaging: Qualitative Findings From 2 Low Resource Settings in Kenya. *Science and Practice.* doi:10.9745/GHSP-D-15-00386

Hessels, J. (2007). Reactie op: Kentering in het Denken over MondialeStrategieën [Response to: A Shift in Thinking about Global Strategies]. *ESB dossier Globaliseren in Nederland, 92*, 59.

Hether, H. J., Huang, G. C., Beck, V., Murphy, S., & Valente, T. W. (2008). Entertainment –education in a media-saturated environment. Examining the impact of single and multiple exposures to breast cancer storylines on two popular medical dramas. *Journal of Health Communication, 13*(8), 808–823. doi:10.1080/10810730802487471 PMID:19051115

Hillbom, E. (2014). Cattle, diamonds and institutions: Main drivers of Botswanas economic development, 1850 to present. *Journal of International Development, 26*(2), 155–176. doi:10.1002/jid.2957

Hindman, M. (2009). *The myth of digital democracy.* Princeton University Press.

Hispian. (2016, June 24). *WhatsApp passes exams.* Retrieved from hisp.org: http://www.hisp.org/wp-content/uploads/2015/11/The-Hispian-18-September.pdf

Hisrich, R., Freeman, E., Standley, A., Yankey, J. A., & Young, D. R. (2000). Entrepreneurship in the not-for-profit sector: The state of the art. *Entrepreneurship*, 321-335.

Hitt, M. A., Hoskisson, R. E., & Kim, H. (1997). International diversification: Effects on innovation and firm performance in product-diversified firms. *Academy of Management Journal, 40*(4), 767–798. doi:10.2307/256948

Hogg, M. A., Terry, D. J., & White, K. M. (1995). A tale of two theories: A critical comparison of identity theory with social identity theory. *Social Psychology Quarterly, 58*(4), 255–269. doi:10.2307/2787127

Holmes, C. (2016, March 29). *Mobilising through Social Media: The Case of South Africa.* Retrieved from http://blogs.lse.ac.uk/africaatlse/2016/02/11/feesmustfall-southafrica-social-media/

Homes, D. (2005). *Communication theory: Media, technology and society.* Sage Publications Inc.

Horst, M., & Miller, D. (2006). *The Cell Phone: An Anthropology of Communication.* Oxford, UK: Oxford International Publishers.

Hossain, M. D. (2015). Manufacturing consent: Framing the liberation war of Bangladesh in the US and UK media. *Journalism, 16*(4), 521–535. doi:10.1177/1464884914524516

Huesmann, R., Moise-Titus, J., Podolski, C., & Eron, L. (2003). Longitudinal relations between childrens exposure to TV violence and their aggressive and violent behavior in young adulthood: 1977–1992. *Developmental Psychology: American Psychological Association*, *39*(2), 201–221. doi:10.1037/0012-1649.39.2.201 PMID:12661882

Hughes, G. (2004). Crime prevention, community safety and crime and disorder reduction. In Student Handbook of Criminal Justice and Criminology. Cavendish Publishing Ltd.

Hughes, N., & Lonie, S. (2007). M-Pesa: Mobile money for the "unbanked" turning cell phones into 24-hour tellers in Kenya. *Innovations*, 66–81.

Human Rights Watch. (2015). *World Report 2015: Nigeria*. Retrieved from https://www.hrw.org/world-report/2015/country-chapters/nigeria

Hwang, E. W., & Cheung, R. (2011). Global epidemiology of hepatitis B virus (HBV) infection. *The Northern. The American Journal of the Medical Sciences*, *4*(1), 7–13.

ICT Empowerment Charter Working Group. (2004). *The ICT Charter*. Retrieved from http://www.workinfo.com/BEE/ICTBEECharterNov04.pdf

ICT-Security-Monitoring-Services-Report. (2016, June 26). *ICT Security Monitoring Services Report*. Retrieved from www.ssa.gov.za: http://www.ssa.gov.za/Portals/0/SSA%20docs/CSIRT/ICT%20Security%20Monitoring%20Services%20Report%2026%20Ma%202016.pdf

Idiagbon, A. (2013). Meaning conception and perception in political campaign adverts – A multimodal discourse approach. *English Language and Literature Studies*, *3*(1), 84–93.

Idris, S. (2015, December 6). Senate vows to go ahead with anti-social media bill despite criticism. *Today*. Retrieved from https://www.today.ng/news/national/48193/senate-vows-to-go-ahead-with-anti-social-media-bill-despite-criticism

Ifeduba, E. (2013). Media entrepreneurship in Nigeria: A study of new patterns, orientations and coping strategies. *Covenant Journal of Communication*, *1*(2), 145–162.

Igwesi, B. N. (2012). Enhancing women participation in national development through a change in the gender system of Nigeria. *Asian Social Science*, *8*(1), 217–223.

Ijeh, N. P., Umukoro, E., & Amune, P. (2015). Use of social media by youths in selected cities of south-south, Nigeria. In D. Wilson (Ed.), *Communication, society and the Nigerian child: Issues and emerging trends in the 21st century* (pp. 61–74). ACCE.

Imre, A. (2012). Adventures in early socialist television edutainment. *Journal of Popular Film and Television*, *40*(3), 119–130. doi:10.1080/01956051.2012.697790

Inooro TV. (2016). *Dagitari* [Doctor]. Author.

International Institute for Sustainable Development. (2013). *A Guidebook to Fossil-fuel Subsidy Reform for Policy-Makers in South East Asia*. Manitoba, Canada: Author.

International Labor Office. (n.d.). *Discrimination at work in Africa*. Retrieved on 10th March, 2014 from http://www.ilo.org/wcmsp5/groups/public/@ed_norm/@declaration/documents/publication/wcms_decl_fs_89_en.pdf

Interpol. (2015). *Cybercrime*. Retrieved from: http://www.interpol.int/Crime-areas/Cybercrime/Cybercrime

IOL. (2016, June 24). *Retrieved June 24, 2016. IOL is on WhatsApp*. Retrieved from iol: http://www.iol.co.za/news/south-africa/iol-is-on-whatsapp-2033092

Iskander, E. (2011). Connecting the national and the virtual: Can Facebook activism remain relevant after Egypt's January 25 uprising? *International Journal of Communication*, *5*, 13–15.

Jaffe, A. B., Trajtenberg, M., & Henderson, R. (1993). Geographic localization of knowledge spillovers as evidenced by patent citations. *The Quarterly Journal of Economics*, 577-598.

Jahannessan, J. (2006). *The Role of the Media in the Cultural Struggle of Gender Transformation in Tanzania* (PhD Thesis). Norwegian University of Science and Technology.

Janniger, K. C., Eastern, S. J., & Hospenthal, R. D. (2014). *Herps Zoster treatment and management*. Retrieved from http://emedicine.medscape.com/article/1132465-treatment

Jenkins, H. (2006, October 20). *Media literacy begins at home*. Retrieved from technologyreview: http://www.technologyreview.com/read_article.aspx?id=13422&ch=infotech&sc=&pg=1

Jenkins, H. (2016, March 28). *Affecting Publics and Social Media: An Interview with Zizi Papacharisi (Part One)*. Retrieved from henryjenkins: http://henryjenkins.org/2015/01/affective-publics-and-social-media-an-interview-with-zizi-papacharissi-part-one.html

Joint United Nations Programme on HIV/AIDS (UNAIDS) and World Health Organisation (WHO). (2006). AIDS Epidemic Update. Geneva: UNAIDS/06.29E.

Joint United Nations Programme on HIV/AIDS (UNAIDS). (2004). *The media and HIV/AIDS: Making a Difference*. Geneva: UNAIDS.

Joint United Nations Programme on HIV/AIDS (UNAIDS). (2007). A Framework for Monitoring and Evaluating HIV Prevention Programmes for Most-At-Risk Populations. Geneva: UNAIDS/07.15E / JC1338E.

Joint United Nations Programme on HIV/AIDS (UNAIDS). (2013). *Global Report: UNIAIDS Report on global AIDS epidemic 2013*. Author.

Jonah, A. A. (2011). *Film as a vehicle for cultural promotion and unity in Nigeria*. Department of mass communication, The Federal Polytechnic, Bida, Nigeria. Retrieved from www.imim- ng.org/downloads/Alice%20Jonahacticle.doc

Jovago. (2014, June 20). *Book an Hotel using WhatsApp, Skype, Viber or Live Chat with Jovago*. Retrieved from Jumia: http://jumia.pr.co/102254-book-an-hotel-using-whatsapp-skype-viber-or-live-chat-with-jovago

Jusoff, K. & Sahimi, N. N. (2009). Television and Media Literacy in Young Children: Issues and Effects in Early Childhood. *International Education Studies*, *3*(2), 151-157.

Kabweza, L. S. (2016, March 29). *Revenge porn laws in Zimbabawe: Parliament considers petition*. Retrieved from Techzim: http://www.techzim.co.zw/2016/03/revenge-porn-laws-zimbabwe-parliament-petition/

Kafewo, S. A. (2008). Using drama for school-based adolescent sexuality education in Zaria, Nigeria. *Reproductive Health Matters*, *16*(31), 203–210. doi:10.1016/S0968-8080(08)31356-1 PMID:18513621

Kahari, L. (2013). The effects of Cell phone use on the study habits of University of Zimbabwe First Year Faculty of Arts Students. *International Journal of Education and Research*, 1–12.

Kainuwa, A., & Yusuf, N. B. M. (2013). Influence of Socio-Economic and Educational Background of Parents on their Children's Education in Nigeria. *International Journal of Scientific and Research Publications*, *3*(10), 1–7.

Kalyango, Y–K. (2013). *Impact of Social Media on Political Mobilisation In East and West Africa*. Retrieved from globalmediajournal: http://www.globalmediajournal.com/openaccess/impact-of-social-media-on-political-mobilization-in-east-and-west-africa.pdf

Karvonen, E. (2004). Introduction. In F. Webster (Ed.), *The information society reader* (pp. 345–350). London: Routledge.

Kasmania, M. F., Sabrana, R., & Ramle, N. (2014). Can Twitter be an effective platform for political discourse in Malaysia? A Study of #PRU13. *Procedia: Social and Behavioral Sciences*, *155*, 348–355. doi:10.1016/j.sbspro.2014.10.304

Katz, J. E., & Aspden, P. (1997). Motivations for and barriers to Internet usage: results of a national public opinion survey. *Internet Research: Electronic Networking Applications and Policy*, *7*(3), 170–88.

Katz, D. (1961). The Functional Approach to the Study of Attitudes. *Public Opinion Quarterly*.

Katz, E., Blumler, J., & Gurevitch, M. (1974). Uses of Mass Communication by the Individual. In W. P. Davidson & Yu. Fredrick (Eds.), *Mass Communication Research: Major Issues and Future Directions* (pp. 11–35). New York: Praeger.

Katz, J. E., & Rice, R. E. (2002). *Social consequences of Internet use: Access, involvement andinteraction*. Cambridge, MA: MIT Press.

Kaul, V. (2013). Journalism in the Age of Digital Technology. *Journal of Communication and Media Technologies.*, *3*(1), 125–143.

Keating, J., Meekers, D., & Adewuyi, A. (2006). Assessing effects of a media campaign on HIV/AIDS awareness and prevention in Nigeria: Results from the VISION Project. *BMC Public Health*, *6*(1), 123. doi:10.1186/1471-2458-6-123 PMID:16672067

Kellow, C. L., & Steeves, H. L. (1998). The role of radio in the Rwandan genocide. *Journal of Communication*, *48*(3), 107–128. doi:10.1111/j.1460-2466.1998.tb02762.x

KickStart. (2013). Retrieved from http://www.kickstart.org/

Kiiti, N., & Mutinda, J. (2011). *Mobile Money Services and Poverty Reduction: A study of Women's Groups in Rural Eastern Kenya*. Academic Press.

Kijeski, T. (1995). Appropriate technology and the Botswana television experience: On the efficacy of waiting. *The Howard Journal of Communications*, *6*(4), 292–305. doi:10.1080/10646179509361704

Kincaid, D. L., Storey, D., & Babalola, S. (2009). *Effect of Communication on HIV Prevention and Living with HIV/AIDS: Key points for discussion*. Baltimore, MD: Johns Hopkins Center for Communication Programs.

Kiprotich, A., & Ong'ondo, C. O. (2013). An Assessment of the Level of Awareness about Children's Rights among Children in Eldoret Municipality, Kenya. *Journal of Emerging Trends in Educational Research and Policy Studies*, *4*(2), 279–287.

Klapper, J. T. (1960). *The Effect of Mass Communication*. New York: The Free Press.

Klapper, J. T. (1960). *The effects of mass communication*. New York: The Free Press.

Klecun, E. (2008). Bringing lost sheep into the fold: Questioning the discourse of the digital divide. *Information Technology & People*, *21*(3), 267–282. doi:10.1108/09593840810896028

Kufin, P., & Sinha, K. A. (2013). Gender discrimination in job Ads: Evidence from China. *The Quarterly Journal of Economics*, *128*(1), 287–336. doi:10.1093/qje/qjs046

Kumar, J. L. (2006). *Fundamentals of Child Rights: Concepts Issues and Challenges*. New Delhi: Mehra Offset Press Delhi.

Labani, S., Zabaleta, C., & Ruiz, L. (2010). *Gender analysis of women's political participation in seven south-east Asian countries: Bangladesh, Cambodia, the Philippines, Indonesia, Sri-Lanka, East Timor and Vietnam (2008-2009)*. Retrieved on 10th March, 2015 from https://en.wikipedia.org/wiki/Gender_inequality_in_Bangladesh

Landemore, H. E. (2012). Why the many are smarter than the few and why it matters. *Journal of Public Deliberation*, *8*(1), 1–12.

Langat, A. C. (2015). Influence of Social Media on Study Habits of Undergraduate Students in Kenyan Universities. *International Journal of Novel Research in Humanity and Social Sciences*, 42-55.

Lasisi, A., & Augoye, J. (2012, January 1). A history of protests against subsidy removal. *The Punch*. p. 16.

Lawrence, D. (2014). *The global cost of cybercrime: More than $400 billion per year.* Retrieved from: http://www.bloomberg.com/bw/articles/2014-06-09/the-global-cost-of-cybercrime-more-than-400-billion-per-year

Lee, J., & Gereffi, G. (2013). *The co-evolution of concentration in mobile phone global value chains and its impact on social upgrading in developing countries*. Capturing the Gains Working Paper 2013/25.

Lee, J., Ryu, H., Mon, L., & Park, S. J. (2013). Citizens' use of Twitter in political information sharing in South Korea. *iConference 2013 Proceedings*, (pp. 351-365). doi:10.9776/13210

Leftwich, A. (1995). Bringing politics back in: Towards a model of the developmental state. *The Journal of Development Studies*, *31*(3), 400–427. doi:10.1080/00220389508422370

Lenhart, A., Lewis, O., & Raine, L. (2001). *Teenager life online*. Retrieved from: http://www.pewinternet.org/2001/06/21/teenage-life-online/

Lennon, M. (2010). *New Report Gauges Emotional Impact of Cybercrime.* Retrieved from http://www.securityweek.com/new-report-gauges-emotional-impact-cybercrime

Lerner, D. (1958). *The passing of traditional society: Modernisation in the Middle East*. Glencoe, IL: The Free Press.

Lesame, Z. (2005). Bridging the digital divide in South Africa. In Z. Lesame (Ed.), *New Media: Technology and policy in developing countries*. Pretoria: Van Schaik.

Levinson, J., & Young, D. (2010). Implicit Gender Bais in the Legal Profession: An Empirical study. *Duke Journal of Gender Law & Policy*, *18*(1), 1–9.

Lewis, K., Gray, K., & Meierhenrich, J. (2014). The structure of online activism. *Sociological Science, 1*, 1-9. DOI: 10.15195/v1.a1

Lievrouw, L. (2002). Determination and contingency in New Media development: Diffusion of Innovations and social shaping of technology perspectives. In L. Lievrouw & S. Livingstone (Eds.), *The handbook of new media* (pp. 183–200). London: Sage.

Lillie, J. (2009). A Critical Analysis of EU Mobile Narratives and the Ideal Citizen of the Mobile World. In *Proceedings of International Communication Association conference*.

Ling, R., & Donner, J. (2009). *Mobile communication*. Cambridge, UK: Polity Press.

Ling, R., & Horst, H. (2011). Mobile Communication in the Global South. *New Media & Society*, *13*(3), 363–374. doi:10.1177/1461444810393899

Lisulo, S. (2016, March). *MTN launches 'Prestige' to raise customer service experience.* Retrieved from Post: http://postzambia.com/news.php?id=16335

Literat, I., & Chen, N. (2014). Communication infrastructure theory and entertainment-education: An integrative model for health communication. *Communication Theory*, *24*(1), 83–103. doi:10.1111/comt.12011

Liu, S-H. (2013). Adolescents' Interpersonal Relationships with Friends, Parents, and Teachers When Using Facebook for Interaction. *Creative Education*, 335-339. Retrieved from Creative Education: http://file.scirp.org/pdf/CE_2013051015134212.pdf

Livingstone, S. (2004). What is media literacy? *Intermedia*, 18-20.

Livingstone, S. (1996). On the continuing problems of media effects research. In J. Curran & M. Gurevitch (Eds.), *Mass Media and Society* (2nd ed.). London: Edward Arnold.

Lodish, H., Berk, A., Zipursky, L., Matsudaira, P., Baltimore, D., & Darnell, J. (2000). *Molecular cell biology* (4th ed.). New York: W. H. Freeman.

Lorimer, R. (1994). *Mass Communication; A Comparative introduction.* Manchester, UK: Manchester University Press.

Luescher, M. T. (2016). *Student power in twenty-first century Africa: The character and role of student organising.* Retrieved from http://scholar.harvard.edu/files/manja_klemencic/files/2016_luescher_-_klemencic_-_chapterfor_r_brooks_final1_0.pdf?m=1454077927

Lusike, L., & Macharia, J. (Eds.). (2016). *Political Influence of the Media in Developing Countries.* IGI Global Publishers.

M'Bayo, R., Amobi, I., & Danjuma, Y. M. (2014). Influence of political campaign advertising on selected voters in Lagos, Nigeria. *KSU Journal of Communication*, (3), 59-76.

Macaluso, N. (2001, August 13). Nearly 60% of homes are online. *E-Commerce Times.*

Macharia, J. W. (2016b). The Portrayal of Men and Women in the Media: The African Landscape. *Journal of Communication and Media Research, 8*(1).

Macharia, J. W. (2016a). *Gendered African Media: The Kenyan Experience. In Political Influence of the Media in Developing Countries.* IGI Global Publishers.

Machin, D. (2013). Introduction: What is multimodal critical discourse studies? *Critical Discourse Studies, 10*(4), 347–355. doi:10.1080/17405904.2013.813770

Macintyre, K., Rutenberg, N., Brown, L., & Karim, A. (2004). Understanding perceptions of HIV risk among adolescents in KwaZulu-Natal. *AIDS and Behavior, 8*(3), 237–250. doi:10.1023/B:AIBE.0000044072.71361.b3 PMID:15475673

Maczewski, M. (2001). Exploring identity through the Internet: Youths experiences online. *Child and Youths Care Forum, 31*(2), 111-129.

Madden, M., & Zickuhr, K. (2011). *Internet and American Life Project: 65% of online adults use social networking sites.* Retrieved from www.pewinternet.org/Reports/2011/Social-Networking-kSites.aspx

Makhoahle, P., & Bagali, T. M. (2012). The Impact of HIV/AIDS awareness campaigns on students who enrolled from 2009-2011 at central University of Technology, Free State. *Interim: Interdisciplinary Journal, 12*(1), 15–25.

Maliki, A. (2010). Attitude of Niger Delta University undergraduates towards HIV/AIDS Campaigns. *Edo Journal of Counselling, 3*(2), 210–219.

Maree, J., Piontak, R., Omwansa, T., Shinyekwa, I., & Njenga, K. (2013). *Developmental uses of mobile phones in Kenya and Uganda.* Manchester, UK: Capturing the Gains.

Mares, M. (1996). Positive effects of television on social behavior: A meta-analysis. *The Annenberg Public Policy Center, 3*, 1–26.

Marianne, P., Majumdar, B., Akhtar-Danesh, N., Boblin, S., & Sherifali, D. 2013. The gendered Knowledge of HIV/AIDS among urban University Students in Southern Ethiopia.*Proceedings of 1st annual International Interdisciplinary Conference*.

Marios, I., & Spyros, H. (2007). International business- A global perspective. Butterworth-Heinemann Publishing Company.

Marshall, P. D. (2004). *New Media Cultures*. New York: Oxford University Press.

Mascheroni, G. V. (2015). "Girls are addicted to likes so they post semi-naked selfies": Peer mediation, normativity and the construction of identity online. *Journal of Psychosocial Research on Cyberspace*, 5.

Maswanya, E., Moji, K., Horiguchi, I., Nagata, K., Aoyagi, K., Honda, S., & Takemoto, T. (1999). Knowledge, risk perception of AIDS and reported sexual behaviour among students in secondary schools and colleges in Tanzania. *Health Education Research: Theory and Practise*, *14*(2), 185–196. doi:10.1093/her/14.2.185 PMID:10387499

Mathenge, D. G. (2011). Mobile Money Transfer in Kenya: An Ethical Perspective. *Research Journal of Finance and Accounting, 2*(1). Retrieved from www.iiste.org

Mavura, A. (2016, March 29). *Empowering women in Ghana and Zimbabwe through Social Media.* Retrieved from social-media-for-development: http://social-media-for-development.org/

Mbakwem-Aniebo, C., Ezekoye, C., & Okonko, I. O. (2012). Knowledge about HIV/AIDS and reported risk behaviours among freshmen of the University of Port Harcourt, Southern Nigeria. *World Applied Sciences Journal*, *16*(8), 1093–1103.

Mbelle, N., Setswe, G., Sifunda, S., Mabaso, M., & Maduna, V. (2014). *HIV and AIDS related Knowledge,attitudes and behaviours of students and staff at South Africa's technical and vocational education and training colleges*. Pretoria: HSRC.

Mbogo, M. (2010). The impact of mobile payments on the success and growth of micro-business: The case of M-Pesa in Kenya. *Journal of Language Technology and Entrepreneurship in Africa*, *2*(1), 182–203.

McDougall, P. P., Covin, J., Robinson, R. Jr, & Herron, L. (1994). The Effects of Industry Growth and Strategic Breadth on New Venture Performance and Strategy Content. *Strategic Management Journal*, *15*(7), 537–553. doi:10.1002/smj.4250150704

McDougall, P. P., & Oviatt, B. M. (1996). New Venture Internationalization, Strategic Change, and Performance: A Follow-Up Study. *Journal of Business Venturing*, *11*(1), 23–40. doi:10.1016/0883-9026(95)00081-X

McDougall, P. P., & Oviatt, B. M. (2000). International Entrepreneurship: The Intersection of Two Research Paths. *Academy of Management Journal*, *43*(5), 902–906. doi:10.2307/1556418

McKenna, K. Y. A., & Bargh, J. A. (2000). Plan 9 from cyberspace: The implications of Internet for personality and social psychology. *Personality and Social Psychology Review*, *4*(1), 57–75. doi:10.1207/S15327957PSPR0401_6

McLuhan, M. (1986). *McLuhanisms*. Available from http://www.marshallmcluhan.com/

McLuhan, M. (1964). *Understanding Media*. London: Routledge.

McPhail, T. L. (2013). *Global communication: theories, stakeholders and trends*. Malden, MA: Wiley Blackwell.

McQuail, D. (1993). Mass Communication Theory: An Introduction (2nd ed.). New Delhi: Sage Publication.

McQuail, D. (1980). *Towards a sociology of Mass Communication*. London: Collier Macmillan Ltd.

McQuail, D. (1982). *Mass Communication Theory: An Introduction*. London: Constable.

McQuail, D. (1998). *Mass Communication Theory. An Introduction* (3rd ed.). London: Sage.

McQuail, D. (2010). *Mass communication theory* (5th ed.). London: Sage Publications.

McQuail, D. (2010). *Mass Communication Theory* (6th ed.). London: Sage.

McQuail, D., Blumler, J. G., & Brown, J. R. (1972). The Television Audience: A Revised Perspective. In *Sociology of Mass Communications*. Harmondsworth, UK: Penguin.

McQuail, D., & Gurevitch, M. (1974). *Explaining Audience Behavior: Three Approaches Considered. In The uses of Mass Communication: Current Perspectives on Gratifications Research*. London: Sage Publication.

McQuail, D., & Windahl, S. (1981). *Communication Model for the Study of Mass Communication*. London: Longman.

McShane, M. D., & Williams, I. I. I. (Eds.). (2007). Youth violence and delinquency: Monster and myths (vols. 1-3). Westport, CT: Praeger Publishers.

Media Council of Kenya. (2005). *Baseline Survey of Media in Kenya*. Author.

Media Council of Kenya. (2011). *The Performance of Vernacular Radio Stations in Kenya Monitoring Report*. Nairobi: Media Council of Kenya.

Media Council of Kenya. (2012). *The Pre-election Coverage by Kenyan Newspapers Monitoring Report*. Nairobi: Media Council of Kenya.

Mehdizadeh, M. (2010). *The theories of media: Current thoughts and critical attitudes*. Tehran: Hamshahri Press.

Mendelsohn, H. (1966). *Mass entertainment*. New Haven, CT: College and University Press.

Menger, C. (1892). *On the origins of money*. Retrieved March 17, 2005, from http://cepa.newschool.edu/het

Meridien Research. (2000, March 8). Shopnow.com unveils PCcharge transaction processing alliance, an innovation in credit card fraud protection. *PR Newswire*.

Mihailidis, P. (2009). *The 5 – A's of media literacy*. Retrieved from https://hofstramass112.files.wordpress.com/2009/09/mihailidis-the-5as-of-media-literacy.pdf

Miller, D., Costa, E., Haynes, N., McDonald, T., Nicolescu, R., Sinanan, J., & Wang, X. (2016). *How the World Changed Social Media*. London: UCL Press. doi:10.14324/111.9781910634493

Ministry of Communications Science and Technology. (2006). *Draft Botswana broadcasting policy*. Gaborone, Botswana: Author.

Mochaba, K., Raffinetti, C., & White, J. (2004). *SADC media law; A handbook for media practitioners; a comparative overview of media law and practice in Botswana, Swaziland and Zambia* (Vol. 2). Johannesburg, South Africa: Konrad Adenauer Foundation.

Moemeka, A. (2000). *Report handbook: An introduction to effective journalism*. Morris Publisher.

Moemeka, A. A. (2000). Reporter's handbook: An Introduction to effective journalism. Morris Publication.

Moen, O., & Servais, P. (2002). Born Global or Gradual Global? Examining the Export Behavior of Small and Medium-Sized Enterprises. *Journal of International Marketing*, *10*(3), 49–72. doi:10.1509/jimk.10.3.49.19540

Mohammed, O. (2016, May 27). *How Tanzanian journalists use WhatsApp to report the news*. Retrieved from ijnet.org: http://adage.com/article/digital/bbc-teams-viber-whatsapp-distribute-content/302906/

Molutsi, P. P., & Holm, J. D. (1990). Developing democracy when civil society is weak: The case of Botswana. *African Affairs*, *89*(356), 323–340.

Moradi, M., Honari, H., Naghshbandi, S., Jabari, N., & Kholouse, P. (2012). Social Functions of Mass Media in Developing Citizen Sports. *International Journal of Academic Research in Business and Social Sciences*, *2*(6), 2222–6990.

Morozov, E. (2011). *The net delusion*. New York: NY Public Affairs.

Mosco, V. (2004). *The digital sublime: myth, power, and cyberspace*. Cambridge, MA: MIT Press.

Mosco, V. (2009). *The political economy of communication: rethinking and renewal*. London: Sage. doi:10.4135/9781446279946

Motsaathebe, G. (2009). Gendered roles, images and behavioural patterns in the soap opera generations. *Journal of Africa Media Studies, 1*(3), 429-435.

Moyer-Guse, E., Jain, P., & Chung, A. H. (2012). Reinforcement or reactance? Examining the effect of an explicit persuasive appeal following an entertainment-education narrative. *Journal of Communication*, *62*(6), 1010–1027. doi:10.1111/j.1460-2466.2012.01680.x

Msonza, N. (2016, March 29). *Internet Freedom and repression in Zimbabwe*. Retrieved from boell.de: https://www.boell.de/sites/default/files/perspectives_africa_juli_2015_gamechanger.pdf

Mulupi, D. (2016, March 29). *Customer care via WhatsApp: Kenyan company finds traction abroad.* Retrieved from howwemadeitinafrica.com: http://www.howwemadeitinafrica.com/customer-care-via-whatsapp-kenyan-company-finds-traction-abroad/

Muncie, J., & McLaughlin, E. (Eds.). (2001). *The problem of crime*. London: Sage Publication Ltd.

Musa, B., Bussell, S., Borodo, M. M., Samaila, A. A., & Femi, O. L. (2015). Prevalence of hepatitis B virus infection in Nigeria, 20002013: A systematic review and meta-analysis. *Nigerian Journal of Clinical Practice*, *18*(2), 163–172. doi:10.4103/1119-3077.151035 PMID:25665986

Musoni. (2012). *Musoni next generation microFinance*. Available at http://www.musoni.eu/

MusvoZimbawe. (2016, March 29). Retrieved from musvozimbabwe.com: https://musvozimbabwe.com/2015/08/daphne-nyasha-mazara-musvo-videoake/

Muthiora, B. (2015). Enabling mobile money policies in Kenya: Fostering a digital financial revolution. In GSMA, 2015.

Mutinta, G., & Govender, K. (2012). The social-environmental determinants of students' sexual risk behavior and HIV prevention at the University of Kwazulu-Natal. *Journal of Human Ecology (Delhi, India)*, *38*, 17–29.

Mzamane, M. V. (2001). *Where there is no vision the people perish: Reflections on the African Renaissance*. Available from www.unisa.edu.au/hawkeinstitute/publications/default.asp-59k-

Nanaghan, B. (2012, January 1). Still on Fuel Subsidy Removal. *The Punch*, p. 16.

National Agency for the Control of AIDS (NACA). (2015a). *National HIV/AIDS Prevention Plan 2014-2015*. Abuja: NACA.

National Agency for the Control of AIDS (NACA). (2015b). *Federal Republic of Nigeria Global AIDS Response: Country Progress Report 2014*. Abuja: NACA.

National Coalition of Affirmative Action. (2009). *National gender policy*. Retrieved on 7, March, 2015 from www.aacoalition.org/national-policy-women

National Crime Prevention Institute (NCPI). (2001). *Understanding crime prevention* (2nd ed.). Woburn, MA: Butterworth-Heinemann.

National Demographic and Health Survey (NDHS). (2013). *National Population Commission*. Abuja: NPC.

National Institute of Health. (2005). Theory at a Glance: A Guide for Health Promotion Practice. Washington, DC: National Institute of Health Publication. No. 05-3896.

National Population Commission. (2009). *Nigeria Demographic and Health Survey 2008*. Abuja: NPC.

National Population Commission. (2013). *Nigeria Demographic and Health Survey 2013*. Abuja: NPC.

Ndlovu, B. (2016, March 29). *Thrilling reactions over Eritrean Law*. Retrieved from The Chronicle: http://www.chronicle.co.zw/thrilling-reactions-over-eritrean-law/

Ndukwe, E. C. A. (2009). *Combating cyber-crime in 21st century*. Paper presented at the 1st annual conference of National assembly anti-money laundering and cyber-crime coalition.

Nehanda Radio. (2013, March 11). Retrieved from nehandaradio.com: http://nehandaradio.com/2013/11/03/why-macheso-willnever-use-whatsapp-or-facebook/

New Zealand Management. (2013, April). Cyber-crime: Cyber threats and data privacy. *Magazine of the Institute of Management*.

Newman, N. D. (2012). Social Media in the Changing Ecology of News. *International Journal of Internet Science*, 6–22.

Ngoa, S. N. (2012). Agenda-setting: A theory in process. In Critique and application of communication theories (pp. 1-25). Ota: Covenant University Press.

Nie, N. H., & Erbring, L. (2000). *Internet and society: A preliminary report*. Stanford Institute for the Quantitative Study of Society (SIQSS), Stanford University, and InterSurvey. Available online at http://www.stanford.edu/group/siqss/

Nigeria Education Fact Sheet. (2012). The United States Embassy. Retrieved from http://photos.state.gov/libraries/nigeria/487468/pdfs/JanuaryEducationFactSheet.pdf

Nigeria Masterweb Citizen News. (2012). *Gender inequality: The Nigerian case*. Retrieved on 2 March, 2015 from www.nigeriamasterweb.com

Nigeria Millennium Development Goals Report 2010. (2010). Government of the Federal Republic of Nigeria.

Nnabuenyi, U. M., & Otinche, S. I. (2015). Culture and the right of the rural women in Nigeria: an overview. *International Journal of Basic, Applied and Innovative Research, 4*(2), 30-43.

Nonoyama, M., Tsurugi, Y., Shirai, C., Ishikawa, Y., & Horiguchi, M. (2005). Influences of Sex-Related Information for STD Prevention. *The Journal of Adolescent Health, 36*(5), 442–445. doi:10.1016/j.jadohealth.2004.01.017 PMID:15837349

Nua.com. (2002). *How many online*? Retrieved from: http://www.nua.ie/surveys/how_many_online/world.html

Nunyenge, R. (2013). The meaning people make of HIV Posters: A study on health improvement at Jirapa District in the upper West Region of Ghana. *Journal of Biology. Agriculture and Healthcare, 12*, 23–34.

Nwabudike, N. (2012). *Stereotyping in Nigerian movies and social interaction among students of Bells University of Technology* (Unpublished undergraduate thesis). Department of Mass Communication, College of Development Studies, Covenant University, Ota, Ogun State, Nigeria.

Nwabueze, C. D. (2014). *Introduction to mass communication: Media ecology in the global village*. Imo: Top Shelve.

Nwachukwu, E. (2012, January 13). Fuel Subsidy: Examining the facts. *The Punch*, p. 18.

Nwadigwe, C. E. (2012). Theatre for development: An alternative programme for reproductive health communication in urban Nigeria. *African Sociological Review*, *16*(2), 102–118.

Nwankwo, E. H., & Okwemba, A. (2002). *How the Kenyan Media Covers the Children's Rights Issues: A Content Analysis*. Paper Presented to International Federation of Journalists (IFJ), Nairobi, Kenya. Retrieved on October 22, 2008 from www.ifj.org/assets/docs/089/021/36f1759- cbfee15.pdf

Nwosu, N. I. (1996). The Politics of Oil Subsidy in Nigeria. *Africa: RivistaTrimestrale Di Studi E Documentazione Dell'istituto Italiano Per L'africa E L'oriente, 51*(1), 80–94. Retrieved from http://www.jstor.org/stable/40761059

Nwulu, N., Adekanbi, A., Oranugo, T., & Adewale, Y. (2010). Television broadcasting in Africa. *Telecommunications Conference (HISTELCON), 2010 Second IEEE Region 8 Conference.*

O'Hara, K. M. (2014). Everyday dwelling with whatsApp.*Proceedings of the 17th ACM conference on Computer supported cooperative work & social computing* (pp. 1131-1142). Baltimore, MD: ACM.

Obasola, K. E. (2013). Gender discrimination and cultural Anachronisms in African Societies: Implication for Girl Child Education. *International Journal of Social Science and Education*, *3*(4), 1027–1034.

Obi, C. (2001). Women's political participation through economic empowerment. Lagos: Human Development Initiatives.

Ochonogor, C. I., Ohiagu, O. P., & Nwammuo, A. (2015). Social media: Connecting or disconnecting Nigeria youths in the 21st century? In D. Wilson (Ed.), *Communication, society and the Nigerian child: Issues and emerging trends in the 21st century* (pp. 1–10). ACCE.

Odekunle, O., & Okuwa, O. (2012). Enhancing Quality Basic Education for the Attainment of Millennium Development Goals in Nigeria: Suggested Policy Interventions. *Journal of Emerging Trends in Educational Research and Policy Studies*, *3*(1), 92–98.

Odimegwu, C., & Okemgbo, C. N. (2008). Mens Perceptions of Masculinities and Sexual Health Risks in Igboland, Nigeria. *International Journal of Mens Health*, *7*(1), 21–40. doi:10.3149/jmh.0701.21

Odusola, A. F. (1997). Poverty in Nigeria: An Eclectic Appraisal. Poverty Alleviation in Nigeria. *Nigerian Economic Society,1997 Annual Conference*. Ibadan: Pat Mag Press.

OECD. (1997). *Globalisation and Small and Medium Enterprises* (Vol. 1). Paris: OECD.

Ogunlade, O. (2015, February 12). *More than ten lucrative ways you can use WhatsApp for business*. Retrieved from Techpoint: https://techpoint.ng/2015/12/02/how-to-use-whatsapp-for-business/

Ojalammi, S. (2011). *Gender and development in Sub-Saharan Africa*. Retrieved on 10th March, 2015 from http://www.bing.com/search?FORM=SK216DF&PC=SK216&q=Ojalammi%2C+S.+(2011).Gender+and+development+in+Sub-Saharan+Africa

Ojebode, A. (2011). Nigeria's Freedom of Information Act: Provisions, strengths, challenges. *African Communication Research*, *4*(2), 267–284.

Ojebode, A. (2013). Ethical dilemma revisited: PBO newspapers and the professional elbowroom of the Nigerian journalist. *Journal of African Media Studies*, *5*(3), 295–312. doi:10.1386/jams.5.3.295_1

Ojekwe, G. I. (2015). *Political advert campaigns and voting behaviour: A study of Akinwunmi Ambode's election ad campaigns in Lagos state*. Retrieved from http://www.inecnigeria.org/wp-content/uploads/2015/07/Conference-Paper-by-Grace-Ojekwe.pdf

Ojiabor, O. (2016, May 17). Finally Senate throws out anti-social media bill. *The Nation.* Retrieved from: http://thenationonlineng.net/finally-senate-throws-anti-social-media-bill/

Ojo, E. (2003). The mass media and the challenges of sustainable democratic values in Nigeria: Possibilities and limitations. *Media Culture & Society*, *25*(1), 62–76.

Okaka, W. (2014) *Effectiveness of ICT for Gender Equality and Women empowerment in small Island Developing States (SIDS).* Third International Conference on small Island Developing States Stakeholders Forum. Retrieved from www.researchgate.net

Okello, D. (2012). *Mobile phones breakout group discussion.* Capturing the gains East African Workshop, Nairobi, Kenya.

Okenwa, S. A. (2000). *The mass media: Uses and regulations*. Enugu: Bismark Publications.

Okogu, B. E. (1995, January01). Issues in Petroleum Product Pricing in Nigeria. *Journal of African Economies*, *4*(3), 378–405. Retrieved from http://jae.oxfordjournals.org/

Okonta, P. I., & Oseji, M. I. (2006). Relationship between knowledge of HIV/AIDS and sexualbehaviour among in school adolescents in Delta State. *Nigerian Journal of Clinical Practice*, *9*(1), 37–39. PMID:16986287

Okorie, N. (2010). Press freedom in Nigeria: Legal constraints and possible solutions. *International Journal of Communication*, *11*(1), 159–173.

Okoro, N. (2004). Constitutional limitations of press freedom in Nigeria: Cases, indices and implication. *An Interdisciplinary Journal of Communication Studies*, *18*(1), 36–70.

Okpe, O. (2005). *Mainstreaming Gender in the African Development Process: a Critic of NEPAD and the Women Question.* Nigeria: BSU Press.

Okpukpara, B. C., & Chukwuone, N. A. (2001). *Child Schooling in Nigeria: The Role of Gender in Urban, Rural, North and South Nigeria*. Centre for Rural Development and Cooperatives.

Okunna, C. S. (2002). *Gender and communication in Nigeria: Is this the twenty-first century?* Department of Mass Communication, Nnamdi Azikiwe University, Akwa, Nigeria. Retrieved from www.portalcommunicacion.com/bcn2002/n_eng/programme/prog_ind?papers/o/pdf/o005se04_okunn.pdf

Okunna, C. (2003). The press, the law and the society: A practical and ethical approach'. *UNILAG Communication Review*, *4*(1), 65–79.

Olaniyan, O. (2011). *The determinants of child schooling in Nigeria (No. RP_217).* African Economic Research Consortium.

Onabajo, O. (2001). Communication Models in Non-Formal Education. *Humanities Review Journal*, *1*(2), 82–88. doi:10.4314/hrj.v1i2.5913

ONS Omnibus Survey. (2000). Retrieved from: http://discover.ukdataservice.ac.uk/Catalogue/?sn=4528&type=Data%20catalogue

Onuzulike, U. (2007). Nollywood: The influence of the Nigerian movie industry on African culture. *The Journal of Human Communication, 10*(3), 231-242.

Onwubiko, C. P. (2012). Empowerment of Nigerian women towards national development. *Journal of Resourcefulness and Distinction*, *2*(1), 68–78.

Onyeizugbe, C. U., & Onwuka, E. M. (2012). Fuel Subsidy Removal as an Imperative for Enhancing Business Development in Nigeria. *VSRD International Journal of Business and Management Research.* Retrieved on May 18, 2013 www.vsrdjournals.com

Oostendorp, M. (2015). Multimodal construction of theidentity of politicians. *Critical Discourse Studies, 12*(1), 39–56. doi:10.1080/17405904.2014.962066

Oppong, A. K., MensahBoafo, I., &Nyamekye, M. (2014). Identifying gender differences in Ghanian university Students' Sexual Practices, Attitudes and Knowledge Regarding HIV. *The Journal of Psychology, 5*(1), 9–18.

Oppong, A. K., & Oti-Boadi, M. (2013). HIV/AIDS knowledge among undergraduate university students: Implication for health education programmes in Ghana. *African Health Sciences, 13*, 270–277. PMID:24235924

Organisation for Economic Co-Operation and Development. (n.d.). *Women's entrepreneurship: issues and policies.* Retrieved from http://www.oecd.org/cfe/smes/31919215.pdf

Osakue, S. O., Kayode, R. A., Marcel, A. O., & Adekunle, A. A. (2009). 'HIV/AIDS and the Broadcast Media in Urban Communities in Edo State. *The Nigeria International Journal of Research and Review, 1*(3), 111–116.

Osofisan, F. (2006). *From Nollywood to Nollyweight? Or reflectons on the possiblities of literature and the burgeoning film industry in Nigeria.* Retrieved from http://www.africultures.com/index.asp

Oso, L. (2012). The burden of history: The press and the 2003 general elections in Nigeria. In S. T. Babalola & T. Azeez (Eds.), *Critical perspectives in language, literature and communication studies.* Obafemi Awolowo University Press.

Oster, J. (2013). Theory and doctrine of 'media freedom' as a legal concept. *Journal of Media Law, 5*, 57-78.

Oteri, O. M., Kibeti, L. P., & Ndungu, E. N. (2015). Mobile Subscription, penetration and coverage trends in Kenya telecommunication sector. *International Journal of Artificial Intelligence, 4*(1).

Otive, I. (2006). *The State Of Education in Nigeria.* A Keynote Address Delivered at a Round Table Organized by Civil Society Action Coalition on EFA.

Otokiti, S. O. (2005). *Methodologies and Mechanics of Computational System.* New Delhi: Sultan Chan and Son.

Owens-Ibie, N. (1989). *A Comparative Study of Mass & Specialized Media Use by Researchers in Nigeria* (Unpublished PhD Thesis). University of Ibadan.

Owolabi, D., & Nnaji, C. I. (2013). The English Language and the Mass Media as Tools for Sustainable Development in Multilingual Nations. *International Journal of Language and Linguistics, 1*(4), 124–130. doi:10.11648/j.ijll.20130104.16

Owoyemi, O., & Olusanya, O. (2014). Gender: A precursor for discriminating against women in paid employment in Nigeria. *American Journal of Business and Management, 3*(1).

Owuamalam, E. (2014). Voter perception of unique selling proposition in Nigeria's presidential political advertisements. *International Journal of International Relation, Media and Mass Communication Studies, 1*(1), 1–16.

Oyero, O. S. & Oyesomi, K. (2014). Perceived influence of television cartoons on Nigerian children's social behaviour. *Estudios em Comunicação*, (17), 91-116.

Oyero, O. (2010). Child Rights and Media Democratization: An Agenda for Realization of the MDGs in Nigeria. *International Journal of Communication*, (11): 202–215.

Oyero, O. S. (2011). *Theories of mass communication. Unpublished lecture notes, Department of Mass Communication.* Ota, Nigeria: College of Development Studies, Covenant University.

Oyesomi, K. O., Oyero, S., & Okorie, N. (2014). Media Advocacy, Development Journalism and Child Right issues in Nigeria. Scholars Journal of Arts, Humanities and Social Sciences, 2(2).

Oyinade, R., Daramola, I., & Lamidi, I. (2013). Media, gender, and conflict: The problem of eradicating stereotyping of women in Nigeria. *Kuwait Chapter of Arabian Journal of Business and Management Review*, *2*(12), 100–114. doi:10.12816/0001275

Packer, L. H. (1968). *The limits of the criminal sanction*. Stanford, CA: Stanford University Press.

Park, P. (2016, January 13). *How many people use social media in Africa*. Retrieved from cnn.com: http://edition.cnn.com/2016/01/13/africa/africa-social-media-consumption/

Peltzer, K., & Promtussananon, S. (2005). HIV/AIDS Knowledge and Sexual Behaviour among Junior Secondary school students in South Africa. *Journal of Social Sciences*, *1*(1), 1–8. doi:10.3844/jssp.2005.1.8

Pennington, J. (2005). Retrieved from www.avert.org/aids-countries

Perrottet, A. &Robie, D. (2011). Pacific media freedom 2011: A status report. *Pacific Journalism Review, 17*, 148-186.

Pettifor, A., Kleinschmidt, I., Levin, J., Ries, H., Macphail, C., & Hlongwa-Madikizela, L. (2005). A community-based study to examine the effect of a youth prevention intervention on young people age 15–24 in South Africa: Results of the baseline study. *Tropical Medicine & International Health*, *10*(10), 971–980. doi:10.1111/j.1365-3156.2005.01483.x PMID:16185231

Pettifor, A., OBrien, K., MacPhail, C., Miller, W. C., & Rees, H. (2009). Early coital debut and associated HIV risk factors among young women and men in South Africa. *International Perspectives on Sexual and Reproductive Health*, *35*(02), 74–82. doi:10.1363/3508209 PMID:19620092

Phippen, A. (2016, March 29). *Sexting: An Exploration of Practises, Attitudes and Influences*. Retrieved from channel4.com: http://www.channel4.com/media/c4-news/pdf/11NSPCCresearch2.pdf

Pilger, J. (Producer). &Lowery, A. (Director). (2001). The New Rulers of the World [Television Documentary]. London: Carlton Television.

Policy and Legal Advocacy Centre. (2016). *What you need to know about the Frivolous Petition Bill 2015*. Retrieved from: http://www.placng.org/new/upload/FrivolousPetitions.pdf

POLITY. (2004). *The Black Economic Empowerment ICT Charter*. Available from www.ictcharter.org.za/-7k-

Ponemon. (2014). *Global report on the cost of cyber-crime*. Retrieved from: http://www.ponemon.org/library/2014-global-report-on-the-cost-of-cyber-crime

Population Council. (2004). *Child Marriage Briefing: Nigeria*. New York, NY: Author.

Porter, M. E. (1990). *The Competitive Advantage of Nations*. New York: Free Press. doi:10.1007/978-1-349-11336-1

Post, J. E., Lawrence, A. T., & Weber, J. (1999). Business and Society: Corporate Strategy, Public Policy, Ethics (9th ed.). Boston: McGraw Hill.

Post, J. E., Lawrence, A. T., & Weber, J. (1999). *Business and Society: Corporate Strategy, Public Policy, Ethics* (9th ed.). Boston: McGraw Hill.

Posts, P. (2016, June 24). *Nairaland Forum*. Retrieved from nairaland.com: http://www.nairaland.com/publisher/posts

Price, E. (2013). Social media and democracy. *Australian Journal of Political Science, 48*(4), 519-527. Retrieved from :10.1080/10361146.2013.846296

Price, R. (2011). *What is the history of entrepreneurship?* Retrieved from www.blog gcase.org/2011/02/04/what-is-the-history-of-entrepreneurship

Prinsloo, C. (2015). *Digital Disruption, Changing the rules of Business for a hyper connected world.* Retrieved from gibs.co.za: https://www.gibs.co.za/aboutus/centres/Business%20Analysis%20And%20Research/Documents/digital%20 disruption_GIBS_small_spreads.pdf

Pyrsopoulous, T. N., & Reddy, R. K. (2015). *Hepatitis B treatment and management.* Retrieved from http://emedicine.mediascape.com/article/177632-treatment

Rahamefy, O., Rivard, M. &Ravaoarinoro, T. 2008. Sexual behaviour and condom use among University students in Madagascar. *Journal of Social Aspects of HIV/AIDS, 5*(1), 28–35.

Rama Naidu, J. D. (2016, March 29). *When Xenophobia Rears its Ugly Head: A Challenge To Responsible and Responsive Governance.* Retrieved from mott.org: http://www.mott.org/files/pubs/WhenXenophobiaRearsItsUglyHead.pdf

Rambe, P. (2013). *Using mobile devices to leverage student access to collaboratively - generated resources: A case of WhatsApp instant messaging at a South African University.* doi:10.2991/icaicte.2013.66

Reddy, P. & Frantz, J. 2011. HIV/AIDS Knowledge, behaviour and beliefs among South African University students. *Journal of Social Aspects of HIV/AIDS, 8*(4), 166–170.

Reeves, H., & Baden, S. (2000). *Gender and development concepts and definitions.* Bridge Institution of Development Studies. Retrieved from http://www.bridge.ids.ac.uk/reports/re55.pdf

Rennie, M. W. (1993). Global Competitiveness: Born Global. *The McKinsey Quarterly, 4*, 45–52.

Report, P. (1989). *Light! Camera! Action! Promoting Family Planning with TV, Video and Film.* Johns Hopkins Bloomberg School of Public Health.

Republic of Botswana. (1968). *National Development Plan (1968-1973).* Gaberone, Botswana: Gaborone Government Printer.

Republic of Kenya. (2001). *The Children's Act, 2001.* National Legislative Bodies / National Authorities Commonwealth Secretariat.

Republic of Kenya. (2006). *Commonwealth Millennium Development Goals Partnerships.* Commonwealth Secretariat.

Republic of South Africa. (2012). *Global AIDS Progress Report.* Pretoria, South Africa: Department of Health.

Research I. C. T. Africa. (2013). *Lifting the Gender Veil on ICT Indicators in Africa.* Accessed on 25 May 2014. http://www.cprsouth.org/wpcontent/uploads/2013/07/Presentation_CPRsouth-8%EF%80%A2CPRafrica-2013_23_08_2013.pdf

Reyes, A. (2015). Building intimacy through linguistic choices, text structure and voices in political discourse. *Language & Communication, 43*, 58–71. doi:10.1016/j.langcom.2015.05.002

Reynolds, P. D. (1997). New and Small Firms in Expanding Markets. *Small Business Economics, 9*(1), 78–84. doi:10.1023/A:1007916217140

Rheingold, H. (1993). *The virtual community: Homesteading on the electronic frontier.* Addison-Wesley.

Riddell, A. (2003). *The Introduction of Free Primary Education in Sub-Saharan Africa.* Background Paper for EFA Global Monitoring Report, 4.

Robinson, W. C. (1967). What is a government subsidy? *National Tax Journal, 20*(1), 86–92. Retrieved from http://www.jstor.org/stable/41792130

Rogers, E. M., & Shoemaker, F. F. (1971). *Communication of Innovations: A cross cultural approach*. New York: Free Press.

Rogers, Y. S. (2015). *Interaction Design: Beyond human-computer interaction* (4th ed.). London: Wiley.

Roossinck, M. J. (2011). The good virus: Viral mutualistic symbioses. *Nature Reviews. Microbiology*, *9*(2), 99–108. doi:10.1038/nrmicro2491 PMID:21200397

Rotich, E. K., Metto, S. K., Siele, L., & Muketha, G. M. (2014). A survey on cyber-crime perpetration and prevention: A review and model for cybercrime prevention. *European Journal of Science and Engineering*, *2*(1), 13–28.

Safaricom. (2012). *Celebrating five years of changing lives*. Available at http://www.safaricom.co.ke/mpesa_timeline/timeline.html

Safaricom. (2013). *M-Pesa agents*. Available at http://www.safaricom.co.ke/personal/m-pesa/m-pesa-agents

Sambe, J. A. (2008). *Introduction to mass communication in Nigeria*. Ibadan: Spectrum Books Limited.

Sandoval, M., & Fuchs, C. (2010). Towards a critical theory of alternative media. *Telematics and Informatics*, *27*(2), 141–150. doi:10.1016/j.tele.2009.06.011

Sani, A. A. (2014). Press freedom: Analyzing the portrayal of Nigerian state on National dailiescover page. *Journal of Yasar University*, *35*(35), 6255–6260. doi:10.19168/jyu.71007

Sani, H. (2001). *Women and national development: The way forward*. Ibadan: Spectrum Books Limited.

Sani, I., Abdullah, M., Ali, A., & Abdullah, F. (2012). Linguistic analysis on the construction of satire in Nigerian political cartoons: The example of newspaper cartoons. *Journal of Media and Communication Studies*, *4*(3), 52–59. doi:10.5897/JMCS11.129

Santana, C., Raj, A., Decker, S., La Marche, A., & Silverman, J. (2006). Masculine Gender Roles Associated with increased sexual Risk and Intimate partner Violence Perpetration among Young Adult men. *Journal of Urban Health: Bulletin of the New York Academy of Medicine*, *83*(4), 575–585. doi:10.1007/s11524-006-9061-6 PMID:16845496

Sanusi, B. O., Omowale, A., & Myke, E. (2015). Changing roles in the Nigerian media industry: A study of multimedia journalism. *Kuwait Chapter of Arabian Journal of Business and Management Review*, *4*(12), 9–17. doi:10.12816/0019000

Sarkar, S. (2015). Gender Stereotyping in Product Promotion: A Dominant Tool of Advertising in Indian Media. *International Journal of Media Journalism and Mass Communications*, *1*(2), 1–12.

Sarri, C. R. (1995). Criminal behavior overview. Encyclopedia of Social Work, 19(1), 637-646.

Sekhar, T., & Hatti, N. (2005). *Discrimination of Female Children in Modern India: From conception through childhood*. Retrieved on 10th March, 2015 from /www.researchgate.net/publication/215875262_Disappearing_Daughters_and_Intensification_of_Gender_Bias_Evidences_from_Two_Village_Studies_in_South_India

Semetko, H. A. (2010). Election campaigns, partisan balance, and the news media. In P. Norris (Ed.), *Public sentinel: News media & governance reform*. Washington, DC: World Bank.

Servaes, J. (1999). *Communication for development: One world multiple cultures*. Cresskill, NJ: Hampton Press.

Seth, M. N. (2007). HIV/*AIDS Mass Media Campaigns*. Indiana University.

Severin, W. J., & Tankard, J. W. (1992). *Communication Theory: Origin and Uses in the Mass Media*. New York: Longman.

Shamabare, R. R. (2012). Are mobile phones the 21st century addiction? *African Journal of Business Management*, 573-577.

Shambare, R. (2014). The Adoption of WhatsApp: Breaking the Vicious Cycle of Technological Poverty in South Africa. *Journal of Economics and Behavioural Sciences*, 542-550.

Shapshak, T. (2016, March 29). *Why WhatsApp is South Africa's Favourite App.* Retrieved from forbes.com: http://www.forbes.com/sites/tobyshapshak/2015/09/04/why-whatsapp-is-south-africas-favourite-app/#68839f8b7939

Shaw, D., Hamm, B., & Knott, D. (2010). Technological change, agenda challenge and social melding: Mass media studies and the four ages of place, class, mass and space. *Journalism Studies*, *1*(1), 57–79. doi:10.1080/146167000361177

Shearer, C. L., Hosterman, S. J., Gillen, M. M., & Lefkowitz, E. S. (2005). Are traditional gender role attitudes associated with risky sexual behavior and condom-related beliefs. *Sex Roles*, *52*(5), 311–324. doi:10.1007/s11199-005-2675-4

Shisana, O., Rehle, T., Simbayi, L. C., Zuma, K., Jooste, S., & Zungu, N. et al.. (2014). *South African National HIV Prevalence, Incidence and Behaviour Survey, 2012.* Cape Town: HSRC Press.

Shriram, V. (2016, March 29). *Build Karma Points on Social Media.* Retrieved from blogs.ucl.ac.uk: http://blogs.ucl.ac.uk/global-social-media/tag/whatsapp/

Shumba, A., Mapfumo, J., & Chademana, J. (2011). The prevalence of concurrent sexual partnerships among students in Institutions of higher Education in Zimbabwe. *Journal of Human Ecology (Delhi, India)*, *34*(1), 53–65.

Signorielli, N. (2001). Television's gender role images and contribution to stereotyping: Past, present and future. In D. Singer & J. Singer (Eds.), *Handbook of children and the media* (pp. 341–358). Thousand Oaks, CA: Sage.

Simone, M., & Fernback, J. (2006). Invisible hands or public spheres? Theoretical foundations for US broadcast policy. *Communication Law and Policy*, *11*(2), 287–313. doi:10.1207/s15326926clp1102_5

Singleton, T. W., & Singleton, A. J. (2010). *Fraud auditing and forensic accounting* (4th ed.). John Wiley & Sons, Inc. doi:10.1002/9781118269183

Sklair, L. (2010). Transnational practices. In G. Ritzer & Z. Atalay (Eds.), *Readings in Globalization. Key concepts and major debates* (pp. 184–212). West Sussex, UK: Wiley-Blackwell.

Smith, L., & Byron, E. (2005). *Is greater decision making power of women associated with reduced gender discrimination in South Asia?* Retrieved on 10th March, 2014 from http://link.springer.com/article/10.1007/s10680-013-9288-3

Smith, M. A., & Kollock, P. (1999). *Communities in cyberspace*. London: Routledge. doi:10.5117/9789056290818

Sobowale, I. (2008). *Scientific Journalism* (2nd ed.). Lagos: Idosan Consult.

Society for Family Health. (2004). *National Behavioural Survey II: Brothel-based sexwork in Nigeria.* Abuja: Society for Family Health.

Sofo, C. A., Ali-Akpajiak, & Pyke. (2003). *Social development and poverty in Nigeria: Measuring poverty in Nigeria.* Oxfam Working Paper.

Soola, E. O. (1988). *An Evaluation Study of The Dissemination of Family Planning Information in Ogun State* (Unpublished PhD Thesis). University of Ibadan.

Spears, R., Lea, M., & Postmes, T. (2000). One side: purview, problems and prospects. In T. Postmes, M. Lea, R. Spears, & S. Reicher (Eds.), *SIDE, issues center stage: recent developments in studies of de-individuation in groups* (pp. 1–16). Amsterdam: KNAW.

Spicy-Snacks. (2016, June 24). Retrieved from Restina: http://restina.net/en/south-africa/randburg/restaurant-cafe/spicy-snacks/295326594006174

Stack, C. (2004). Self esteem and psychological impact of employment. *Social Psychology Quarterly*, *49*(6), 67–72.

Statistita. (2013). *Percentage of online adults who have fallen victim to cyber-crime as of August 2013*. Retrieved from: http://www.statista.com/statistics/294684/online-adult-cyber-crime-victimization/

Statistita. (2014). *What types of cyber-crime have you encountered personally during the last 12 months*? Retrieved from: http://www.statista.com/statistics/429689/internet-users-experiences-with-cyber-crime-germany/

Statistita. (2015). *Average costs of cyber-crime in selected countries as of June 2014 (in million U.S. dollars)*. Retrieved from http://www.statista.com/statistics/293274/average-cyber-crime-costs-to-companies-in-selected-countries/

Steenblik, R. (n.d.). *A Subsidy Primer. Global Subsidies Initiative of the IISD*. Retrieved September 25, 2015 from https://www.iisd.org/gsi/subsidy-primer/

Suggs, L. S., McIntyre, C., Warburton, W., Henderson, S., & Howitt, P. (2015). *Communicating Health Messages: A Framework To Increase The Effectiveness Of Health Communication Globally*. Report of the WISH Communicating Complex, Institute of Global Health Innovation, Imperial College London.

Szabo, R., & Short, R. V. (2000). How does male circumcision protect against HIV infection? *British Medical Journal*, *320*(7249), 1592–1594. doi:10.1136/bmj.320.7249.1592 PMID:10845974

Tagoe, M., & Aggor, R. A. (2009). Knowledge, behaviour, perceptions and attitudes of University of Ghana students towards HIV/AIDS: What does behavioural surveillance survey tell us? *Journal of Health and Human Services Administration*, *32*(1), 51–84. PMID:19558033

Tallman, S., & Li, J. T. (1996). Effects of international diversity and product diversity on the performance of multinational firms. *Academy of Management Journal*, *39*(1), 179–196. doi:10.2307/256635

Taylor, I. (2003). As good as it gets? Botswanas democratic development. *Journal of Contemporary African Studies*, *21*(2), 215–231. doi:10.1080/02589000305446

Tech@work. (n.d.). *CNBC Africa*. Retrieved from http://www.cnbcafrica.com/tv/shows/tech-@-work/

Terry, P. E., Mhloyi, M., Masvaure, T., & Adlis, S. (2006). An examination of knowledge, attitudes and practises related to HIV/AIDS prevention in Zimbabwe university students: Comparing intervention programme participants and non-participants. *International Journal of Infectious Diseases*, *10*(1), 38–46. doi:10.1016/j.ijid.2004.10.007 PMID:16183316

The, U. C. L. A. CCP Internet Report. (2001). *Surveying the digital future*. Retrieved from: http://www.worldinternetproject.net/_files/_Published/_oldis/ucla-internet-2001.pdf

Thompson, T. L., & Zerbinos, E. (1995). Gender roles in animated cartoons: Has the picture changed in 20 years. *Sex Roles*, *32*(9/10), 651–673. doi:10.1007/BF01544217

Titcomb, J. (2016, March 29). *WhatsApp reaches one billion users*. Retrieved from Telegraph: http://www.telegraph.co.uk/technology/2016/02/02/whatsapp-reaches-one-billion-users/

Toholde, L. Z., & Olaniran, S. (2015). Cultural imperialism and national image: The role of the media. In J. E. Aliede (Ed.), *Studies in communication, mass media and society: Discourses and debates* (pp. 38–52). Lagos: PR-MIX.

Trottier, D. (2015). Theorising Social Media, Politics and the State. In Social Media, Politics, and the State. Protests, Revolutions, Riots, Crime and Policing in the Age of Facebook, Twitter and Youtube (pp. 3-38). New York: Routledge.

Tsegyu, S., & Danjuma, Y. M. (2014). Impact of social media networks on the academic performance of mass communication students of Ibrahim Badamasi Babangida University, Lapai Nigeria. *KSU Journal of Mass Communication*, *3*, 47–58.

Tubbs, S., & Moss, S. (2008). *Human communication: Principles and contexts*. Boston: Mc Graw Hill.

Tufte, E. R. (2001). *The visual display of quantitative information* (2nd ed.). Cheshire, CT: Graphics Press.

Turkle, S. (1995). *Life on the screen: identity in the Age of the Internet*. New York: Simon and Schuster.

Tuurosong, D. (2014). The Social Media Scourge among Unversity Students: A study of the University for Development Studies in, Ghana. *Journal of Asian Development Studies*, 62-74.

Tuwor, T., & Sossoh, M. A. (2008). Gender discrimination and education in West Africa: Strategies for maintaining girls in school. *International Journal of Inclusive Education*, *12*(4), 363–379. doi:10.1080/13603110601183115

Udende, P., & Ugande, G. B. (2015). The puzzle of media violence and actual exhibition of violence by Nigerian youths. In D. Wilson (Ed.), *Communication, society and the Nigerian child: Issues and emerging trends in the 21st century* (pp. 355–370). Nigeria: ACCE.

Udeze, S. E., Nwodu, L. C., & Nwodu, E. G. (2015). Social media and global news flow: What has changed? In J. E. Aliede (Ed.), *Studies in communication, mass media and society: Discourses and debates* (pp. 247–254). Lagos: PR-MIX.

Udeze, S., & Uwem, A. (2013). The influence of political advertising on Nigerian electorate. *Journal of Communication*, *4*(1), 49–55.

Udofa, I. J. (2011). Right to freedom of expression and the law of defamation in Nigeria. *International Journal of Advanced Legal Studies and Governance*, *2*(1), 75–84.

Ufuophu-Biri, E. (2006). Post-Mortem of the Nigerian press freedom. *An Interdisciplinary Journal of Communication Studies*, *20*(4), 143–157.

Ugangu, W. (2012). *Normative media theory and the rethinking of the role of the Kenyan media in a changing social economic context* (Doctoral dissertation). University of South Africa. Retrieved from http://uir. unisa.ac.za/bitstream/handle/10500/8606/thesis_ugangu_w.pdf?sequence=1

Ugwuanyi, E. (2015, July1). What next for NNPC after board dissolution? *Nation (New York, N.Y.)*, 2–3.

Ugwulebo, B. (2011). Female gender in professional education. *Marketing and Management of Professional Survey Education.* Retrieved from www.fig.net/pub/fig2011

Ukoha, K. E. (2013). *Media violence and violent behaviour of Nigeria youth: Intervention strategies.* Nigeria Educational Research and Development Council. Retrieved from www.web.a.ebscohost.com

Ukonu, M. (2004). Issues in the Nigerian press laws versus press freedom. *An Interdisciplinary Journal of Communication Studies*, *18*(1), 14–28.

Ukonu, O. I., & Tafamel, A. E. (2011). Problems, challenges and prospects of female Entrepreneurs in Gwagwalada, Abuja. *Africa Research Review*, *5*(3), 226–246. doi:10.4314/afrrev.v5i3.67354

Umeh, C. C. (1989). The advent and growth of television broadcasting in Nigeria: Its political and educational overtone. *Africa Media Review*, *3*(2), 54–66. PMID:12342789

Umoru, H., & Erunke, J. (2016, May 18). Senate throws out anti-social media bill. *Vanguard Newspaper.* Retrieved from http://www.vanguardngr.com/2016/05/senate-throws-anti-social-media-bill-2/http://www.vanguardngr.com/2016/05/senate-throws-anti-social-media-bill-2/

UNAIDS. (2013). UNAIDS Report on the Global AIDS epidemic update, 2013. Geneva: Switzerland: UNAIDS.

UNDP. (n.d.). *United Nations Development Program*. Retrieved from www.undp.org/info21

UNESCO. (2003C). *Gender Issues in information society*. Paris: UNESCO.

UNICEF. (2015). *Strategic communication for behaviour and social change in South Asia*. Retrieved from http://www.unicef.org/rosa/Strategic_Communication_for_Behaviour_and_Social_Change.pdf

United Nations. (2012). *Community mobilization, outreach and mass media*. Retrieved from http://www.endvawnow.org/en/modules/view/3-campaigns.html

United Nations. (2014). *Equality for women in progress for all women watch*. Retrieved on 5th March, 2015 from http://www.un.org/womenwatch/future/iwd

United Nations. (2014). Measuring ICT and gender: An assessment. New York: UN.

University of Rochester Medical Center. (2015). *Is it a virus or a bacterium? Know the difference*. Retrieved from http://www.urmc.rochester.edu./encyclopedia/content.aspx

UNODC. (2013). *Comprehensive study on cyber-crime—draft*. Vienna, Austria: UNODC.

Valentine, G., & Holloway, S. L. (2002). *Cyberkids? Exploring children's identities and social networks in on-line and off-line worlds*. Retrieved from: http://knowledgepublic.pbworks.com/f/cyberkids.pdf

Van Loon, J. (2008). *Media technology: Critical perspectives*. New York: Open University Press.

Van Staden, A., & Badenhorst, G. (2009). Reviewing gender and cultural factors associated with HIV/AIDS among university students in the South African context. *Curationais*, *32*(4), 19–28.

Vermeulen, J. (2016, March 29). *The data cost of WhatsApp voice calling in SA*. Retrieved from http://mybroadband.co.za/news/smartphones/122106-the-data-cost-of-whatsapp-voice-calling-in-sa.html

Vesnic-Alujevic, L. (2012). Political participation and web 2.0 in Europe: A case study of Facebook. *Public Relations Review*, *389*(3), 466–470. doi:10.1016/j.pubrev.2012.01.010

Vigna, S. D., Enikolopov, R., Mironova, V., Petrova, M., & Zhuravskaya, E. (2011). Cross-border media and nationalism: Evidence from Serbian radio in Croatia. Cambridge, MA: National Bureau of Economic Research Working Paper Series. National Bureau of Economic Research. Retrieved from http://www.nber.org/papers/w16989.pdf?new_window=1

Vilanilam, J. V. (2009). *Development communication in practice: India and the Millenium Development Goals*. Thousand Oaks, CA: Sage. doi:10.4135/9788132104148

Vircoulon, T. (2016, June 24). *Burundi turns to WhatsApp as political turmoil brings media blackout*. Retrieved from The Guardian: http://www.theguardian.com/global-development/2016/jun/14/burundi-turns-to-whatsapp-as-political-turmoil-brings-media-blackout

Vodaworld. (n.d.). *Vodacom*. Retrieved from http://www.vodacom.com/vodacom/vodacom-world

Wakunuma, K. (2012). Mobile phones for development: Are we in danger of losing sight of the bigger Picture. *Standpoint*, 8.

Waldman, L. (2016, March 29). *Sexual and Reproductive Health Rights and Information and Communications Technologies: A Policy Review and Case Study from South Africa*. Retrieved from http://opendocs.ids.ac.uk/opendocs/bitstream/handle/123456789/5787/ER113_SexualandReproductiveHealth.pdf.pdf;jsessionid=5DA3C2417198E54224C0B0B8812D75BC?sequence=1

Wallace, P. (1999). *The psychology of the Internet*. Cambridge, UK: Cambridge University Press. doi:10.1017/CBO9780511581670

Wamala, C. (2010). *Does IT count? Complexities between access to and use of information technologies among Uganda's farmers* (PhD thesis). Luleå University of Technology, Sweden.

Wandibba, S., Nangendo, S. M., & Mulemi, B. A. (2014). *Gender empowerment and access to financial services in Machakos County*. Academic Press.

Wang, J. (2014). Criticising images: Critical discourse analysis of visual semiosis in picture news. *Critical Arts (South-North Cultural Media Studies)*, *28*(2), 264-286. doi: 10.1080/02560046.2014.906344

Wangui, J. (2014). *Government Goes Big on Social Media*. Nairobi: Business and Tech.

Warnock, K., Schoemaker, E., & Wilson, M. (2007). *At the heart of change. The role of communication in sustainable development*. Retrieved from http://panos.org.uk/wp-content/files/2007/09/The-Case-for-Communication-in-Sustainable-Development.pdf

Wellman, B., & Haythornwaite, C. (Eds.). (2002). *The Internet in everyday life*. Oxford, UK: Blackwell. doi:10.1002/9780470774298

White, L. (2001). *Reconsidering cultural imperialism theory*. Retrieved from http://www.tbsjournal.com

Widodo, T. G. A., Sahadewo, S. U. S., & Chaerriyah, M. (2012). Cambodia's Electricity Sector in the Context of Regional Electricity Market Integration. In Y. Wu, X. Shi, & F. Kimura (Eds.), *Energy Market Integration in East Asia: Theories, Electricity Sector and Subsidies* (pp. 173–206). Jakarta: ERIA.

Wilcox, D. (1975). *Mass media in black Africa: Philosophy and Control*. New York: Praeger.

Williams, M. (2006). *Virtual criminal: Crime, deviance and regulation online*. New York: Routledge.

Wilson, M. (2002). Understanding the international ICT and development discourse: Assumptions and implications. *The South African Journal of Information and Communication*, *3*(1).

Winston, B., & Walton, P. (1996). Netscape: Virtually free. *Index on Censorship, 25*(168).

Winston, B. (1998). *Media technology and society: A history from the telegraph to the Internet*. London: Routledge.

Wirtz, B. (2008). *What is social learning theory?* Retrieved 30 April, 2010 from http://www,xyhd.tv/2008/uncategorised/what-is-social-learning-theory

Wood, J. (1994) Gender Media: The influence of media on views of gender. In *Communication, gender and gender*. Thomson Learning.

Wood, J. (2011). *Communication, gender and culture*. Belmont: Thomson Wadsworth.

World Bank Group Strategy. (2012-15). *ICT for greater development impact*. Washington, DC: Author.

World Bank Group. (2006). *2006 Information and communications for development. Global trends and policies*. Washington, DC: World Bank.

World Bank. (2003a). *Engendering ICT: Ensuring gender equality in ICT for development*. Washington, DC: Author.

World Bank. (2003b). *ICT and MDGs: A World Bank Group Perspective*. Author.

World Bank. (2005). *Gender –responsive social analysis: A guidance note. Incorporating social dimension into bank-supported projects.* Retrieved form http://web.worldbank.org/archive/website01028/WEB/IMAGES/GENDERGU.PD

World Health Organisation (WHO). (2011). *The sexual and reproductive health of younger adolescents: Research issues in developing countries.* Geneva, Switzerland: WHO Press.

World Health Organisation. (2015). *Ebola virus disease: Fact sheet.* Retrieved from http://who.int/mediacentre/factsheets/fs103/en/

Worthington, N. (2011). Gender discourse and Ubuntu Media Philosophy. *Journalism Studies, 12*(5), 608–623. doi:10.1080/1461670X.2010.545205

Wright, C. R. (1979). Sociology of mass communication. *Annual Review of Sociology, 5*(1), 193–247. doi:10.1146/annurev.so.05.080179.001205

Yankah, E., & Aggleton, P. (2008). Effects and effectiveness of life skills education for HIV prevention in young people. *AIDS Education and Prevention, 20*(6), 465–485. doi:10.1521/aeap.2008.20.6.465 PMID:19072524

Yar, M. (2009). Computer crime control as industry: Virtual insecurity and the market for private policing. In Technologies of insecurity: The surveillance of everyday life. Routledge-Cavendish.

Yassin-Kassab, R., El Fattah, A. A., Soueif, A., Barghouti, M., Lalami, L., Shehadeh, R., . . . Haddad, J. (2016, January 23). 'I was terribly wrong' - writers look back at the Arab spring five years on'. *The Guardian.* Retrieved from https://www.theguardian.com/books/2016/jan/23/arab-spring-five-years-on-writers-look-back

Yazdi, C. A., Aschbacher, K., Arvantaj, A., Naser, H. M., Abdollahi, E., Asadi, A., … Moghadam, A. K. (2006). Knowledge, attitudes and sources of information regarding HIV/AIDS in Iranian adolescents. *AIDS Care, 18*(8), 1004-10.

Yeboah, J. &. (2014). The Impact of WhatsApp Messenger Usage on Students Performance in Tertiary Institutions in Ghana. *Journal of Education and Practice*, 157-164.

Young, L. E. (2016, March 29). *Mobilization under threat.* Retrieved from laurenelyssayoung.com: http://www.laurenelyssayoung.com/wp-content/uploads/2016/02/Young-Party-Emotions-Experiment-Sep-2015.pdf

Youngs, G. (2007). Virtual communication, globalization and new public sphere issues. In G. Youngs (Ed.), *Global political economy in the information age: Power and inequality* (pp. 77–91). New York: Routledge.

Zaffiro, J. (2000). Broadcasting reform and democratization in Botswana. *Africa Today, 47*(1), 87–102. doi:10.1353/at.2000.0023

Zahariadis, N. (2008). *State subsidies in the global economy.* New York: Palgrave Macmillan. doi:10.1057/9780230610514

Zahra, S. A., Ireland, R. D., & Hitt, M. A. (2000). International expansion by new venture firms: International diversity, mode of market entry, technological learning and performance. *Academy of Management Journal, 43*(5), 925–950. doi:10.2307/1556420

Zanamwe, N. R. (n.d.). Use of Social Networking Technologies in Higher Education in Zimbabwe: A learner's perspective. *International Journal of Computer and Information Technology*, 8-18.

Zhang, W., & Kramarae, C. (2014). SlutWalk on connected screens: Multiple framings of a social media discussion. *Journal of Pragmatics, 73*, 66–81. doi:10.1016/j.pragma.2014.07.008

Zhang, Y., & Leung, L. (2015). A review of social networking service (SNS) research in communication journals from 2006 to 2011. *New Media & Society, 17*(7), 1007–1024. doi:10.1177/1461444813520477

Zimbabwean, Z. (2016, March 29). *Indecent Pictures of Undressed ZRP Officer go Viral!* Retrieved from http://iharare.co.zw/constable-chiwanza-sagas/

Zimbardo, P. G. (1999). *Stanford prison experiment.* Retrieved fromhttp://www.prisonexp.org/

Zonta Organisation. (n.d.). *Mass communication with a purpose: Global partnership on edutainment for social change.* Retrieved from https://www.zonta.org/Portals/0/PDFs/04c_Edutainment%20Description.pdf

Zoonen, L. (1991). Feminist perspectives on the media. In J. Curran & M. Gurevitch (Eds.), *Mass Media and society.* London: Arnold Ltd.

About the Contributors

Okorie Nelson is one of the notable lecturers in the Department of Mass Communication, Covenant University. Okorie has more than eight (8) years teaching experience at the higher institution level. In addition, Okorie has obtained a Ph.D degree in the field of Mass Communication. Currently, Okorie has 29 publications comprising of 24 Journal articles, 3 Chapters in peer reviewed books, and 2 published books. Okorie's research goal is aimed at creating knowledge via employing innovative, leading edge teaching and learning methods that promote integrated, life-applicable and life-transforming education relevant to the context of human capacity building.

Babatunde Raphael Ojebuyi, PhD. is a full-time lecturer in the Department of Communication and Language Arts, University of Ibadan, Nigeria. Ojebuyi's teaching and research focus is on media studies in the areas of media theories, media and ethics, and new media. He has written books and published well researched articles in reputable local and international journals.

* * *

Taiwo Olubunmi Abioye holds a Ph.D degree in English with background in Stylistics, Applied Linguistics, and English Literacy. A Resource Person for NTI /MDGs Project Retraining Workshop for Teachers of English and Executive Board member and Secretary, ISA:RC 25 (2010-2014), she chairs the Women Development and Human Security research cluster of her university. She is widely published in international journals and is the immediate past Deputy Vice-Chancellor, Covenant University, Nigeria.

Adeyemi Aderogba won the Bureau for External Aid Scholarship award to study Journalism in the Belorussian State University, Minsk, Belarus in 1984. He returned to Nigeria in 1996 after completing his doctorate degree programme from the same University. He initially lectured at the Delta State University, Lagos Study Centre (now Lagos State University, Anthony Village Campus) from 2001 to 2002 on part-time basis and Bowen University from 2002 to 2007. While at Bowen University, Dr. Adeyemi was privileged to be part of the team that packaged the department's curricula. Between 2006 and 2007, he was appointed the Acting Head of Department of Human Communications. He also headed the team that carried out a comprehensive review of the curriculum of the Department. He joined Covenant University in September 2007 as a lecturer at the Department of Mass Communication, College of Human Development where, among other responsibilities, he served as the coordinator of the department's post graduate programme. He was the secretary to the University's Internship Committee. Major duty of the Committee was to ensure linkages of the University's young academics with industries, for the purpose of enhancement of their work experiences. Dr Adeyemi Aderogba moved to Osun State University, Osogbo, in 2010. He was the pioneered Acting Head of the Department of Communication Studies of the University. Dr. Adeyemi is privileged to have served in several committees of all the universities where he had worked. Currently he is a member of several committees in Osun State University.

Samuel O. Adeyeye holds a BA in Linguistics and Communication Studies from Osun State University, Nigeria. Following his interests and proficiency with Information Technology and computers, he became Mozilla Campus Ambassador for Osun State University, a position he held from 2011 until his graduation in 2015. Due to his skills and drive, he later became an approved Mozillian in 2015. Presently he's working on the Mozilla translation of two Nigerian indigenous languages namely Yoruba and Hausa at his primary place of assignment where he's doing his National Youth Service Corp.

Jegede Ajibade is a graduate of the University of Ibadan Nigeria is a lecturer in the Department of Sociology, Covenant University, Ota, Ogun State, Nigeria. Prior to joining the academic profession, he worked under the Nigerian Army as a soldier and an instructor in the Nigerian Army Intelligence school, Apapa, Lagos. After his disengagement from the Army in 2006, he obtained a Ph.D. degree in Sociology (specialization in criminology/technological crimes) from Covenant University. He has widely published and married with four children.

Esther Ajiboye is a doctoral researcher at Covenant University. Her research interests include Critical Discourse Analysis, (New) Media Discourse, Crisis Discourse, Language in Society, as well as Language and Identity. She is driven by a passion to make meaningful contributions to discourses on social issues such as inequalities and crises. Her publications appear in reputable local and international journals.

Moses Akanbi is a Demographer and a Statistician with over 9 years of professional experience as a University lecturer. He also teaches Statistics at undergraduate and Post-graduate levels Lecturer in the department of Economics and Development Studies in Covenant University. He had both Bachelor of Science(Bsc)and Masters' degree in Demography and Social Statistics from Obafemi Awolowo University,Ile-Ife,Osun State,Nigeria and finally bagged his PhD in Demography and Social Statistics at Covenant University,Ota,Ogun state,Nigeria.

Nneoma Anozie got her first degree in Mass Comunication from Chukwuemeka Odumegwu Ojukwu University, Nigeria and is currently running her masters programme in the same institution. She is cheerful, loves to teach and to lend a helping hand. She enjoys learning, has a penchant for languages and attempts 6 of them. She has interest in communication for development both at individual and national level.

Aretha Oluwakemi Asakitikpi is an alumnus of Brown University's Institute of Advanced Research, USA; a resource person to the Council for the Development of Social Science Research in Africa (CODESRIA) headquartered in Senegal; and has collaborated with academics and researchers at Humboldt University, Germany on the theme of leisure and development. She has taught in a number of universities in Nigeria as well as South Africa. Her broad research interests are in gender, media and cultural studies; history and development of which she has published extensively in books and journals that have national as well as international spread. Her interests in symbolic representations has resulted in research publications in the area of text and image creations using online and print media sources.

Paul Bigala is a Post-Doctoral Research fellow at the North West University. He has written several papers especially in the area of HIV/AIDS together with recent research work on the same subject. he has taught and supervised both under and Post graduate students at the same campus/university.

Ngozi Chilaka is an aspiring researcher and author with special interest in Nollywood as a vibrant film industry.

Ugonna Kelly Ejimofor is a versatile expert in business and IT management. She has positioned herself to encounter growing request for quality in the field of IT and business development. Ugonna has imparted business knowledge and IT development in various several Organizations since she graduated from Covenant University in 2013.

Gbenga Elegbe is currently a lecturer of applied communication and broadcasting in the Department of Communication and Language Arts, University of Ibadan with a wide knowledge in scripting for Advertising, Broadcast Media Production and Strategic Communication for Community Development, Advertising Campaign, Interpersonal communication and Group Cohesion.

Juliet Wambui Macharia (PhD) is an Associate Professor of Communication Studies. She is the former Dean School of Business, Karatina University. She holds a Masters degree, a Bachelor of Education degree, PGD in Public Relations and PGD in Personnel Management. She is a published author of articles in journals, chapters in books and an IGI-Global author. She has 26 years teaching experience both at undergraduate and post graduate university levels. She joined Karatina in 2015 as an Associate professor. She has occupied several administrative positions, among them Head of Department, Communication Studies, Moi University and Head of Department of Human Resource Development at Karatina University. She has attended many international and local conferences, seminars and workshops. She is a published author of many articles in peer reviewed journals and several chapters in books. She is an editor of several peer reviewed Journals. She has conducted sponsored and self sponsored research. Her research interests are: Gender, communication, Media, politics, Culture and Public Relations.

Seamogano Mosanako (PhD) is a Lecturer in the Department of Media Studies, University of Botswana. Mosanako teaches theoretical and practical courses in Media Studies, specialising in broadcasting and media theory. She teaches courses in news writing and production, television production, radio production, media in Botswana, media representation, and theories of mass communication. She is one of the pioneers of the national television Btv, where she rose through the ranks to the position of News Editor. Her research interests are in the areas of broadcasting, journalism, television, media policy, development communication, and media in Botswana. Mosanako has been engaged in a number of media training projects ranging from news writing to media relations. She has been involved in consultancies in development of communication strategies for reputable organisations in Botswana.

Mercy Ogbari is a Lecturer in the Department of Business Management, Covenant University, Ota, (Nigeria) where she teaches Management Principles, Business and Strategic Management, Research Methodology and Entrepreneurship courses. She obtained a Bachelor of Science degree in Cooperative Economics and Management from the Nnamdi Azikiwe University Awka. She also holds a Master of Science degree as well as a Ph.D in Business Management & Entrepreneurship from Covenant University Ota, Nigeria. Her research interests include Entrepreneurship, Innovation, General Business, management Science and Economic Development. She has published articles in several national & international journals.

Oyewole Oladapo is a Research Assistant and Doctoral Student in the Department of Communication and Language Arts, University of Ibadan. His areas of interest are New Media and Political Communication.

Gbenga Olanipekun is a fledgling academic. He graduated from the Osun State University as the best student in his class and as a college scholar. Presently, he is a writer, a speaker of diverse fields; sort after in various walks of life due to his fresh perspective on issues. He is an author of two unpublished works. He is a social media crusader on tailored academic curriculum. He is presently processing admission to University to do his Sc in communications. He has had the opportunity of working in a couple of companies at managerial level. He is presently the MD of Melandrie and La Purcher.

Segun Omidiora is a lecturer in Literature and the Chaplain Covenant University, Nigeria with interest in Gender Studies, African Folklore, African-American Literature, Bible and Literature, Christian Narrative Literature, New Writings and Eschatological Literature. He has years of experience working with young persons aged 7-25. He is currently on his PhD programme.

Kehinde Opeyemi is a researcher and lecturer at the Department of Mass Communication, Covenant University, Ota, Ogun state, Nigeria. Her responsibilities as a lecturer include teaching, advising, project supervision, research and community activities. Her research interests are on Gender, media, political communication and development communication. She possesses a strong positive character and determination to contribute to scholarship in her area of discipline. She has published in reputable international and local journals and contributed in book chapters. She also possesses the ability to manage multiple projects simultaneously and efficiently, and to work smartly and resourcefully in a fast-paced environment.

Tunji Oyedepo is a teacher and researcher with interest in the use of new media technology in Public Relations practice in Nigeria.

Olusola Oyero, PhD, is an Associate Professor in the Department of Mass Communication, Covenant University, Ota, Nigeria. His research interests include children's media, development communication and media and democratic governance. He has written several articles in international and national journal. He is currently the postgraduate chair for the department of Mass Communication, Covenant University.

Brian Pindayi is a lecturer of Journalism and Communication Studies at Rusangu University. His research interests include social media, political economy of the media, political mobilisation, cultural studies and health communication.

Anusharani Sewchurran is a lecturer at the Department of Media & Cultural Studies - University of KwaZulu-Natal, Durban. She recently completed a PhD in the field of telecommunications in South Africa. Her interests range from the regulation of media and new media technology to religion and the mediation of religious identity via social media. She enjoys cooking a lot.

Barnabas Suleiman is a professional academic with expertise in statistical analysis of financial data, Project management and resource planning. He holds a Bsc in Sociology from Covenant University and a Msc in Economics (Econs) and Business from the University of Hull, UK. He is currently undergoing a PhD degree at Covenant University, Ota, Nigeria.

Suleimanu Usaini is a teacher, researcher, media personality and documentary filmmaker. His research interests span across Stereotyping in the media especially film, entertainment media, Nollywood film industry and its influence on the audience.

Index

B

C

D

E

F

G

H

I

K

L

M

N

P

www.ingramcontent.com/pod-product-compliance
Lightning Source LLC
LaVergne TN
LVHW062144200926
848484LV00007B/1855